HEINEMANN [illegible]
for
EDEXCEL AS AND A-LEVEL

Pure Mathematics 1

Geoff Mannall Michael Kenwood

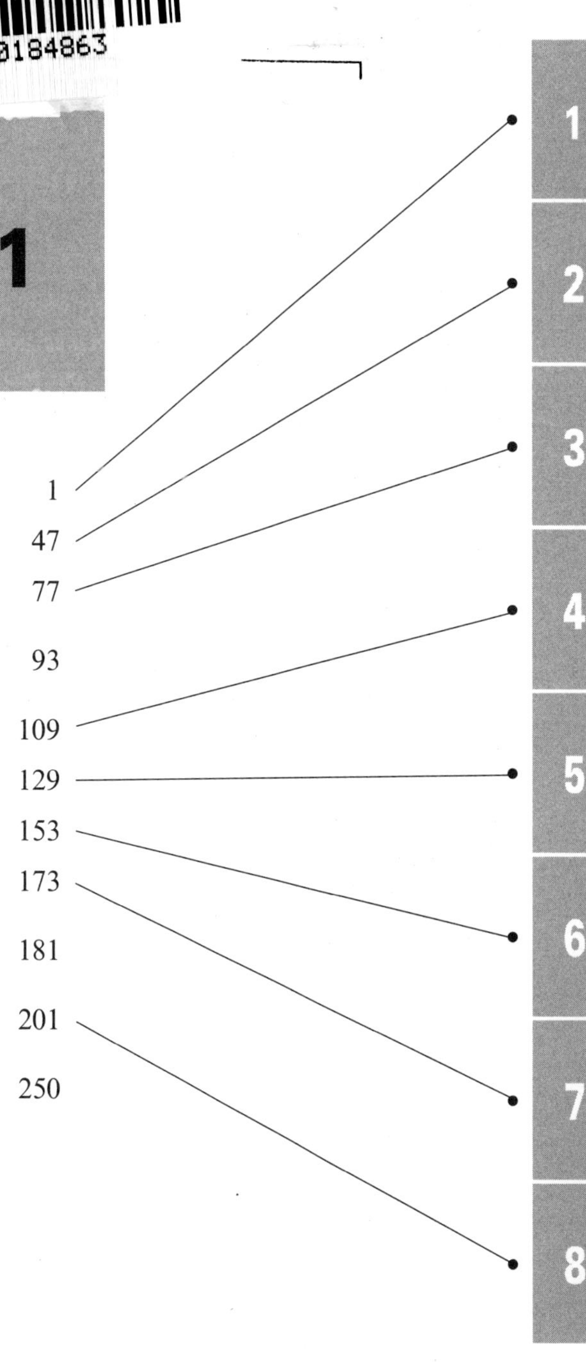

Heinemann Educational Publishers
Halley Court, Jordan Hill, Oxford OX2 8EJ
A Part of Harcourt Education Limited

Heinemann is the registered trademark of
Harcourt Education Limited

First published 2000

03 04 05 12 11 10

ISBN 0 435 51088 6

Cover design by Gecko Limited

Original design by Geoffrey Wadsley; additional design work by Jim Turner

Typeset and illustrated by Tech-Set Limited, Gateshead, Tyne and Wear

Printed in Great Britain by The Bath Press, Bath

Acknowledgements:

The publisher's and authors' thanks are due to Edexcel for permission to reproduce questions from past examination papers. These are marked with an [E].

The answers have been provided by the authors and are not the responsibility of the examining board.

About this book

This book is designed to provide you with the best preparation possible for your Edexcel P1 exam. The series authors are senior examiners and exam moderators themselves and have a good understanding of Edexcel's requirements.

Finding your way around

To help to find your way around when you are studying and revising use the:

- **edge marks** (shown on the front page) – these help you to get to the right chapter quickly;
- **contents list** – this lists the headings that identify key syllabus ideas covered in the book so you can turn straight to them;
- **index** – if you need to find a topic the **bold** number shows where to find the main entry on a topic.

Remembering key ideas

We have provided clear explanations of the key ideas and techniques you need throughout the book. Key ideas you need to remember are listed in a **summary of key points** at the end of each chapter and marked like this in the chapters:

■ **At the point of inflexion C, $\mathbf{f''(x) = 0}$**

Exercises and exam questions

In this book questions are carefully graded so they increase in difficulty and gradually bring you up to exam standard.

- **past exam questions** are marked with an [E];
- **review exercises** on pages 93 and 181 help you practise answering questions from several areas of mathematics at once, as in the real exam;
- **exam style practice paper** – this is designed to help you prepare for the exam itself;
- **answers** are included at the end of the book – use them to check your work.

Essential skills

This new edition of P1 includes a new **Essential skills** chapter beginning on page 201. This chapter covers all prerequisite material for the P1 course, providing worked examples and revision exercises.

Essential skills will be invaluable for students who have previously studied Intermediate tier GCSE and will also provide useful revision for candidates who have completed Higher tier GCSE.

Contents

8 Essential skills

Algebra

1

Algebra is the area of mathematics that uses symbols to represent numbers and to make generalisations about relationships between them.

In order to succeed at an advanced level in mathematics you need to have confidence and accuracy in using algebra in many situations. This chapter introduces some further terminology of algebra and builds on the basic knowledge you should already have from your GCSE work. The essential skills chapter (chapter 8) provides revision of this material. Your ability to process algebraic expressions and to solve equations will be greatly enhanced by regular practice.

1.1 Laws of indices for rational exponents

The rules of indices:

$$x^a \times x^b = x^{a+b}$$

and

$$(x^a)^b = x^{ab}$$

which you have already used in situations where a and b are positive integers, can be extended to include all rational exponents like this.

Example 1

What does x^0 mean?

Consider $5^4 \div 5^4$.

Using $x^a \div x^b = x^{a-b}$ you have:

$$5^4 \div 5^4 = 5^{4-4} = 5^0$$

But

$$5^4 \div 5^4 = \frac{5^4}{5^4} = 1$$

So:

$$5^0 = 1$$

In general,

$$x^n \div x^n = x^{n-n} = x^0$$

$$x^n \div x^n = \frac{x^n}{x^n} = 1$$

■ **So $x^0 = 1$**

Example 2 **Negative indices**

Consider $2^3 \div 2^7 = 2^{3-7} = 2^{-4}$

But: $$2^3 \div 2^7 = \frac{\cancel{2} \times \cancel{2} \times \cancel{2}}{\cancel{2} \times \cancel{2} \times \cancel{2} \times 2 \times 2 \times 2 \times 2} = \frac{1}{2^4}$$

So: $$2^{-4} = \frac{1}{2^4}$$

■ **In general** $x^{-n} = \dfrac{1}{x^n}$

Example 3 **Fractional indices**

Consider $$x^{\frac{1}{2}} \times x^{\frac{1}{2}} = x^{\frac{1}{2}+\frac{1}{2}} = x^1 = x$$

and $$x^{\frac{1}{3}} \times x^{\frac{1}{3}} \times x^{\frac{1}{3}} = x^{\frac{1}{3}+\frac{1}{3}+\frac{1}{3}} = x^1 = x$$

Since $\sqrt{x} \times \sqrt{x} = x$ also, you can conclude that $x^{\frac{1}{2}} = \sqrt{x}$, the square root of x, and similarly $x^{\frac{1}{3}} = \sqrt[3]{x}$, the cube root of x.

■ **In general** $x^{\frac{1}{n}} = \sqrt[n]{x}$, **the** n**th root of** x.

Notice also that $x^{\frac{1}{3}} \times x^{\frac{1}{3}} = x^{\frac{1}{3}+\frac{1}{3}} = x^{\frac{2}{3}}$.

Now as you have just seen, $x^{\frac{1}{3}}$ is the cube root of x.

So $x^{\frac{2}{3}} = (x^2)^{\frac{1}{3}}$ is the same as $\sqrt[3]{x^2}$.

Also, $x^{\frac{2}{3}} = (x^{\frac{1}{3}})^2$ is the same as $(\sqrt[3]{x})^2$.

■ **In general** $x^{\frac{m}{n}} = (\sqrt[n]{x})^m = \sqrt[n]{x^m}$

Example 4

Evaluate (a) $9^{\frac{1}{2}}$ (b) $125^{\frac{1}{3}}$ (c) $81^{\frac{3}{2}}$ (d) $36^{-\frac{3}{2}}$.

(a) $9^{\frac{1}{2}} = \sqrt{9} = 3$

(b) $125^{\frac{1}{3}} = \sqrt[3]{125} = 5$

(c) $81^{\frac{3}{2}} = (81^{\frac{1}{2}})^3 = (\sqrt{81})^3 = 9^3 = 729$

(d) $36^{-\frac{3}{2}} = \dfrac{1}{36^{\frac{3}{2}}} = \dfrac{1}{(\sqrt{36})^3} = \dfrac{1}{6^3} = \dfrac{1}{216}$

Exercise 1A

Evaluate:

1 $4^{\frac{1}{2}}$	**2** $16^{\frac{1}{2}}$	**3** $8^{\frac{1}{3}}$	**4** $243^{\frac{1}{5}}$	**5** $243^{\frac{3}{5}}$
6 $8^{-\frac{2}{3}}$	**7** 10^0	**8** 10^{-2}	**9** $(-7)^{-3}$	**10** $(6\frac{1}{4})^{\frac{1}{2}}$
11 $(1\frac{7}{9})^{\frac{1}{2}}$	**12** $(\frac{4}{9})^{-\frac{1}{2}}$	**13** $(\frac{8}{27})^{\frac{2}{3}}$	**14** $(-27)^{\frac{2}{3}}$	**15** $(\frac{5}{6})^0$
16 $(\frac{5}{6})^{-1}$	**17** $(\frac{5}{6})^{-2}$	**18** $(15\frac{5}{8})^{\frac{2}{3}}$	**19** $(\frac{16}{25})^{-\frac{1}{2}}$	**20** $(2\frac{7}{81})^{\frac{1}{2}}$
21 $125^{-\frac{2}{3}}$	**22** $1296^{\frac{1}{4}}$	**23** $(\frac{512}{343})^{-\frac{2}{3}}$	**24** $(3\frac{1}{16})^{\frac{3}{2}}$	

1.2 Using and manipulating surds

Some numbers which contain a root sign can be evaluated exactly, for example, $\sqrt{4} = 2$, $\sqrt{100} = 10$, $\sqrt[3]{27} = 3$, $\sqrt[5]{\frac{1}{32}} = \frac{1}{2}$. They can be written as exact numbers.

This is because each of these numbers is **rational**. However, some numbers which contain a root sign cannot be evaluated exactly because they are **irrational** – they cannot be written exactly as a decimal. Hence $\sqrt{3} = 1.73205\ldots$ and $\sqrt[3]{7} = 1.9129\ldots$ These decimals go on for ever without repeating. Such numbers are called **surds**.

The answers to many mathematical problems appear more elegant than would otherwise be the case if they are given as surds. Also, when an answer to a problem is given as a surd, it is the *exact* answer to the problem. If the answer to a problem is irrational and is given in decimal form then it can only be given to a specified number of decimal places and so it can only be an approximation to the correct answer. For both these reasons many mathematical questions require the answer to be given as a surd. So you need to learn how to manipulate surds.

Manipulating surds

Here are the rules for the manipulation of surds:

- $\sqrt{ab} = \sqrt{a} \cdot \sqrt{b}$
- $\sqrt{\frac{a}{b}} = \frac{\sqrt{a}}{\sqrt{b}}$
- $a\sqrt{b} + c\sqrt{b} = (a + c)\sqrt{b}$
- $a\sqrt{b} - c\sqrt{b} = (a - c)\sqrt{b}$

Notice that $\sqrt{a} + \sqrt{b}$ is *not* the same as $\sqrt{(a + b)}$.

For example, $\sqrt{25} + \sqrt{144} = 5 + 12 = 17$, which is not the same as

$$\sqrt{(25 + 144)} = \sqrt{169} = 13$$

Example 5

Simplify $\sqrt{80}$.

$$\sqrt{80} = \sqrt{(16 \times 5)} = \sqrt{16} \times \sqrt{5} = 4\sqrt{5}$$

Example 6

Simplify $\frac{\sqrt{63}}{3}$.

$$\frac{\sqrt{63}}{3} = \frac{\sqrt{(9 \times 7)}}{3} = \frac{\sqrt{9} \times \sqrt{7}}{3} = \frac{3 \times \sqrt{7}}{3} = \sqrt{7}$$

Example 7

Simplify $\sqrt{75} + 2\sqrt{48} - 5\sqrt{12}$.

$$\begin{aligned}
&\sqrt{75} + 2\sqrt{48} - 5\sqrt{12} \\
&= \sqrt{(25 \times 3)} + 2\sqrt{(16 \times 3)} - 5\sqrt{(4 \times 3)} \\
&= (\sqrt{25} \times \sqrt{3}) + 2(\sqrt{16} \times \sqrt{3}) - 5(\sqrt{4} \times \sqrt{3}) \\
&= 5\sqrt{3} + 8\sqrt{3} - 10\sqrt{3} \\
&= 3\sqrt{3}
\end{aligned}$$

Rationalising the denominator

When the denominator of a fraction is a surd, it is normal to try to remove the surd from the denominator. This process is called 'rationalising the denominator'. Before you can do this you need to learn three more rules:

- **If a fraction is in the form $\frac{1}{\sqrt{a}}$, then multiply both the top and bottom of the fraction by $\sqrt{a}$.**
- **If a fraction is the form $\frac{1}{a + \sqrt{b}}$, then multiply both the top and the bottom of the fraction by $(a - \sqrt{b})$.**
- **If a fraction is in the form $\frac{1}{a - \sqrt{b}}$, then multiply both the top and bottom of the fraction by $(a + \sqrt{b})$.**

The following examples demonstrate how to rationalise the denominator by applying these three rules.

Example 8

Rationalise the denominator of $\frac{1}{\sqrt{3}}$.

$$\frac{1}{\sqrt{3}} = \frac{1 \times \sqrt{3}}{\sqrt{3} \times \sqrt{3}} = \frac{\sqrt{3}}{3}$$

Example 9

Rationalise the denominator of $\frac{1}{2 - \sqrt{3}}$

$$\begin{aligned}
\frac{1}{2 - \sqrt{3}} = \frac{1 \times (2 + \sqrt{3})}{(2 - \sqrt{3})(2 + \sqrt{3})} &= \frac{2 + \sqrt{3}}{4 - 3} \\
&= 2 + \sqrt{3}
\end{aligned}$$

Example 10

Rationalise the denominator of $\frac{\sqrt{3}-\sqrt{2}}{\sqrt{3}+\sqrt{2}}$.

$$\frac{\sqrt{3}-\sqrt{2}}{\sqrt{3}+\sqrt{2}} = \frac{(\sqrt{3}-\sqrt{2})(\sqrt{3}-\sqrt{2})}{(\sqrt{3}+\sqrt{2})(\sqrt{3}-\sqrt{2})}$$

$$= \frac{3 - 2\sqrt{2}\sqrt{3} + 2}{3 - 2}$$

$$= 5 - 2\sqrt{6}$$

Exercise 1B

Simplify:

1 $\sqrt{27}$ **2** $\sqrt{45}$ **3** $\sqrt{162}$

4 $\sqrt{48}$ **5** $\sqrt{75}$ **6** $\sqrt{147}$

7 $\sqrt{567}$ **8** $\sqrt{112}$ **9** $\frac{\sqrt{12}}{2}$

10 $\frac{\sqrt{98}}{7}$ **11** $\frac{\sqrt{18}}{\sqrt{2}}$ **12** $\frac{\sqrt{27}}{\sqrt{3}}$

13 $\sqrt{12} + 3\sqrt{75}$ **14** $\sqrt{200} + \sqrt{18} - 2\sqrt{72}$

15 $\sqrt{20} + 2\sqrt{45} - 3\sqrt{80}$ **16** $5\sqrt{6} - \sqrt{24} + \sqrt{294}$

17 $\sqrt{63} - 2\sqrt{28} + \sqrt{175}$

Rationalise the denominators:

18 $\frac{1}{\sqrt{2}}$ **19** $\frac{1}{\sqrt{7}}$ **20** $\frac{7}{\sqrt{5}}$

21 $\frac{\sqrt{2}}{3\sqrt{3}}$ **22** $\frac{\sqrt{8}}{\sqrt{32}}$ **23** $\frac{\sqrt{5}}{\sqrt{45}}$

24 $\frac{\sqrt{3}}{\sqrt{21}}$ **25** $\frac{\sqrt{11}}{\sqrt{132}}$ **26** $\frac{1}{1-\sqrt{2}}$

27 $\frac{1}{\sqrt{3}-1}$ **28** $\frac{2}{\sqrt{5}-1}$ **29** $\frac{3}{4-\sqrt{7}}$

30 $\frac{1}{3-\sqrt{7}}$ **31** $\frac{1}{\sqrt{11}-4}$ **32** $\frac{1}{\sqrt{7}-\sqrt{3}}$

33 $\frac{2}{\sqrt{11}-\sqrt{5}}$ **34** $\frac{7}{\sqrt{13}-\sqrt{3}}$ **35** $\frac{7}{2+\sqrt{7}}$

36 $\frac{2-\sqrt{3}}{\sqrt{11}-4}$ **37** $\frac{7\sqrt{2}}{\sqrt{8}-\sqrt{7}}$ **38** $\frac{\sqrt{5}-\sqrt{3}}{\sqrt{5}+\sqrt{3}}$

39 $\frac{\sqrt{13}-\sqrt{7}}{\sqrt{13}+\sqrt{7}}$ **40** $\frac{\sqrt{43}-\sqrt{23}}{\sqrt{43}+\sqrt{23}}$

1.3 Processing polynomials

A **polynomial** is an algebraic expression that is a sum of a number of terms. The name comes from the Greek words *poly* (many) and *nomen* (names or terms). Here is an example:

$$2x^3 + 7x^2 + 5x$$

These are each called **terms**

The most general form of a polynomial is written as:

$$a_n x^n + a_{n-1} x^{n-1} + \ldots + a_2 x^2 + a_1 x + a_0$$

This is called a polynomial in x of **degree** n, meaning the highest power of x is n. In a polynomial, n must be a positive integer (written $n \in \mathbb{Z}^+$) and a_n must not be zero ($a_n \neq 0$). So $2x^3 + 7x^2 + 5x + 2$ is a polynomial of degree 3.

Also $5x^6 + 2x^3 + 4x + 1$ is a polynomial of degree 6, since the expression could be written as

$$5x^6 + 0x^5 + 0x^4 + 2x^3 + 0x^2 + 4x + 1.$$

$2.3x^4 - 1.2x^2 + 7$ is a polynomial of degree 4, but

$$5x^3 + 6x - \frac{2}{x} - \frac{6}{x^5}$$

is *not* a polynomial, as it includes terms such as

$$\frac{2}{x} \text{ and } \frac{6}{x^5}$$

The number a_n is called the **coefficient** of x^n and can be positive or negative.

Adding and subtracting polynomials

To add or subtract two polynomials write one expression underneath the other and align corresponding terms – those in which x is raised to the same power. Leave gaps for missing terms.

Example 11

Add $3x^2 + 2x + 5$ to $6x^3 + x + 7$.

Write this as

$$\begin{array}{r} 3x^2 + 2x + 5 \\ 6x^3 \qquad\quad + \ x + 7 \\ \hline 6x^3 + 3x^2 + 3x + 12 \end{array}$$

Example 12
Add $5x^7 - 2x^3 - 3$ to $6x^6 + 5x^5 - 3x^3 + 2$.

Write this as:

$$\begin{array}{r} 5x^7 \qquad\qquad\qquad - 2x^3 - 3 \\ 6x^6 + 5x^5 - 3x^3 + 2 \\ \hline 5x^7 + 6x^6 + 5x^5 - 5x^3 - 1 \end{array}$$

Example 13
Subtract $2x^3 - 7x + 3$ from $4x^4 - 2x^2 + 3x - 2$.

Write this as:

$$\begin{array}{r} 4x^4 \qquad\quad - 2x^2 + \ 3x - 2 \\ 2x^3 \qquad\quad - \ 7x + 3 \\ \hline 4x^4 - 2x^3 - 2x^2 + 10x - 5 \end{array}$$

When adding or subtracting polynomials it is usually easier to start with each polynomial in its own set of brackets, then remove the brackets and add or subtract the corresponding terms. If you use this method remember that a + sign in front of a bracket leaves the signs inside the bracket unaltered when the bracket is removed, but a − sign in front of the bracket changes each of the signs inside the bracket when the bracket is removed.

Example 14
Simplify $(2x^2 - 3x + 2) + (-3x^2 + 2x - 6)$.

$$\begin{aligned} &(2x^2 - 3x + 2) + (-3x^2 + 2x - 6) \\ &= 2x^2 - 3x + 2 - 3x^2 + 2x - 6 \\ &= -x^2 - x - 4 \end{aligned}$$

Example 15
Add $(3x^2 - 2)$, $(6x^3 - 2x^2 + 1)$, $(4x^2 - 2x + 3)$ and $(6x - 7)$.

$$\begin{aligned} &(3x^2 - 2) + (6x^3 - 2x^2 + 1) + (4x^2 - 2x + 3) + (6x - 7) \\ &= 3x^2 - 2 + 6x^3 - 2x^2 + 1 + 4x^2 - 2x + 3 + 6x - 7 \\ &= 6x^3 + 5x^2 + 4x - 5 \end{aligned}$$

Example 16
Subtract $(8x^4 - 2x^2 - 3x)$ from $(6x^3 - 7x + 2)$.

$$\begin{aligned} &(6x^3 - 7x + 2) - (8x^4 - 2x^2 - 3x) \\ &= 6x^3 - 7x + 2 - 8x^4 + 2x^2 + 3x \\ &= -8x^4 + 6x^3 + 2x^2 - 4x + 2 \end{aligned}$$

Example 17
Simplify $(2x^3 + 6x - 2) - (3x^2 - 4x + 7) - (6x - 3) + (2x^2 + 5x)$.

Removing the brackets:

$$2x^3 + 6x - 2 - 3x^2 + 4x - 7 - 6x + 3 + 2x^2 + 5x$$
$$= 2x^3 - x^2 + 9x - 6$$

Multiplying one polynomial by another

You can use similar methods to multiply two polynomials.

Example 18
Multiply $x^3 - 2x + 4$ by $3x - 7$.

Write this as:		x^3		$-\ 2x$	$+\ 4$
				$3x$	$-\ 7$
Multiplying by $3x$:	$3x^4$		$-\ 6x^2$	$+\ 12x$	
Multiplying by -7:		$-\ 7x^3$		$+\ 14x$	$-\ 28$
Adding:	$3x^4$	$-\ 7x^3$	$-\ 6x^2$	$+\ 26x$	$-\ 28$

Example 19
Multiply $2x^5 + 7x^3 + 2$ by $8x^3 - 2x - 3$.

Write:			$2x^5$		$+\ 7x^3$		$+\ 2$
					$8x^3$	$-\ 2x$	$-\ 3$
Multiplying by $8x^3$:	$16x^8$	$+\ 56x^6$			$+\ 16x^3$		
Multiplying by $-2x$:		$-4x^6$		$-\ 14x^4$		$-\ 4x$	
Multiplying by -3:			$-6x^5$		$-\ 21x^3$		$-\ 6$
Adding:	$16x^8$	$+\ 52x^6$	$-\ 6x^5$	$-\ 14x^4$	$-\ 5x^3$	$-\ 4x$	$-\ 6$

As with addition and subtraction, when multiplying polynomials it is usually quicker to start with each polynomial in its own set of brackets, then remove the brackets and collect corresponding terms. Once again, remember that a + sign outside the bracket leaves the signs inside the bracket unaltered and a − sign outside the bracket changes all the signs inside the bracket.

Example 20
Multiply $(2x^2 - 3x + 4)$ by $(3x^3 - x + 1)$.

$$(2x^2 - 3x + 4)(3x^3 - x + 1)$$
$$= 2x^2(3x^3 - x + 1) - 3x(3x^3 - x + 1) + 4(3x^3 - x + 1)$$
$$= 6x^5 - 2x^3 + 2x^2 - 9x^4 + 3x^2 - 3x + 12x^3 - 4x + 4$$
$$= 6x^5 - 9x^4 + 10x^3 + 5x^2 - 7x + 4$$

Example 21

Multiply $(5x^2 - 2x + 3)$ by $(2x^3 - 4x^2 + 2x - 3)$.

$$(5x^2 - 2x + 3)(2x^3 - 4x^2 + 2x - 3)$$
$$= 5x^2(2x^3 - 4x^2 + 2x - 3) - 2x(2x^3 - 4x^2 + 2x - 3) + 3(2x^3 - 4x^2 + 2x - 3)$$
$$= 10x^5 - 20x^4 + 10x^3 - 15x^2 - 4x^4 + 8x^3 - 4x^2 + 6x + 6x^3 - 12x^2 + 6x - 9$$
$$= 10x^5 - 24x^4 + 24x^3 - 31x^2 + 12x - 9$$

Exercise 1C

1 Add $(5x^7 + 2x^5 + 3x^2)$ and $(6x^7 + 2x^6 + 3x^5 + 2x)$

2 Add $(3x^5 - 2x^2 + 6x)$ and $(4x^4 - 2x^3 + 3x^2 + 2)$

3 Add $(2x^3 - 3x + 1)$, $(2x^2 + x - 6)$ and $(2x - 3)$

4 Add $(8x^4 - 3x^2 + 6x + 1)$, $(5x^3 - 4x^2 - 2x + 2)$ and $(-2x^4 + 6x - 3)$

5 Simplify $(5x^6 - 2x^3 + 7x) + (3x^5 - 2x^4 + 3x^3 + 7)$

6 Simplify $(-3x^4 + 2x^2 + 7x + 2) + (8x^3 - 3x^2 + 2x - 5)$

7 Simplify $(5x^7 - 6x^3 + 7) + (-6x^6 + 2x^3 - 9) + (5x^4 - 2x - 3)$

8 Simplify $(2x^2 - 3) + (3x^2 + 2x - 1) + (-2x^3 - 2x^2 + 2x) + (6x - 3)$

9 Subtract $(4x^5 - 2x^3 + 7)$ from $(5x^5 + 7x - 2)$

10 Subtract $(3x^3 + 7x^2 + 2)$ from $(6x^3 + 7x + 3)$

11 Subtract $(2x^5 - 3x^3 - x - 2)$ from $(6x^3 - 9x^2 + 2x - 3)$

12 Subtract $(-2x^3 - 4x^2 - 3x + 2)$ from $(-3x^5 + 2x^3 - 4x^2 + 6x - 7)$

13 Simplify $(6x^4 + 3x^2 - 2x + 9) - (3x^3 - 7x^2 - 6x + 7)$

14 Simplify $(5x^3 - 2x + 1) - (2x^3 - 7x + 3)$

15 Simplify $(6x^5 - 2x + 2) - (3x^3 - 2x - 5) - (7x + 4)$

16 Simplify $(4x^6 - 2x + 1) + (3x^3 + 7x - 4) - (-3x^6 - 2x + 7)$

17 Simplify $(8x^4 - 2x^2 + 7) - (6x^3 - 3x^2 - 1) + (6x + 4) - (3x^2 - 2x - 3)$

18 Multiply $(3x^2 + 2x + 4)$ by $(4x^3 + 7x + 6)$

19 Multiply $(2x^3 - 6x - 3)$ by $(5x^4 + 2x^2 - x)$

20 Multiply $(4x^2 + 2x - 1)$ by $(6x^2 - x + 7)$

21 Multiply $(3x^3 - 2x - 6)$ by $(2x^2 - 3x + 1)$

22 Simplify $(8x^5 - 2x^4 + 3x - 7)(2x^3 - 3x^2 - 9x + 6)$

23 Simplify $(2x^4 - 7x + 6)(x^3 - 2x^2 + 7x - 3)$

24 Simplify $(3x^2 - 2x - 4)(8x^5 - 3x^3 - 7x + 3)$

25 Simplify $(-7x + 2x^2 - 6x^3)(-x^4 + 6x^2 - 2x - 5)$

1.4 Factorising polynomials

From the work done so far you should be able to show that:

$(2x^2 - 1)(3x + 4)$ is equivalent to $6x^3 + 8x^2 - 3x - 4$.

Because these two expressions are equal for *all* values of x (not just for some values of x as in the case of equations) we say that:

$(2x^2 - 1)(3x + 4)$ **is identically equal to** $6x^3 + 8x^2 - 3x - 4$

$(2x^2 - 1)(3x + 4) \equiv 6x^3 + 8x^2 - 3x - 4$ is an **identity**. The identity symbol is $\equiv$.

You should by now be able to check that the following are identities (that is, they are true for all values of x and y).

$$(x + y)^2 \equiv (x + y)(x + y) \equiv x^2 + 2xy + y^2$$

$$(x - y)^2 \equiv (x - y)(x - y) \equiv x^2 - 2xy + y^2$$

$$(x - y)(x + y) \equiv x^2 - y^2$$

$$(x + y)^3 \equiv (x + y)(x + y)(x + y) \equiv x^3 + 3x^2y + 3xy^2 + y^3$$

$$(x - y)^3 \equiv (x - y)(x - y)(x - y) \equiv x^3 - 3x^2y + 3xy^2 - y^3$$

The **factors** of each expression are shown on the left-hand side. Multiplying the factors together gives the full expression. So $(x - y)$ and $(x + y)$ are the factors of $x^2 - y^2$. The process of taking an expression and writing it as a product of its factors is called **factorisation**. So if you factorise $x^2 - y^2$ you get $(x - y)(x + y)$.

The simplest type of factor is the **common factor**. As its name suggests, it is a factor which is common to each term in a polynomial.

Example 22
Factorise $4x^3 + 7x$.

Each term in the polynomial contains the factor x so $4x^3 + 7x \equiv x(4x^2 + 7)$.

Example 23
Factorise $3x^3 + 6x^2 - 12x$.

Each term in the polynomial contains the factor $3x$ so $3x^3 + 6x^2 - 12x \equiv 3x(x^2 + 2x - 4)$.

Many other polynomials can be factorised using the five identities given above.

Example 24

Factorise $4x^2 + 12x + 9$.

If you write $4x^2 + 12x + 9$ as $(2x)^2 + 2(2x)(3) + 3^2$, then you can compare it with the identity $X^2 + 2XY + Y^2 \equiv (X + Y)^2$. In this case $X = 2x$ and $Y = 3$.

So: $$4x^2 + 12x + 9 \equiv (2x + 3)^2$$

Example 25

Factorise $4x^2 - 1$.

If $4x^2 - 1$ is written as $(2x)^2 - 1^2$ then it may be compared with the identity $X^2 - Y^2 \equiv (X - Y)(X + Y)$. Here, $X = 2x$ and $Y = 1$.

So $$4x^2 - 1 \equiv (2x - 1)(2x + 1)$$

Example 26

Factorise $x^3 - 12x^2 + 48x - 64$.

Write $x^3 - 12x^2 + 48x - 64$ as $x^3 - 3x^2(4) + 3x(4)^2 - (4)^3$.

You can then compare it with the standard identity

$$(X - Y)^3 \equiv X^3 - 3X^2Y + 3XY^2 - Y^3$$

where $X = x$ and $Y = 4$.

So: $$x^3 - 12x^2 + 48x - 64 \equiv (x - 4)^3$$

A polynomial of the form $ax^2 + bx + c$ is called a **trinomial** as it contains three terms. Many trinomials can be factorised. But not many can be factorised as a perfect square using

$$(x + y)^2 \equiv x^2 + 2xy + y^2 \text{ or } (x - y)^2 \equiv x^2 - 2xy + y^2$$

These are special cases.

However, you should be able to show that:

$$(x + a)(x + b) \equiv x^2 + (a + b)x + ab$$

Polynomials of degree two which contain three terms (i.e. trinomials) and which have the coefficient of x^2 as 1 can often be factorised using this identity.

Example 27

Factorise $x^2 + 5x + 6$.

If you compare $x^2 + 5x + 6$ with $x^2 + (a + b)x + ab$, you can see that $ab = 6$. If you assume that a and b are integers then either $a = 6$ and $b = 1$, or $a = 1$ and $b = 6$, or $a = 2$ and $b = 3$, or $a = 3$ and $b = 2$. If you take $a = 2$ and $b = 3$ then the middle term, $(a + b)x$, becomes $5x$, as required.

So: $$x^2 + 5x + 6 \equiv (x + 2)(x + 3)$$

Example 28
Factorise $x^2 + 3x - 28$.

Compare $x^2 + 3x - 28$ with $x^2 + (a + b)x + ab$, and you see that $ab = -28$. So $a = -28$, $b = 1$ or $a = 28$, $b = -1$, or $a = -2$, $b = 14$ or $a = 2$, $b = -14$ or $a = -4$, $b = 7$ or $a = 4$, $b = -7$. If you choose $a = -4$, $b = 7$ you find that $(a + b)x$ equates to $3x$ as required.

So: $$x^2 + 3x - 28 \equiv (x - 4)(x + 7)$$

Trinomials that need to be factorised often have a coefficient of x^2 that is not 1. In such cases the identity

$$(ax + b)(cx + d) \equiv acx^2 + (ad + bc)x + bd$$

can be used.

Example 29
Factorise $3x^2 - 4x - 4$.

Comparing $3x^2 - 4x - 4$ with $acx^2 + (ad + bc)x + bd$ you find that if $a = 1$ and $c = 3$ (to give $ac = 3$) and if $b = -2$ and $d = 2$ (to give $bd = -4$) then the middle term $(ad + bc)x$ is, in this case,

$$[(1 \times 2) + (-2 \times 3)]x = -4x$$

as required.

So: $$3x^2 - 4x - 4 \equiv (x - 2)(3x + 2)$$

Example 30
Factorise $15x^2 - 11x - 12$.

Compare $15x^2 - 11x - 12$ with $acx^2 + (ad + bc)x + bd$.

Then $ac = 15$ and $bd = -12$.

If $a = 3$ and $c = 5$ (to give $ac = 15$), and if $b = -4$ and $d = 3$ (to give $bd = -12$) then $(ad + bc)x$ becomes

$$[(3 \times 3) + (-4 \times 5)]x = -11x$$

as required.

So: $$15x^2 - 11x - 12 \equiv (3x - 4)(5x + 3)$$

It is extremely important that you have a clear, confident strategy for factorising polynomials and particularly trinomials. The following exercise is intended to give you plenty of opportunity for practice, which is the sure route towards building your own successful strategy.

Exercise 1D

Factorise completely:

1 $2x^2 - 3x^3$ **2** $5y^5 + 2y^2 + 7y$ **3** $2x^2 + 18x$
4 $40y - 8$ **5** $6xy + 18x$ **6** $5xy^3 + 15x^2y$
7 $2x - 6y + 10xy$ **8** $x^3 + 3x^2 - 6x$ **9** $6pq^2 + 9p^2q$
10 $2c(c + d) + 4cd$ **11** $x^2 - 36$ **12** $y^2 - 81$
13 $100 - b^2$ **14** $8a^2 - 18b^2$ **15** $18 - 50b^2$
16 $a^2 - a^2b^2$ **17** $2x^2 - 18$ **18** $3x^2 - 27$
19 $17a^2 - 68$ **20** $48 - 147c^2$ **21** $x^2 + 6x + 9$
22 $x^2 - 8x + 16$ **23** $x^2 + 14x + 49$ **24** $x^2 - 18x + 81$
25 $3x^2 - 36x + 108$ **26** $9x^2 + 30x + 25$
27 $4x^2 - 28x + 49$ **28** $45x^2 - 30x + 5$
29 $6x^2 + 12x + 2$ **30** $25x^2 - 30x + 9$
31 $25x^2 + 20x + 4$ **32** $9x^2 - 42x + 49$
33 $x^3 + 6x^2 + 12x + 8$ **34** $x^3 - 9x^2 + 27x - 27$
35 $8x^3 + 12x^2 + 6x + 1$ **36** $27x^3 - 27x^2 + 9x - 1$
37 $8x^3 + 36x^2 + 54x + 27$ **38** $8x^3 - 60x^2 + 150x - 125$
39 $x^3y + 6x^2y + 12xy + 8y$ **40** $3x^3 - 27x^2 + 81x - 81$
41 $x^2 + 9x + 14$ **42** $x^2 - 8x + 15$ **43** $x^2 - x - 12$
44 $x^2 + 13x + 12$ **45** $x^2 - 9x - 10$ **46** $x^2 - 3x + 2$
47 $x^2 + 15x + 56$ **48** $x^2 + 4x - 12$ **49** $x^2 + 7x + 10$
50 $x^2 - 3x - 40$ **51** $x^2 - 7x + 12$ **52** $x^2 - 14x + 45$
53 $x^2 + 11x + 18$ **54** $x^2 - 10x + 21$ **55** $x^2 + 9x + 20$
56 $x^2 - 17x + 30$ **57** $x^2 - 22x + 120$ **58** $x^2 + 3x - 180$
59 $x^2 - x - 110$ **60** $x^2 + 11x - 620$ **61** $3x^2 - 4x + 1$
62 $2x^2 - x - 1$ **63** $8x^2 + 6x + 1$ **64** $4x^2 + 3x - 1$
65 $14x^2 + 5x - 1$ **66** $14x^2 - 5x - 1$ **67** $18x^2 - 3x - 1$
68 $18x^2 + 9x + 1$ **69** $2x^2 - 24x + 70$ **70** $3x^2 - 6x - 144$
71 $2x^2 + 7x + 6$ **72** $2x^2 + 3x - 9$ **73** $3x^2 + 7x + 2$
74 $5x^2 - 14x - 3$ **75** $3y^2 + 13y + 14$ **76** $2y^2 - 9y - 5$
77 $7y^2 - 30y + 8$ **78** $5y^2 - 27y + 10$ **79** $3y^2 - 10y - 25$
80 $3y^2 + 4y - 4$ **81** $4x^2 + 8x + 3$ **82** $6x^2 - x - 2$
83 $8x^2 + 14x - 15$ **84** $10x^2 - 23x + 12$ **85** $12x^2 - 5x - 3$
86 $10x^2 + 11x - 35$ **87** $12x^2 - 11x + 2$ **88** $9x^2 + 21x + 10$
89 $4x^2 + 8x - 21$ **90** $35x^2 + x - 6$ **91** $x^3 - 4x^2 - 21x$
92 $6x - 7x^2 + 2x^3$ **93** $10x^5 + 9x^4 + 2x^3$ **94** $12 - 4x - 40x^2$
95 $6x^2y + 19xy - 7y$ **96** $3x^4 + 4x^2 + 1$ **97** $6y^2 - 5y - 6$
98 $6y^2 + 13y + 6$ **99** $28x^2 - 19x - 20$ **100** $4x^4 - 5x^2 + 1$

1.5 Identities

In your work on factorising polynomials, you met the word **identity**. Identities are true for all values of the variable and, in particular, a factorised form of a polynomial is identically equal to its expanded form for every value of the variable. As a very simple illustration you could say:

the **equation** $2x + 4 = 0$ is true for just one value of x;
that is, $x = -2$, often called the **solution** of the equation;
the **identity** $2x + 4 \equiv 2(x + 2)$ is true for *all values of x*.

As identities are true for all values of the variable, or variables, you can equate the coefficients of respective terms when required. The following examples show this method.

Example 31

Find the constants A and B such that $A(x + 2) + B(x + 1) \equiv x$.

The left-hand side can be written as

$$\begin{aligned} &Ax + 2A + Bx + B \\ &= Ax + Bx + 2A + B \\ &= (A + B)x + (2A + B) \end{aligned}$$

So: $$(A + B)x + (2A + B) \equiv x$$

Equate coefficients of x:

$$A + B = 1 \qquad (1)$$

Equate the constant terms:

$$2A + B = 0 \qquad (2)$$

There are now two equations in A and B which can be solved simultaneously.

For a reminder about solving equations simultaneously see page 211.

Subtract (1) from (2):

$$A = -1$$

Substitute the value of A in (1):

$$\begin{aligned} -1 + B &= 1 \\ B &= 2 \end{aligned}$$

Example 32

Find the values of A and B for which

$$A(4 - x) + B(x - 7) \equiv 30 - 6x$$

The left-hand side can be rewritten:

$$\begin{aligned} &4A - Ax + Bx - 7B \\ &= (-A + B)x + (4A - 7B) \end{aligned}$$

So we have: $$(-A + B)x + (4A - 7B) \equiv -6x + 30$$

Equate coefficients of x:

$$-A + B = -6 \quad (1)$$

Equate the constant terms:

$$4A - 7B = 30 \quad (2)$$

Multiply (1) by 4:

$$-4A + 4B = -24 \quad (3)$$

Add (3) and (2):

$$-3B = 6$$
$$B = -2$$

Substitute the value of B into (1):

$$-A - 2 = -6$$
$$A = 4$$

Example 33

Find the values of A, B and C such that

$$A(x^2 + 4) + (x - 2)(Bx + C) \equiv 7x^2 - x + 14$$

Rewrite the left-hand side (LHS) as a polynomial in descending powers of x:

$$\text{LHS} = Ax^2 + 4A + Bx^2 - 2Bx + Cx - 2C$$
$$= (A + B)x^2 + (-2B + C)x + (4A - 2C)$$

So: $\quad (A + B)x^2 + (-2B + C)x + (4A - 2C) \equiv 7x^2 - x + 14$

Equate coefficients of x^2:

$$A + B = 7 \quad (1)$$

Equate coefficients of x:

$$-2B + C = -1 \quad (2)$$

Equate the constant terms:

$$4A - 2C = 14 \quad (3)$$

You have three equations in three unknowns to be solved simultaneously. Substitute the expression for A given in (1) into equation (3) so that you have two equations in B and C which you can solve in the usual way.

From (1): $\quad A = 7 - B$

Substitute in (3):

$$4(7 - B) - 2C = 14$$
$$28 - 4B - 2C = 14$$
$$-4B - 2C = -14$$
$$4B + 2C = 14$$
$$2B + C = 7 \quad (4)$$

Add (2) and (4): $$2C = 6$$
$$C = 3$$

Substitute in (2): $$-2B + 3 = -1$$
$$-2B = -4$$
$$B = 2$$

Substitute in (1): $$A + 2 = 7$$
$$A = 5$$

Example 34

Find the constants A, B and C such that

$$A(x+2)(x+3) + B(x+1)(x+3) + C(x+1)(x+2) \equiv 4x + 6$$

Rearrange the left-hand side as a polynomial in descending powers of x:

$$A(x^2 + 5x + 6) + B(x^2 + 4x + 3) + C(x^2 + 3x + 2)$$
$$= (A + B + C)x^2 + (5A + 4B + 3C)x + (6A + 3B + 2C)$$

So: $(A + B + C)x^2 + (5A + 4B + 3C)x + (6A + 3B + 2C) \equiv 4x + 6$

Equate coefficients of x^2:

$$A + B + C = 0 \qquad (1)$$

Equate coefficients of x:

$$5A + 4B + 3C = 4 \qquad (2)$$

Equate constant terms:

$$6A + 3B + 2C = 6 \qquad (3)$$

From (1): $$A = -B - C$$

Substitute into (2) and (3):

$$5(-B - C) + 4B + 3C = 4$$
$$-5B - 5C + 4B + 3C = 4$$
$$-B - 2C = 4 \qquad (4)$$

$$6(-B - C) + 3B + 2C = 6$$
$$-6B - 6C + 3B + 2C = 6$$
$$-3B - 4C = 6 \qquad (5)$$

Multiply (4) by 2: $$-2B - 4C = 8 \qquad (6)$$

Subtract (6) from (5):

$$-B = -2$$
$$B = 2$$

Substitute for B in (6):

$$-4 - 4C = 8$$
$$4C = -12$$
$$C = -3$$

Substitute for B and C in (1):

$$A = -2 + 3$$
$$A = 1$$

Another method used in the processing of identities is to substitute particular values of the variable into the identity. This is allowed, since an identity is true for *all* values of the variable. Your aim is to choose values to substitute that will make some of A, B and C disappear from the identity. This will make it easier to find the others.

Example 35

Find the constants A and B such that $A(x + 2) + B(x + 1) \equiv x$.

Substitute $x = -1$ since this makes the bracket $(x + 1)$ zero.

$$A(-1 + 2) + B(-1 + 1) = -1$$
$$A = -1$$

Substitute $x = -2$ and the bracket $(x + 2)$ becomes zero:

$$A(-2 + 2) + B(-2 + 1) = -2$$
$$-B = -2$$
$$B = 2 \text{ as in example 31}$$

Example 36

Find the values of A and B for which

$$A(4 - x) + B(x - 7) \equiv 30 - 6x$$

Substitute $x = 4$:

$$A(4 - 4) + B(4 - 7) = 30 - 24$$
$$-3B = 6$$
$$B = -2$$

Substitute $x = 7$:

$$A(4 - 7) + B(7 - 7) = 30 - 42$$
$$-3A = -12$$
$$A = 4 \text{ as in example 32}$$

Example 37

Find the values of A, B and C such that

$$A(x+2)(x+3) + B(x+1)(x+3) + C(x+1)(x+2) \equiv 4x+6$$

Substitute $x = -1$:

$$A(-1+2)(-1+3) + B(-1+1)(-1+3) + C(-1+1)(-1+2) = -4+6$$

$$A(1)(2) + B(0)(2) + C(0)(1) = 2$$

$$2A = 2$$

$$A = 1$$

Substitute $x = -2$:

$$A(-2+2)(-2+3) + B(-2+1)(-2+3) + C(-2+1)(-2+2) = -8+6$$

$$A(0)(1) + B(-1)(1) + C(-1)(0) = -2$$

$$-B = -2$$

$$B = 2$$

Substitute $x = -3$:

$$A(-3+2)(-3+3) + B(-3+1)(-3+3) + C(-3+1)(-3+2) = -12+6$$

$$A(-1)(0) + B(-2)(0) + C(-2)(-1) = -6$$

$$2C = -6$$

$$C = -3 \text{ as in example 34}$$

You will gather from the above examples that, in general, it is much easier to solve problems like this by substitution than by equating coefficients. However, sometimes the method of substitution will not work completely. For example, if one of the brackets in the identity is, say, (x^2+9) there is no real value of x that will make this bracket zero. For if $x^2+9=0$ then $x^2=-9$ and a negative number has no real square roots. Under these circumstances it is usually easier to use the method of substitution as far as you can and then finish off the problem by equating coefficients.

Example 38

Find the values of A, B and C such that

$$A(x^2+4) + (x-2)(Bx+C) \equiv 7x^2 - x + 14$$

Substitute $x = 2$:

$$A(4+4) + (2-2)(2B+C) = 28 - 2 + 14$$

$$8A = 40$$

$$A = 5$$

Since there is no real value of x that will make (x^2+4) equate to zero, revert to equating coefficients:

$$A(x^2+4)+(x-2)(Bx+C)$$
$$=Ax^2+4A+Bx^2-2Bx+Cx-2C$$
$$=(A+B)x^2+(-2B+C)x+(4A-2C)$$

But since $A=5$ the LHS of the identity is:

$$(5+B)x^2+(-2B+C)x+(20-2C)$$

So: $\quad (5+B)x^2+(-2B+C)x+(20-2C)\equiv 7x^2-x+14$

Equate coefficients of x^2:

$$5+B=7$$
$$B=2$$

Equate the constant terms:

$$20-2C=14$$
$$C=3$$

Another approach would be to take two other values of x and form equations that can be solved to find B and C.

For $x=0$, $\quad 4A-2C=14$

So: $\quad 20-2C=14$

and: $\quad C=3$

For $x=1$, $\quad 5A-B-C=20$

So: $\quad 25-B-3=20$

and: $\quad B=2$

You should appreciate that the two approaches are essentially equivalent.

Example 39

Find the values of A, B and C for which

$$A(2x^2+1)+(2x-3)(Bx+C)\equiv 11x$$

Substitute $x=1\frac{1}{2}$:

$$A(\tfrac{18}{4}+1)+(3-3)(1\tfrac{1}{2}B+C)=16\tfrac{1}{2}$$
$$5\tfrac{1}{2}A=16\tfrac{1}{2}$$
$$A=3$$

Since there is no real number that will make $(2x^2+1)$ equate to zero, revert to the method of equating coefficients.

$$\begin{aligned} &A(2x^2 + 1) + (2x - 3)(Bx + C) \\ &= 2Ax^2 + A + 2Bx^2 - 3Bx + 2Cx - 3C \\ &= (2A + 2B)x^2 + (-3B + 2C)x + (A - 3C) \end{aligned}$$

Since $A = 3$, this can be written:

$$(6 + 2B)x^2 + (-3B + 2C)x + (3 - 3C)$$

So: $$(6 + 2B)x^2 + (-3B + 2C)x + (3 - 3C) \equiv 11x$$

Equate coefficients of x^2:

$$6 + 2B = 0$$
$$B = -3$$

Equate the constant terms:

$$3 - 3C = 0$$
$$C = 1$$

Exercise 1E

1 Each of the following is either
(i) true for all values of x, or
(ii) true for some values of x, or
(iii) true for no values of x.
Distinguish which is which and explain why.
(a) $(3x - 1)^2 = 9x^2 - 1$
(b) $(3x - 1)^2 = 25$
(c) $(3x - 1)^2 = 9x^2 - 6x + 1$
(d) $(2x + 3)^2 = x^2 - 4$
(e) $x^2 + (x + 2)^2 = (x + 1)^2$

In questions **2** to **21** find the values of the constants A, B, and C:

2 $A(x + 3) + B(x + 2) \equiv 4x + 9$
3 $A(x + 5) + B(x + 2) \equiv x + 8$
4 $A(x + 2) + B(x - 1) \equiv 6x + 3$
5 $A(x + 4) + B(x + 2) \equiv x + 12$
6 $A(x + 1) + B(x - 3) \equiv 8x + 16$
7 $A(x + 3)(x + 4) + B(x + 2)(x + 4) + C(x + 2)(x + 3) \equiv 6x^2 + 34x + 46$
8 $A(x + 1)(x + 3) + B(x - 1)(x + 1) + C(x - 1)(x + 3) \equiv 6x + 2$
9 $A(x + 2)(x + 3) + B(x + 1)(x + 3) + C(x + 1)(x + 2) \equiv 4x + 6$
10 $A(x + 2)(x - 1) + B(2x + 1)(x - 1) + C(2x + 1)(x + 2) \equiv 4x^2 - 17x - 14$
11 $A(x - 1)(x + 4) + B(2x + 1)(x + 4) + C(2x + 1)(x - 1) \equiv 12x^2 + 59x - 26$

12 $(Ax+B)(x+1)+C(x^2+3)\equiv x-3$

13 $(Ax+B)(x^2+3)+Cx^2(5x-2)\equiv -13x^3+18x^2+6x+36$

14 $A(x^2+x+3)+(Bx+C)(2x+1)\equiv x^2-x+2$

15 $A(x-2)^2+B(x+1)(x-2)+C(x+1)\equiv 3$

16 $A(x^2+4)+(x-2)(Bx+C)\equiv 4x^2-7x+22$

17 $A(x^2+x+2)+(Bx+C)x\equiv -x^2-2x+4$

18 $A(x^2+x+1)+(Bx+C)(2x-5)\equiv 7x^2-7x+3$

19 $A(x^2+1)+(Bx+C)(2x+3)\equiv 8x^2+4x+1$

20 $A(x^2+5)+(Bx+C)(x-3)\equiv 3x^2+5x+28$

21 $(Ax+B)(3x-4)+C(x^2+x+1)\equiv x^2-12x+6$

22 Find the values of A, B and C for which

$$4x^2-12x+25\equiv A(x+B)^2+C$$

Hence find the minimum value of $4x^2-12x+25$.

23 Find the values of A, B and C for which

$$3x^2+18x-5\equiv A(x+B)^2+C$$

Hence find the minimum value of $3x^2+18x-5$.

24 Find the values of A, B and C for which

$$16+4x-x^2\equiv A-(B-Cx)^2$$

Hence find the maximum value of $16+4x-x^2$.

25 Find the values of A and B for which

$$9x^2+30x+A\equiv(3x+B)^2$$

1.6 Algebraic division

Not many years ago questions such as 'divide 12 603 by 14' were commonplace in mathematics lessons, not only in secondary schools but also in primary schools. The method that was used was called long division. Students used to practise questions using long division for weeks on end!

Today most students have an electronic calculator that produces an answer to such questions in a fraction of a second and the method of long division has become largely redundant. However, most calculators today are still unable to divide one algebraic expression by another and so the method of long division has to be used in these circumstances. It is a technique with which you therefore need to become familiar.

The method of long division is best explained by using an example.

Example 40

Divide $2x^4 - 9x^3 + 13x^2 - 17x + 15$ by $x - 3$.

You first need to ask, 'If I divide the first term of $2x^4 - 9x^3 + 13x^2 - 17x + 15$ by the first term of $x - 3$, what will be the answer?' In this case, $2x^4$ divided by x gives $2x^3$. Now multiply this answer of $2x^3$ by $x - 3$ to give $2x^4 - 6x^3$.

For a reminder about manipulating indices see page 202.

The solution thus far is written:

$$\begin{array}{rl} & 2x^3 \\ x-3 & \overline{)2x^4 - 9x^3 + 13x^2 - 17x + 15} \\ & \;\; 2x^4 - 6x^3 \end{array}$$

Notice that $2x^3$, the term you obtain when you divide $2x^4$ by x, is written immediately above $2x^4$. Notice also that the terms of $2x^4 - 6x^3$ are written immediately underneath the corresponding terms of $2x^4 - 9x^3 + 13x^2 - 17x + 15$. In other words, $2x^4$ is written underneath $2x^4$ and $-6x^3$ is written underneath $-9x^3$.

You now subtract $2x^4 - 6x^3$ from $2x^4 - 9x^3$.

So:

$$\begin{aligned} & 2x^4 - 9x^3 - (2x^4 - 6x^3) \\ &= 2x^4 - 9x^3 - 2x^4 + 6x^3 \\ &= -3x^3 \end{aligned}$$

The solution now looks like this:

$$\begin{array}{rl} & 2x^3 \\ x-3 & \overline{)2x^4 - 9x^3 + 13x^2 - 17x + 15} \\ & \;\; \underline{2x^4 - 6x^3} \\ & \qquad -3x^3 \end{array}$$

At this stage copy the next term of $2x^4 - 9x^3 + 13x^2 - 17x + 15$ that does not have anything written directly underneath it, next to $-3x^3$. In this case the $13x^2$ needs to be copied:

$$\begin{array}{rl} & 2x^3 \\ x-3 & \overline{)2x^4 - 9x^3 + 13x^2 - 17x + 15} \\ & \;\; \underline{2x^4 - 6x^3} \\ & \qquad -3x^3 + 13x^2 \end{array}$$

Now repeat the whole process from the beginning. So ask, 'If I divide the first term of $-3x^3 + 13x^2$ by the first term of $x - 3$, what will be the answer?' This time the answer is $-3x^2$, which you write next to $2x^3$. You then multiply $-3x^2$ by $x - 3$ to give $-3x^3 + 9x^2$.

Write these two terms underneath the corresponding terms of $-3x^3 + 13x^2$ and subtract:

$$\begin{aligned} & -3x^3 + 13x^2 - (-3x^3 + 9x^2) \\ &= -3x^3 + 13x^2 + 3x^3 - 9x^2 \\ &= 4x^2 \end{aligned}$$

The solution now looks like this:

$$\begin{array}{r@{\,}l} & \underline{2x^3 - 3x^2 } \\ x - 3 \,\big) & 2x^4 - 9x^3 + 13x^2 - 17x + 15 \\ & \underline{2x^4 - 6x^3} \\ & -3x^3 + 13x^2 \\ & \underline{-3x^3 + 9x^2} \\ & 4x^2 \end{array}$$

Copy the next term from $2x^4 - 9x^3 + 13x^2 - 17x + 15$ that has nothing written directly underneath it next to $4x^2$, and repeat the process.

$$\begin{array}{r@{\,}l} & \underline{2x^3 - 3x^2 + 4x } \\ x - 3 \,\big) & 2x^4 - 9x^3 + 13x^2 - 17x + 15 \\ & \underline{2x^4 - 6x^3} \\ & -3x^3 + 13x^2 \\ & \underline{-3x^3 + 9x^2} \\ & 4x^2 - 17x \\ & \underline{4x^2 - 12x} \\ & - 5x \end{array}$$

Repeat the process once more:

$$\begin{array}{r@{\,}l} & \underline{2x^3 - 3x^2 + 4x - 5 } \\ x - 3 \,\big) & 2x^4 - 9x^3 + 13x^2 - 17x + 15 \\ & \underline{2x^4 - 6x^3} \\ & - 3x^3 + 13x^2 \\ & \underline{- 3x^3 + 9x^2} \\ & 4x^2 - 17x \\ & \underline{4x^2 - 12x} \\ & - 5x + 15 \\ & \underline{- 5x + 15} \end{array}$$

There are no more terms to copy so the solution is finished. Thus $2x^4 - 9x^3 + 13x^2 - 17x + 15$ divided by $x - 3$ gives a result of $2x^3 - 3x^2 + 4x - 5$. This can easily be checked by multiplication.

Example 41

Divide $6x^3 + x^2 + 13x + 7$ by $2x + 1$.

$$\begin{array}{rl} & 3x^2 - x + 7 \\ 2x + 1 & \overline{)6x^3 + x^2 + 13x + 7} \end{array}$$

$2x$ multiplied by $3x^2 = 6x^3$:	$2x + 1\,\overline{)6x^3 + x^2 + 13x + 7}$
Multiply $2x + 1$ by $3x^2$:	$\underline{6x^3 + 3x^2}$
Subtract and copy $13x$:	$-2x^2 + 13x$
Multiply $2x + 1$ by $-x$:	$\underline{-2x^2 - x}$
Subtract and copy 7:	$14x + 7$
Multiply $2x + 1$ by 7:	$\underline{14x + 7}$

The answer is $3x^2 - x + 7$.

Sometimes the polynomial that you are given to divide has one or more terms in x missing. For example, $3x^5 + 2x^3 + 7x - 6$ has the terms in x^4 and x^2 missing. In questions involving long division these must be added and a zero placed in front of them. So the polynomial above would be written $3x^5 + 0x^4 + 2x^3 + 0x^2 + 7x - 6$.

Example 42

Divide $-4x^4 - 5x^2 + 5x + 4$ by $2x + 1$.

Write $-4x^4 - 5x^2 + 5x + 4$ as

$$-4x^4 + 0x^3 - 5x^2 + 5x + 4$$

$$\begin{array}{rl} & -2x^3 + x^2 - 3x + 4 \\ 2x + 1 & \overline{)-4x^4 + 0x^3 - 5x^2 + 5x + 4} \end{array}$$

$2x$ multiplied by $-2x^3 = -4x^4$:	$2x + 1\,\overline{)-4x^4 + 0x^3 - 5x^2 + 5x + 4}$
Multiply $2x + 1$ by $-2x^3$:	$\underline{-4x^4 - 2x^3}$
Subtract and copy $-5x^2$:	$2x^3 - 5x^2$
Multiply $2x + 1$ by x^2:	$\underline{2x^3 + x^2}$
Subtract and copy $5x$:	$-6x^2 + 5x$
Multiply $2x + 1$ by $-3x$:	$\underline{-6x^2 - 3x}$
Subtract and copy 4:	$8x + 4$
Multiply $2x + 1$ by 4:	$\underline{8x + 4}$

The answer is $-2x^3 + x^2 - 3x + 4$.

Example 43

Divide $-3x^4 + 8x^3 - 10x^2 + 18x - 14$ by $-x + 2$.

	$3x^3 - 2x^2 + 6x - 6$
$-x$ multiplied by $3x^3 = -3x^4$:	$-x + 2\ \overline{)-3x^4 + 8x^3 - 10x^2 + 18x - 14}$
Multiply $-x + 2$ by $3x^3$:	$\underline{-3x^4 + 6x^3}$
Subtract and copy $-10x^2$:	$2x^3 - 10x^2$
Multiply $-x + 2$ by $-2x^2$:	$\underline{2x^3 - 4x^2}$
Subtract and copy $18x$:	$-6x^2 + 18x$
Multiply $-x + 2$ by $6x$:	$\underline{-6x^2 + 12x}$
Subtract and copy -14:	$6x - 14$
Multiply $-x + 2$ by -6:	$\underline{6x - 12}$
Subtract:	-2

Since there is no other term to copy next to -2 you cannot go any further. So $-x + 2$ divided into $-3x^4 + 8x^3 - 10x^2 + 18x - 14$ gives an answer of $3x^3 - 2x^2 + 6x - 6$ (called the **quotient**) and leaves a **remainder** of -2. (This is a similar situation to dividing 39 by 5. The number 5 divided into 39 gives a quotient of 7 and leaves a remainder of 4. That is, $39 = (7 \times 5) + 4$.) This result is written as

$$-3x^4 + 8x^3 - 10x^2 + 18x - 14 = (3x^3 - 2x^2 + 6x - 6)(-x + 2) - 2$$

Example 44

Divide $3x^5 - 8x^4 + 8x^3 - 11x^2 + 15x - 7$ by $x^2 - 2x + 1$.

	$3x^3 - 2x^2 + x - 7$
x^2 multiplied by $3x^3 = 3x^5$:	$x^2 - 2x + 1\ \overline{)3x^5 - 8x^4 + 8x^3 - 11x^2 + 15x - 7}$
Multiply $x^2 - 2x + 1$ by $3x^3$:	$\underline{3x^5 - 6x^4 + 3x^3}$
Subtract and copy $-11x^2$:	$-2x^4 + 5x^3 - 11x^2$
Multiply $x^2 - 2x + 1$ by $-2x^2$:	$\underline{-2x^4 + 4x^3 - 2x^2}$
Subtract and copy $15x$:	$x^3 - 9x^2 + 15x$
Multiply $x^2 - 2x + 1$ by x:	$\underline{x^3 - 2x^2 + x}$
Subtract and copy -7:	$-7x^2 + 14x - 7$
Multiply $x^2 - 2x + 1$ by -7:	$\underline{-7x^2 + 14x - 7}$

Example 45

Divide $-3x^4 + 28x^3 - 43x^2 + 16x + 21$ by $-x^2 + 7x + 4$.

$-x^2$ multiplied by $3x^2 = -3x^4$:

Multiply $-x^2 + 7x + 4$ by $3x^2$:

Subtract and copy $16x$:

Multiply $-x^2 + 7x + 4$ by $-7x$:

Subtract and copy 21:

Multiply $-x^2 + 7x + 4$ by 6:

Subtract:

$$
\begin{array}{r}
3x^2 - 7x + 6 \\
-x^2 + 7x + 4 \;\overline{)\,-3x^4 + 28x^3 - 43x^2 + 16x + 21} \\
\underline{-3x^4 + 21x^3 + 12x^2} \qquad\qquad\quad \\
7x^3 - 55x^2 + 16x \qquad \\
\underline{7x^3 - 49x^2 - 28x} \qquad \\
-6x^2 + 44x + 21 \\
\underline{-6x^2 + 42x + 24} \\
2x - 3
\end{array}
$$

So the answer when $-3x^4 + 28x^3 - 43x^2 + 16x + 21$ is divided by $-x^2 + 7x + 4$ is $3x^2 - 7x + 6$ with a remainder of $2x - 3$. This can be expressed as

$$-3x^4 + 28x^3 - 43x^2 + 16x + 21 \equiv (-x^2 + 7x + 4)(3x^2 - 7x + 6) + (2x - 3)$$

where $3x^2 - 7x + 6$ is the quotient and $2x - 3$ is the remainder.

Exercise 1F

Divide:

1. $2x^3 - 3x^2 - 3x + 2$ by $x - 2$
2. $2x^3 + 3x^2 - 1$ by $2x - 1$
3. $2x^3 - 11x^2 + 12x - 35$ by $x - 5$
4. $-3x^4 + 11x^3 - 13x^2 + 26x - 15$ by $x - 3$
5. $-18x^3 + 33x^2 - 29x + 10$ by $-3x + 2$
6. $2x^4 + 12x^3 + 14x^2 - 8x$ by $x + 4$
7. $4x^3 + 4x^2 - x + 1$ by $2x + 1$
8. $15x^4 - x^3 + 7x^2 + 5$ by $3x + 1$
9. $3x^4 + 19x^3 - 25x^2 - 57x + 130$ by $-x - 7$
10. $-8x^4 + 24x^3 - 12x^2 + 17x - 26$ by $-2x + 5$
11. $3x^5 + 4x^4 + x^2 + 3x + 1$ by $x^2 + 2x + 1$
12. $-2x^4 + 13x^3 - 3x^2 + 37x + 35$ by $2x^2 - x + 7$
13. $-2x^4 + 3x^3 - 4x^2 - 24x + 7$ by $x^2 - 3x + 7$
14. $2x^5 + 6x^4 + 11x^3 - 11x^2 + 5x - 7$ by $2x^2 + 1$
15. $-6x^5 - 12x^4 + 25x^3 - 10x^2 - 14x + 12$ by $-3x^2 + 2$
16. $3x^4 - 4x^3 + 12x^2 - 6x + 11$ by $x^2 + 2$
17. $x^5 + x^4 + 10x^2 + 10x + 7$ by $x^2 + 2x$
18. $-6x^5 + 3x^4 - 4x^3 + 20x^2 - 3x + 10$ by $3x^2 + 2$
19. $4x^4 - 4x^3 - 4x^2 + 6x - 7$ by $2x^2 + 2x - 1$
20. $-3x^5 + 3x^4 - 19x^3 - 3x^2 + 17x - 9$ by $-x^2 + x - 7$

1.7 The factor theorem

Consider the trinomial $3x^2 - 2x - 1$ which can be denoted by $f(x)$ so that:

$$f(x) \equiv 3x^2 - 2x - 1$$

This notation is extremely useful when you are considering particular values which the trinomial takes for given values of x. For example:

$$f(2) = 3(2)^2 - 2(2) - 1 = 12 - 4 - 1 = 7$$
$$f(-1) = 3(-1)^2 - 2(-1) - 1 = 3 + 2 - 1 = 4$$
$$f(0) = 0 - 0 - 1 = -1$$

In particular notice that:

$$f(1) = 3(1^2) - 2(1) - 1 = 3 - 2 - 1 = 0$$

and also that you can factorise $f(x)$ so that:

$$(3x^2 - 2x - 1) = (3x + 1)(x - 1)$$

That is, $f(1) = 0$ and $x - 1$ is a factor of $f(x)$. This is a particular illustration of the **factor theorem** which states that:

- **If $f(x)$ is a polynomial and $f(a) = 0$, then $x - a$ is a factor of $f(x)$.**

This important theorem is used widely in factorising cubic and higher power polynomials. Consider this statement '$(x - a)$ is a factor of the polynomial $f(x)$'. It follows that $f(x)$ can be written as $g(x)(x - a)$ where $g(x)$ is a polynomial whose degree is one less than $f(x)$.

Further it is clear that since $g(x)(x - a) \equiv f(x)$ then

$$g(a)(a - a) = f(a) = 0$$

The argument is reversible and for a polynomial $f(x)$, if $f(a) = 0$, then you can say at once that $x - a$ is a factor of $f(x)$.

- **If $f(x)$ is a polynomial and $f\left(\frac{b}{a}\right) = 0$ then $(ax - b)$ is a factor of $f(x)$.**

Example 46
Factorise $x^3 - 2x^2 - x + 2$.

Let $f(x) \equiv x^3 - 2x^2 - x + 2$.

Put $x = 1$: $\quad f(1) = 1 - 2 - 1 + 2 = 0$

By the factor theorem $x - 1$ is a factor of $f(x)$.

Now consider the identity

$$x^3 - 2x^2 - x + 2 \equiv (x - 1)(Ax^2 + Bx + C)$$

$$\begin{aligned} x^3 - 2x^2 - x + 2 &\equiv x(Ax^2 + Bx + C) - 1(Ax^2 + Bx + C) \\ &= Ax^3 + Bx^2 + Cx - Ax^2 - Bx - C \\ &= Ax^3 + Bx^2 - Ax^2 + Cx - Bx - C \end{aligned}$$

Equating the x^3 coefficients gives:

$$1 = A$$

Equating the x^2 coefficients gives:

$$-2 = B - A$$

Since $A = 1$:

$$-2 = B - 1 \Rightarrow B = -1$$

Equating the x coefficients:

$$-1 = C - B$$

and since $B = -1$ then:

$$C = -2$$

You can also consider the constant term, where the result $C = -2$ is confirmed.

So:
$$\begin{aligned} f(x) &\equiv (x - 1)(x^2 - x - 2) \\ &\equiv (x - 1)(x + 1)(x - 2) \end{aligned}$$

because $x^2 - x - 2 \equiv (x + 1)(x - 2)$.

In this example, you could have found the three factors by repeated application of the factor theorem for $x = 1$, -1 and 2 but this approach breaks down if the quadratic factor does not in turn factorise.

Example 47

Given that $f(x) \equiv x^3 - x + 6$ has a factor $x + a$, find the value of a and factorise $f(x)$ completely.

By repeated trial you find that:

$$f(-2) = (-2)^3 - (-2) + 6 = -8 + 2 + 6 = 0$$

Therefore $x + 2$ is a factor, and $a = 2$.

Take

$$\begin{aligned} x^3 - x + 6 &\equiv (x + 2)(Ax^2 + Bx + C) \\ &= Ax^3 + Bx^2 + Cx + 2Ax^2 + 2Bx + 2C \\ &= Ax^3 + (B + 2A)x^2 + (C + 2B)x + 2C \end{aligned}$$

Equating coefficients of x^3 gives $A = 1$

Equating constant terms gives $2C = 6 \Rightarrow C = 3$

Equating coefficients of x^2 gives $B + 2A = 0$

So: $B + 2 = 0 \Rightarrow B = -2$

That is:

$$x^3 - x + 6 \equiv (x + 2)(x^2 - 2x + 3)$$

and the quadratic $x^2 - 2x + 3$ does not factorise into real factors. (See section 1.4).

You could have found the quadratic factor by long division like this:

$$\begin{array}{r|l} & x^2 - 2x + 3 \\ \hline x + 2 & x^3 \qquad\quad - x + 6 \\ & \underline{x^3 + 2x^2} \\ & \quad -2x^2 - x \\ & \quad \underline{-2x^2 - 4x} \\ & \qquad\qquad 3x + 6 \\ & \qquad\qquad \underline{3x + 6} \end{array}$$

Either method, using identities or using long division, is equally valid. As multiplication is, in general, easier than division you can *check* your answer by either method by confirming that:

$$\begin{aligned}(x + 2)(x^2 - 2x + 3) &= x(x^2 - 2x + 3) + 2(x^2 - 2x + 3) \\ &= x^3 - 2x^2 + 3x + 2x^2 - 4x + 6 \\ &= x^3 - x + 6, \text{ as required.}\end{aligned}$$

Exercise 1G

Each of the following expressions has a factor $(x + p)$. Find a value of p and hence factorise the expression completely.

1 $x^3 - 6x^2 + 11x - 6$

2 $x^3 + 2x^2 - x - 2$

3 $x^3 + x^2 - 4x - 4$

4 $x^3 + 3x^2 - 4x - 12$

5 $x^3 - 7x - 6$

6 Show that $(2x - 1)$ is a factor of $2x^3 + x^2 + x - 1$ and find the quadratic factor.

7 Show that $3x^3 - 2x^2 + 3x - 2$ has $3x - 2$ as a factor and find the quadratic factor.

Factorise completely:

8 $2x^3 - 3x^2 - 11x + 6$

9 $4x^3 - 20x^2 + 13x + 12$

10 $3x^3 - 8x^2 + 25x - 14$

1.8 Quadratic functions and their graphs

In mathematics you will meet many situations in which each member of one set of numbers is related to a member of another set of numbers by some well understood rule.

For example, the rule $y = 3x + 2$ relates the set of numbers $x = 0$, 1, 2, 3 ... to the set $y = 2, 5, 8, 11, \ldots$ respectively.

The rule $y = x^2 - 2x + 3$ relates the set of numbers $x = -1, 0, 1$, 2 ... to the set of numbers $y = 6, 3, 2, 3, \ldots$ respectively.

In cases like this, y is said to be a **function** of x and you will be studying functions more generally in Book P2.

The rule $y = ax^2 + bx + c$, where a, b, c are constants and $a \neq 0$, often written as $f(x) \equiv ax^2 + bx + c$, is the **quadratic function**. The x-set of values is called the **domain** of f and the y-set is called the **range** of f. It is usual for the domain to be the set of all real numbers, denoted by $\mathbb{R}$.

Either by using a graphical calculator or by plotting values of $[x, f(x)]$ for a quadratic function, you can confirm that, for specific values of a, b and c, the curve with equation $y = ax^2 + bx + c$ will be one of the forms shown on page 31. For a less than zero, which we write $a < 0$, the curve will be of shape $\bigcap$ because the x^2 term will be negative and will dominate the other terms when x is large positive or large negative.

Similarly, for a greater than zero, which we write $a > 0$, the curve will be of shape $\bigcup$. In all cases, the curve of the quadratic function is called a **parabola**.

The second fact you will need to investigate when dealing with a quadratic function curve is whether the curve intersects, touches or misses the x-axis. The condition for this is whether $b^2 - 4ac$, called the **discriminant** of $ax^2 + bx + c$, is positive, zero or negative. If $b^2 > 4ac$ the curve intersects the x-axis, if $b^2 = 4ac$ the curve touches the x-axis and if $b^2 < 4ac$ the curve misses the x-axis. The proof of this will be given in section 1.10.

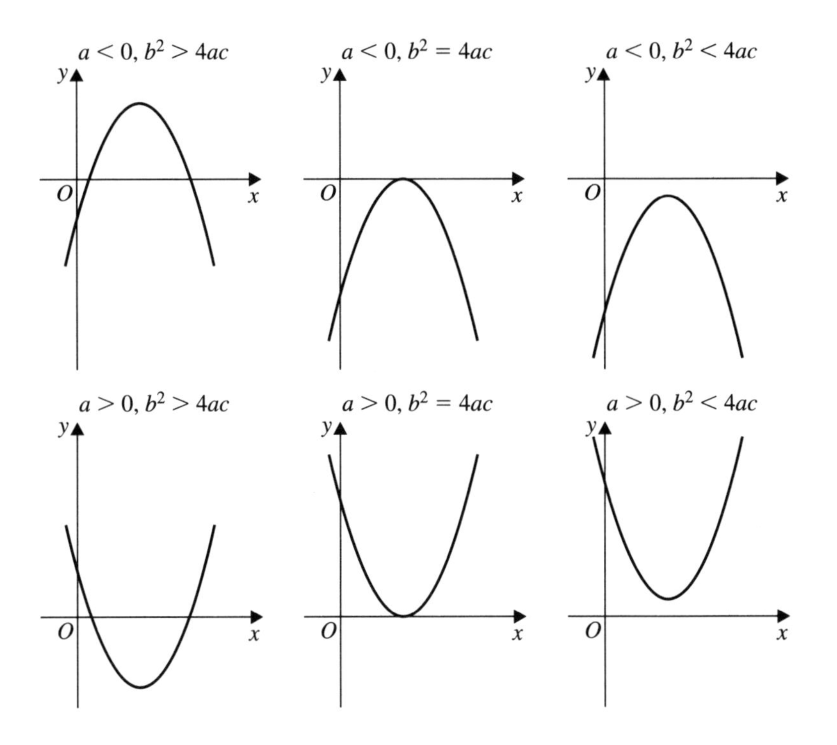

Example 48

Sketch the curves with equation:

(a) $y = x^2 - 4x + 3$
(b) $y = -x^2 + 4x - 5$

indicating the coordinates of any points where the curves meet the coordinate axes.

(a) Here $a = 1$, $b = -4$ and $c = 3$.

Also: $$x^2 - 4x + 3 \equiv (x - 3)(x - 1)$$

For $x = 1$ and $x = 3$ it follows that $y = 0$ and for $x = 0$, $y = 3$. Also $a > 0$, so it is a $\cup$ curve. The discriminant

$$b^2 - 4ac = (-4)^2 - 4(1)(3) = 16 - 12 > 0$$

In other words $b^2 > 4ac$ so the curve intersects the x-axis. The sketch of the curve looks like this:

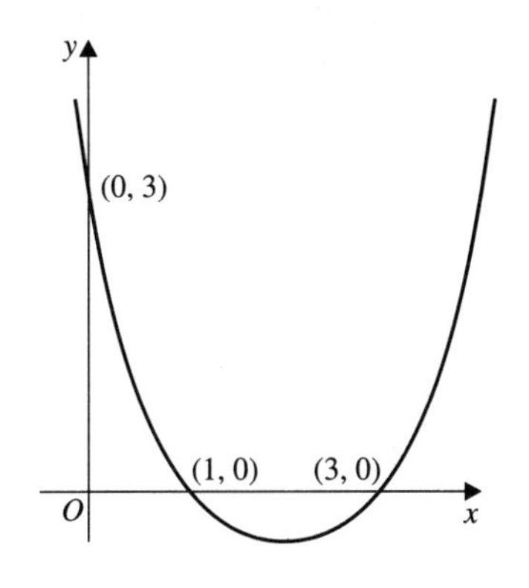

(b) Here $a = -1$, $b = 4$, $c = -5$.
For $x = 0$, $y = -5$.
Also $a < 0$ and

$$b^2 - 4ac = 16 - 4(-1)(-5)$$
$$= 16 - 20 = -4 < 0$$

In other words $b^2 < 4ac$, so it is a $\frown$ curve which does not intersect the x-axis and looks like this:

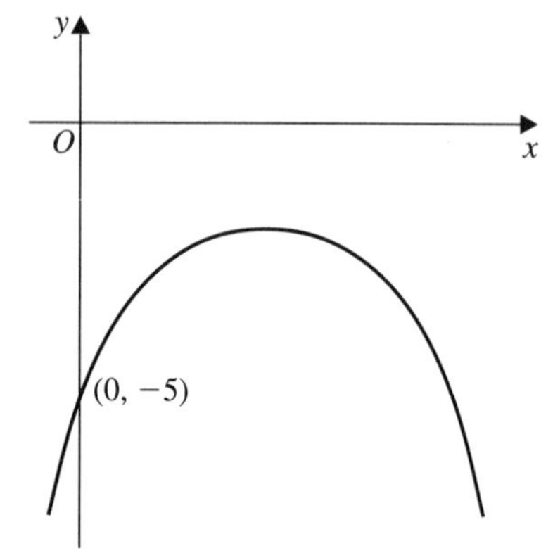

1.9 Completing the square of a quadratic function

The function $x^2 - 2ax + a^2$ can be written as $(x - a)^2$, so this quadratic function is called a **perfect square**. A function such as $x^2 + Ax$ requires a further term to *complete the square*. This term is $\frac{A^2}{4}$.

Then:
$$x^2 + Ax \equiv x^2 + Ax + \frac{A^2}{4} - \frac{A^2}{4}$$
$$= \left(x + \frac{A}{2}\right)^2 - \frac{A^2}{4}$$

For quadratic expressions which have an x^2 coefficient of 1, you add the square of half the coefficient of x in order to complete the square. Remember to subtract this term as well, in order to keep the identity valid.

In completed square form, the expression $x^2 + Ax$ is

$$\left(x + \frac{A}{2}\right)^2 - \frac{A^2}{4}$$

For expressions whose x^2 coefficient is not 1, an extra step is needed at the start of the process as illustrated in the following example.

Example 49

Express $3x^2 + 12x + 5$ in the form $A(x + B)^2 + C$, where the constants A, B and C are to be found. Hence sketch the graph of $y = 3x^2 + 12x + 5$ showing the coordinates of the points where the curve meets the axes and the turning point (i.e. the maximum or minimum point).

$$3x^2 + 12x + 5 = 3\left[x^2 + 4x + \tfrac{5}{3}\right]$$

Now $x^2 + 4x + \frac{5}{3}$, has an x^2 coefficient of 1, so take half the coefficient of x and square it, then add this and subtract it from the expression. Since $\left(\frac{4}{2}\right)^2 = 4$, then:

$$3x^2 + 12x + 5 = 3\left[x^2 + 4x + 4 - 4 + \frac{5}{3}\right]$$

$$= 3\left[(x + 2)^2 - \frac{7}{3}\right]$$

$$= 3(x + 2)^2 - 7$$

So you have $3x^2 + 12x + 5$ in the form:

$$A(x + B)^2 + C \equiv 3(x + 2)^2 - 7$$

where $A = 3$, $B = 2$ and $C = -7$, as required.

Notice carefully the steps required to complete the process: first extract 3 (the coefficient of x^2), complete the square on $x^2 + 4x$, remembering to *add* and to *subtract* 4 to maintain equality, and then finally write the result in the form required.

By writing $3x^2 + 12x + 5$ in the completed square form $3(x + 2)^2 - 7$, you are able to make several important deductions about the graph of $y = 3(x + 2)^2 - 7$ at once. These are:

(i) when $x = 0$, $y = 5$

(ii) when $y = 0$,

$$3(x + 2)^2 - 7 = 0$$

$$3(x + 2)^2 = 7$$

$$(x + 2)^2 = \frac{7}{3}$$

$$x + 2 = \pm\sqrt{\frac{7}{3}}$$

$$x = -3.53 \text{ or } -0.47 \text{ (2 decimal places)}$$

(iii) The minimum value of $3(x + 2)^2 - 7$ occurs where the first term is zero; that is, where $x = -2$. So the minimum value is -7. This is because a square term like $3(x + 2)^2$ is always greater than or equal to zero.

Using these facts, the graph of $y = 3x^2 + 12x + 5$ looks like this:

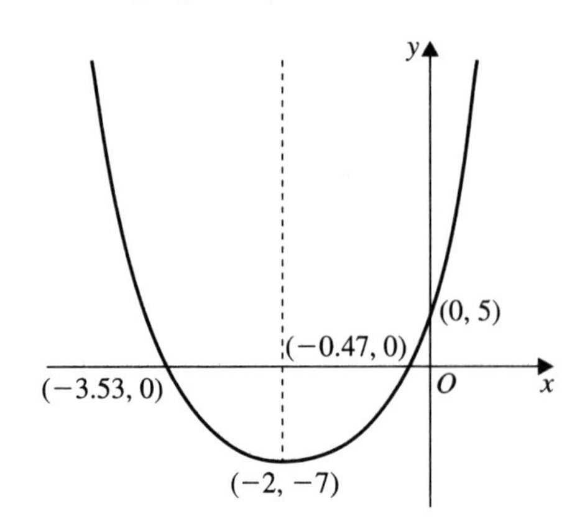

Notice also that the range of the function f, given by $f(x) \equiv 3x^2 + 12x + 5$, $x \in \mathbb{R}$, is the set of values taken by y, that is, $y \geqslant -7$.

Completing the square for $ax^2 + bx + c$

To do this, take the same steps as shown in example 49:

$$ax^2 + bx + c = a\left[x^2 + \frac{b}{a}x + \frac{c}{a}\right]$$

The coefficient of x is $\frac{b}{a}$, so half this squared is $\left(\frac{b}{2a}\right)^2$. You must both add and subtract this.

So: $$ax^2 + bx + c = a\left[x^2 + \frac{b}{a}x + \left(\frac{b}{2a}\right)^2 - \left(\frac{b}{2a}\right)^2 + \frac{c}{a}\right]$$

But since $x^2 + \frac{b}{a}x + \left(\frac{b}{2a}\right)^2 = \left(x + \frac{b}{2a}\right)^2$ then:

$$\begin{aligned} ax^2 + bx + c &= a\left[\left(x + \frac{b}{2a}\right)^2 - \frac{b^2}{4a^2} + \frac{c}{a}\right] \\ &= a\left[\left(x + \frac{b}{2a}\right)^2 - \left(\frac{b^2}{4a^2} - \frac{c}{a}\right)\right] \\ &= a\left[\left(x + \frac{b}{2a}\right)^2 - \left(\frac{b^2 - 4ac}{4a^2}\right)\right] \end{aligned}$$

So, in completed square form, you have:

$$ax^2 + bx + c \equiv a\left[\left(x + \frac{b}{2a}\right)^2 - \left(\frac{b^2 - 4ac}{4a^2}\right)\right]$$

Again, notice the steps involved:
(i) Extract the factor a, because this is the coefficient of x^2.
(ii) Complete the square on $x^2 + \frac{b}{a}$.
(iii) Remember to both add $\frac{b^2}{4a^2}$ and subtract it to maintain the identity.
(iv) Tidy up to final form.

Example 50

Use the process of completing the square to find:

(a) the least value of $x^2 - 3x + 4$
(b) the greatest value of $-4x^2 + 6x - 1$.

(a) $$x^2 - 3x + 4 = x^2 - 3x + \left(\frac{-3}{2}\right)^2 - \left(\frac{-3}{2}\right)^2 + 4$$

$$= \left(x - \frac{3}{2}\right)^2 - \frac{9}{4} + 4$$

$$= \left(x - \frac{3}{2}\right)^2 + \frac{7}{4}$$

The least value of $x^2 - 3x + 4$ occurs when $\left(x - \frac{3}{2}\right)^2 = 0$, that is, where $x = \frac{3}{2}$. So the least value is $\frac{7}{4}$.

(b) $$-4x^2 + 6x - 1 = -4\left[x^2 - \frac{6}{4}x + \frac{1}{4}\right]$$

$$= -4\left[x^2 - \frac{3}{2}x + \frac{1}{4}\right]$$

$$= -4\left[x^2 - \frac{3}{2}x + \left(\frac{-3}{4}\right)^2 - \left(\frac{-3}{4}\right)^2 + \frac{1}{4}\right]$$

$$= -4\left[\left(x - \frac{3}{4}\right)^2 - \frac{9}{16} + \frac{1}{4}\right]$$

$$= -4\left[\left(x - \frac{3}{4}\right)^2 - \frac{5}{16}\right]$$

$$= -4\left(x - \frac{3}{4}\right)^2 + \frac{5}{4}$$

Since $-4\left(x - \frac{3}{4}\right)^2$ is never positive and is zero when $x = \frac{3}{4}$, the greatest value of $-4x^2 + 6x - 1$ is $\frac{5}{4}$ and occurs when $x = \frac{3}{4}$.

Exercise 1H

In questions 1–6 copy and complete the identities.

1 $x^2 - 8x + \quad \equiv (x - \quad)^2$

2 $x^2 + 10x + \quad \equiv (x + \quad)^2$

3 $x^2 + 5x + \quad \equiv (x + \quad)^2$

4 $25x^2 - 10x + \quad \equiv (5x - \quad)^2$

5 $3x^2 - 12x + \quad \equiv 3(x - \quad)^2$

6 $7x^2 + 14x + \quad \equiv 7(x + \quad)^2$

In questions 7–12 express the quadratic function given in the form $A(x + B)^2 + C$, where the constants A, B and C are to be found.

7 $4x^2 - 16x + 1$

8 $-3x^2 + 5x$

9 $2x^2 + 3x + 4$

10 $-x^2 - x + 1$

11 $4x^2 - 3x - 5$

12 $5x^2 + 9x - 3$

In questions 13–20, sketch the curve with the given equation. On your sketch give the coordinates of the turning point and of any points where the curve meets the coordinate axes.

13 $y = x^2 - 6x + 9$

14 $y = x^2 - 3x - 18$

15 $y = -x^2 - 4x - 5$

16 $y = -x^2 + 25$

17 $y = -4x^2 + 12x - 9$

18 $y = 3x^2 - 4x - 5$

19 $y = 5x^2 - 15x$

20 $y = 3x^2 - 6x + 19$

21 Use the method of completing the square to find the maximum value of $5 - 2x - 3x^2$ and the value of x when it occurs. [E]

22 The expression $3x^2 + px + 6$, where p is a constant, has $x + 2$ as a factor. Find the other factors and sketch the curve with equation $y = 3x^2 + px + 6$.

1.10 Solving quadratic equations

Any equation in x which can be written in the form

$$ax^2 + bx + c = 0, \quad a \neq 0$$

with the highest power being x^2, is called a **quadratic equation** or quadratic for short. It is an equation of the second degree. Quadratic equations have two solutions or roots, which are sometimes equal.

The starting point when solving a quadratic equation is to *always* put everything on the left-hand side of the equation and to have zero on the right-hand side, so that you have an equation of the form $ax^2 + bx + c = 0$. When you have done this there are three common ways of solving the quadratic equation: by factorisation, by completing the square and by using a standard formula.

Solving quadratic equations by factorisation

This is usually the quickest way of solving a quadratic. (Page 40 shows you how to find out whether a quadratic will factorise or not.)

After moving all the terms to the left-hand side of the equation you factorise the left-hand side.

Example 51

Solve $x^2 - 4 = 0$.

$$x^2 - 4 = 0$$

Factorise the left-hand side:

$$(x - 2)(x + 2) = 0$$

So either $\quad x - 2 = 0$ and $x = 2$

or $\quad x + 2 = 0$ and $x = -2$

Example 52

Solve $3x^2 + 4x = 0$.

$$3x^2 + 4x = 0$$

So: $\quad x(3x + 4) = 0$

Either $\quad x = 0$

or $\quad 3x + 4 = 0$ and $x = -\frac{4}{3} = -1\frac{1}{3}$

Example 53

Solve $x^2 - 5x - 14 = 0$.

$$x^2 - 5x - 14 = 0$$

So: $\quad (x + 2)(x - 7) = 0$

Either $\quad x + 2 = 0$ and $x = -2$

or $\quad x - 7 = 0$ and $x = 7$

Example 54

Solve $2x^2 - x - 3 = 0$.

$$2x^2 - x - 3 \equiv (2x - 3)(x + 1)$$

So either $\quad 2x - 3 = 0$ or $x + 1 = 0$

That is: $\quad x = \frac{3}{2}$ or $x = -1$

Example 55

Solve $12x^2 - 23x + 10 = 0$.

$$12x^2 - 23x + 10 \equiv (3x - 2)(4x - 5)$$

So either $\qquad 3x - 2 = 0$ or $4x - 5 = 0$

That is: $\qquad x = \frac{2}{3}$ or $x = \frac{5}{4}$

Solving quadratic equations by completing the square

You know from section 1.9 that any quadratic expression such as $ax^2 + bx + c$ can be expressed in the form

$$a\left[\left(x + \frac{b}{2a}\right)^2 - \left(\frac{b^2 - 4ac}{4a^2}\right)\right]$$

Once this process has been completed, the quadratic equation $ax^2 + bx + c = 0$ is easily solved.

Consider, for example, the quadratic equation

$$x^2 - 4x + 2 = 0$$

$$x^2 - 4x + 2 \equiv x^2 - 4x + 4 - 4 + 2 = (x - 2)^2 - 2$$

So $x^2 - 4x + 2 = 0$ is the same as saying

$$(x - 2)^2 - 2 = 0$$

That is: $\qquad (x - 2)^2 = 2$

Taking the square roots gives:

$$x - 2 = \pm\sqrt{2}$$

So: $\qquad x = 2 + \sqrt{2}$ or $x = 2 - \sqrt{2}$

are the solutions of the equation. If accuracy to 2 decimal places is required then the answers, called **roots**, are 3.41 and 0.59.

Example 56

Solve $2x^2 + 8x - 5 = 0$, giving the roots to 2 decimal places.

$$2x^2 + 8x - 5 \equiv 2\left(x^2 + 4x - \tfrac{5}{2}\right)$$

$$= 2\left[x^2 + 4x + 4 - 4 - \tfrac{5}{2}\right]$$

$$= 2\left[(x + 2)^2 - \tfrac{13}{2}\right]$$

So the equation $2x^2 + 8x - 5 = 0$ is the same equation as $2\left[(x + 2)^2 - \frac{13}{2}\right] = 0$ which, dividing by 2, can be written as $(x + 2)^2 - \frac{13}{2} = 0$.

That is: $(x+2)^2 = \frac{13}{2}$

so: $x + 2 = \pm\sqrt{\frac{13}{2}}$

$$x = -2 \pm \sqrt{\tfrac{13}{2}}$$

$$= 0.55 \text{ or } -4.55$$

Solving quadratic equations by using the formula

Since the equation $ax^2 + bx + c = 0$ can also be written in the form $a\left[\left(x + \frac{b}{2a}\right)^2 - \left(\frac{b^2 - 4ac}{4a^2}\right)\right] = 0$ then:

$$\left(x + \frac{b}{2a}\right)^2 - \left(\frac{b^2 - 4ac}{4a^2}\right) = 0$$

So: $$\left(x + \frac{b}{2a}\right)^2 = \frac{b^2 - 4ac}{4a^2}$$

By taking the square root of both sides you obtain:

$$x + \frac{b}{2a} = \pm\frac{\sqrt{(b^2 - 4ac)}}{2a}$$

Thus: $$x = \frac{-b \pm \sqrt{(b^2 - 4ac)}}{2a}$$

and these are the solutions of the equation $ax^2 + bx + c = 0$.

Notice that if $b^2 - 4ac < 0$, no real square root can be found and so the roots of the equation $ax^2 + bx + c = 0$ are not real. In this case, the curve with equation $y = ax^2 + bx + c$ does not intersect the x-axis (see section 1.8).

- **The solutions of $ax^2 + bx + c = 0$ are**

$$x = \frac{-b \pm \sqrt{(b^2 - 4ac)}}{2a}$$

This is often known as the **quadratic formula** and *you should memorise it.*

Example 57

Solve the equation $5x^2 - 11x + 4 = 0$ giving the roots to 2 decimal places. This is the general equation $ax^2 + bx + c = 0$ with $a = 5$, $b = -11$ and $c = 4$.

Substituting these values into the formula:

$$
\begin{aligned}
x &= \frac{-b \pm \sqrt{(b^2 - 4ac)}}{2a} \\
&= \frac{11 \pm \sqrt{[121 - (4 \times 5 \times 4)]}}{10} \\
&= \frac{11 \pm \sqrt{41}}{10} \\
&= \frac{11 \pm 6.403}{10} \\
&= 1.74\,(2\,\text{d.p.}) \text{ or } 0.46\,(2\,\text{d.p.})
\end{aligned}
$$

How to check whether a quadratic can be factorised

As you were told earlier, the expression $(b^2 - 4ac)$ is called the **discriminant. If the discriminant has an exact square root then the quadratic equation can be factorised into two linear factors**. You will save yourself a lot of time and energy if you use the method of factorisation whenever you can.

Exercise 1I

Solve the following quadratic equations by the method of factorisation:

1 $x^2 - 49 = 0$

2 $4x^2 - 25 = 0$

3 $x^2 = 144$

4 $x^2 + 5x = 0$

5 $x^2 - 7x = 0$

6 $x^2 - x - 2 = 0$

7 $x^2 + x - 6 = 0$

8 $x^2 + 7x + 10 = 0$

9 $x^2 - 9x + 20 = 0$

10 $x^2 + 6x + 9 = 0$

11 $x^2 + 12x + 32 = 0$

12 $x^2 - 9x - 10 = 0$

13 $x^2 - x - 12 = 0$

14 $x^2 + 4x - 12 = 0$

15 $x^2 - 11x - 12 = 0$

16 $x^2 + 13x + 12 = 0$

17 $2x^2 - 3x - 2 = 0$

18 $2x^2 - x - 1 = 0$

19 $2x^2 + 5x + 2 = 0$

20 $3x^2 - 7x + 2 = 0$

21 $3x^2 + 19x - 14 = 0$

22 $3x^2 - 13x - 10 = 0$

23 $12x^2 - 7x + 1 = 0$

24 $6x^2 - 5x - 6 = 0$

25 $8x^2 - 22x + 15 = 0$

26 $28x^2 - 51x - 27 = 0$

27 $15x^2 + 22x + 8 = 0$

28 $20x^2 - 31x - 9 = 0$

Solve the following quadratic equations by (i) completing the square, (ii) using the formula, giving your answers to 2 decimal places:

29 $x^2 - 2x - 1 = 0$

30 $x^2 + 8x + 6 = 0$

31 $x^2 - 3x - 5 = 0$

32 $10x^2 - 2x - 3 = 0$

33 $7x^2 + 6x + 1 = 0$

34 $4x^2 - 3x - 2 = 0$

35 $3x^2 - 5x + 1 = 0$

36 $5x^2 + 3x - 3 = 0$

37 $3x^2 - 4x = 2$

38 $(2x - 3)^2 = 2x$

1.11 Solving two simultaneous equations, one linear and one quadratic

When you can solve two linear simultaneous equations and can solve quadratic equations, you can also solve two simultaneous equations of which one is linear and the other quadratic. To solve these make either x or y the subject of the linear equation, substitute this into the quadratic equation and then solve the quadratic equation you are left with.

Notice that two simultaneous equations, one linear and one quadratic, have, in general, two pairs of solutions. Each pair involves a value of x and a value of y.

Example 58

Solve these simultaneous equations:

$$x - 2y = 7$$
$$x^2 + 4y^2 = 37$$

Taking the linear equation and making x the subject gives $x = 7 + 2y$.

Substituting into the second equation gives

$$(7 + 2y)^2 + 4y^2 = 37$$
$$49 + 28y + 4y^2 + 4y^2 - 37 = 0$$
$$8y^2 + 28y + 12 = 0$$

Divide by 4: $2y^2 + 7y + 3 = 0$

Solve the quadratic: $(2y + 1)(y + 3) = 0$

So either $2y + 1 = 0 \Rightarrow y = -\frac{1}{2}$

or $y + 3 = 0 \Rightarrow y = -3$

Now substitute these values of y into the linear equation to find the corresponding values of x.

When $y = -\frac{1}{2}$, $x = 7 - 1 = 6$

When $y = -3$, $x = 7 - 6 = 1$

So the solutions are:

$$\left\{\begin{matrix} x = 6 \\ y = -\frac{1}{2} \end{matrix}\right\} \text{and} \left\{\begin{matrix} x = 1 \\ y = -3 \end{matrix}\right\}$$

Notice that to specify the solutions correctly you need to show which value of x is paired with which value of y. An easy way of doing this is by using braces { }. Alternatively, the solutions can be written in brackets, for example: $(6, -\frac{1}{2})$, $(1, -3)$.

Exercise 1J

Solve the simultaneous equations:

1 $x + y = 1$
$16x^2 + y^2 = 65$

2 $2x + y = 1$
$x^2 + y^2 = 1$

3 $y - x = 2$
$2x^2 + 3xy + y^2 = 8$

4 $x - 3y = 10$
$x^2 + y^2 = 20$

5 $x + y = 9$
$x^2 - 3xy + 2y^2 = 0$

6 $x = 2y$
$x^2 + 3xy = 10$

7 $y = x^2 + 3$
$y = 4x$

8 $u - v = 3$
$u^2 + v^2 = 89$

9 $x + 2y = -3$
$x^2 - 2x + 3y^2 = 11$

10 $y - x = 4$
$2x^2 + xy + y^2 = 8$

1.12 Solving linear and quadratic inequalities in one variable

Linear inequalities are similar to linear equations except that the $=$ sign in the equation is replaced by one of four inequality signs:

$>$ is greater than

$\geqslant$ is greater than or equal to

$<$ is less than

$\leqslant$ is less than or equal to

So 'four is greater than 1' can be written $4 > 1$, and 'minus 3 is less than 10' can be written $-3 < 10$, and so on.

The rules for manipulating linear inequalities are, with one major exception, the same as those used for manipulating linear equations. So if $x > y$ then

$$x \pm c > y \pm c$$

and $\qquad mx > my$ if m is positive

and $\qquad \dfrac{x}{m} > \dfrac{y}{m}$ if m is positive

The one major exception comes if you try to multiply or divide both sides of a linear inequality by a negative number. In these cases the inequality sign has to be *reversed*. That is, if $x > y$ then

$$nx < ny \text{ if } n \text{ is negative}$$

and $\qquad \dfrac{x}{n} < \dfrac{y}{n}$ if n is negative

The validity of this rule can easily be seen from two examples:

(i) $7 > -4$, i.e. 'seven is greater than minus four'.

But if you multiply by -2, then $-14 < 8$, i.e. 'minus fourteen is less than eight'.

(ii) $-15 < 3$, i.e. 'minus fifteen is less than three'.

But if you divide by -3 then $5 > -1$, i.e. 'five is greater than minus one'.

Example 59

Find the set of values for which $8 + 3x \geqslant 23$.

$$\begin{aligned} 8 + 3x &\geqslant 23 \\ 3x &\geqslant 23 - 8 \\ 3x &\geqslant 15 \\ x &\geqslant 5 \end{aligned}$$

You can show this solution diagrammatically on a number line:

−2 −1 0 1 2 3 4 5 6 7 8 9 10 11

The heavy dot at 5 indicates that 5 is included in the solution.

Example 60

Find the set of values for which $15 - 7x < 24 - 4x$.

$$\begin{aligned} 15 - 7x &< 24 - 4x \\ -7x + 4x &< 24 - 15 \\ -3x &< 9 \\ 3x &> -9 \\ x &> -3 \end{aligned}$$

This solution is shown on the number line like this:

−6 −5 −4 −3 −2 −1 0 1 2 3 4 5 6

In this case, the circle at -3 indicates that -3 is *not* included in the solution.

You can also manipulate **quadratic inequalities** to find the set of values for which they are true. However, manipulating a quadratic inequality is somewhat more complex than manipulating a linear inequality. You must examine either the graph of the quadratic function or the signs of the factors of the quadratic function.

Example 61

Find the set of values for which $x^2 - 5x + 4 > 0$.

$$x^2 - 5x + 4 > 0$$

factorises to:

$$(x - 4)(x - 1) > 0$$

A sketch of the quadratic function $y = x^2 - 5x + 4$ looks like this:

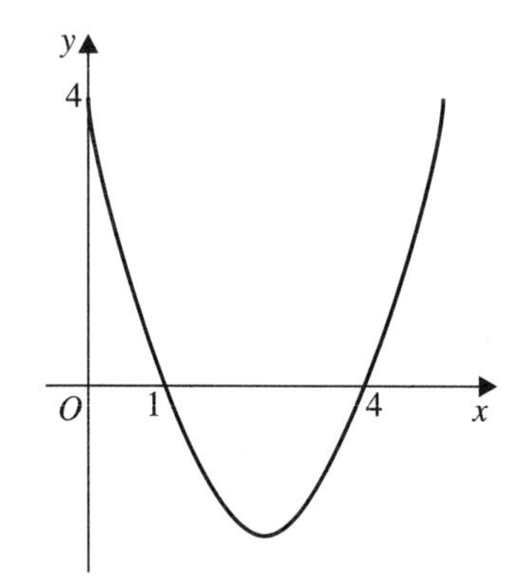

From the graph you can see that $(x^2 - 5x + 4)$ is greater than zero (i.e. it lies above the x-axis) if $x < 1$ or $x > 4$ and so this is the solution.

The alternative way of obtaining the solution is to look at the signs of the factors in the regions into which the graph is divided by the critical values. The **critical values** are the values of x where the graph cuts the x-axis; that is, they are the roots of the quadratic equation $(x - 4)(x - 1) = 0$. Since they are the roots of the corresponding quadratic equation they can be found, if necessary, without sketching the graph of the quadratic function.

Draw up a table which gives the *sign* of the factors in each of the three regions:

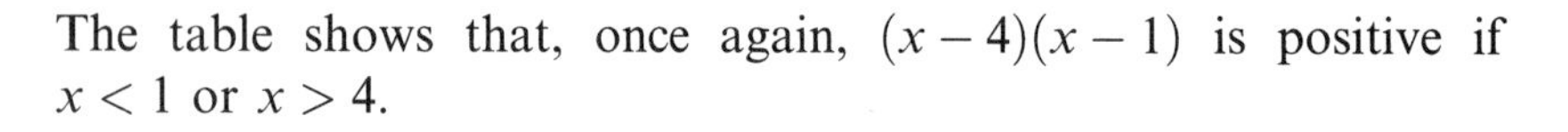

x	$x < 1$	$1 < x < 4$	$x > 4$
$x - 4$	$-ive$	$-ive$	$+ive$
$x - 1$	$-ive$	$+ive$	$+ive$
$(x - 4)(x - 1)$	$+ive$	$-ive$	$+ive$

The table shows that, once again, $(x - 4)(x - 1)$ is positive if $x < 1$ or $x > 4$.

Example 62

Find the set of values for which $2x^2 - 7x - 15 \leqslant 0$.

$$2x^2 - 7x - 15 \leqslant 0$$
$$(2x + 3)(x - 5) \leqslant 0$$

The critical values are thus $-\frac{3}{2}$ and 5, and so the three regions are $x \leqslant -\frac{3}{2}$, $-\frac{3}{2} \leqslant x \leqslant 5$, $x \geqslant 5$.

x	$x \leqslant -\frac{3}{2}$	$-\frac{3}{2} \leqslant x \leqslant 5$	$x \geqslant 5$
$2x + 3$	$-ive$	$+ive$	$+ive$
$x - 5$	$-ive$	$-ive$	$+ive$
$(2x + 3)(x - 5)$	$+ive$	$-ive$	$+ive$

From the table you can see that $(2x + 3)(x - 5) \leqslant 0$ for all values of x in the set $-\frac{3}{2} \leqslant x \leqslant 5$.

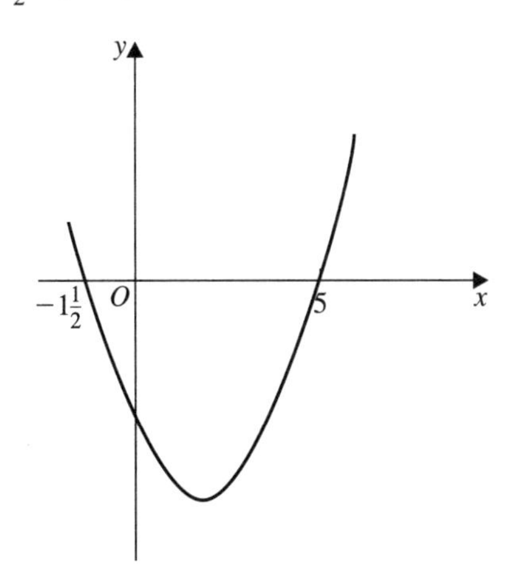

The sketch of the curve $y = (2x + 3)(x - 5)$ confirms that the function is negative or zero (i.e. lies below or crosses the x-axis) for $-1\frac{1}{2} \leqslant x \leqslant 5$.

Exercise 1K

Find the set of values for which:

1 $3x + 4 > 7$

2 $7x - 2 < 19$

3 $3 > 4x + 11$

4 $26 < 8 - 9x$

5 $6x - 7 + 2x - 9 \geqslant 0$

6 $9x - 17 + 6x - 8 \leqslant 20$

7 $5x - 30 < x - 4$

8 $17 - 3x \geqslant 9x + 45$

9 $4 + 5x < 8 - 11x$

10 $7x - 5 + 4x \geqslant 20 - 6x + 26$

11 $2(3x - 7) + 3 > 13 - 2x$

12 $3(2x - 7) > 5(6 - x) + 4$

13 $2(x - 7) + 4(3 - 2x) \leqslant -26$

14 $3(x - 7) + 2(2x - 9) < 2(12 - x)$

15 $8(2x - 4) - 9x \leqslant 3$

16 $2(3x-7)+3(x+1) \geqslant 3(10-x)+1$

17 $-7-4(x-3) > 23+2x$

18 $5(2x+7)+20 \leqslant 3(x+9)-1$

19 $3(5x+7)+2(x-9) < 10(6-x)+3(1-3x)$

20 $-15+7(2x+3) \geqslant 6x+8(2-x)$

Find the set of values for which:

21 $x^2+7x+10 \geqslant 0$

22 $x^2-5x+6 \leqslant 0$

23 $x^2+x-12 > 0$

24 $x^2-9x+18 < 0$

25 $x^2-7x-18 \leqslant 0$

26 $x^2+11x+28 > 0$

27 $2x^2-11x+12 > 0$

28 $3x^2+2x-8 > 0$

29 $3x^2-19x-14 < 0$

30 $2x^2-13x+21 > 0$

31 $2x^2+13x+20 \leqslant 0$

32 $5x^2-4x-9 > 0$

33 $3x^2+34x+63 \geqslant 0$

34 $4x^2-23x+15 < 0$

35 $3x^2-23x+30 > 0$

36 $3x^2-4x-2 \leqslant 2$

37 $2x^2+x-6 < 0$

38 $12x^2-5x-2 \leqslant 0$

39 $21x^2+5x-6 \geqslant 0$

40 $20+7x-6x^2 > 0$

SUMMARY OF KEY POINTS

1 $a^m \times a^n = a^{m+n}$

2 $(a^m)^n = a^{mn}$

3 $a^m \div a^n = a^{m-n}$

4 $a^0 = 1$

5 $a^{-m} = \dfrac{1}{a^m}$

6 $a^{\frac{1}{m}} = \sqrt[m]{a}$

7 $a^{\frac{m}{n}} = \sqrt[n]{a^m}$

8 If $f(x)$ is a polynomial and $f(a)=0$, then $(x-a)$ is a factor of $f(x)$.

9 If $f(x)$ is a polynomial and $f\left(\dfrac{b}{a}\right)=0$ then $(ax-b)$ is a factor of $f(x)$.

10 The quadratic equation $ax^2+bx+c=0$ $(a \neq 0)$ has roots

$$x = \frac{-b \pm \sqrt{(b^2-4ac)}}{2a}$$

Trigonometry

2

2.1 Radian measure

In GCSE mathematics courses angles are usually measured in units called degrees, where 360° is a complete circle. There is no real need for a complete revolution to be divided into 360 equal parts – this is just a convention. A revolution could just as well be divided into 100 parts, 1000 parts or any other number of equal parts and a suitable name given to the new unit.

In the study of trigonometry and in mathematics generally you will often find it useful to measure the size of an angle in units called **radians**.

- **One radian is the angle subtended at the centre by the arc of a circle whose length is equal to the radius of the circle.**

'The angle subtended by the arc of a circle' just means the angle between the radii that join each end of the arc to the centre of the circle.

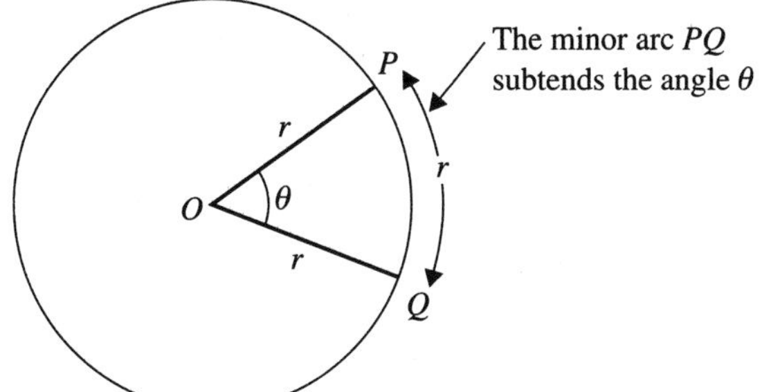

In the diagram the radius of the circle is r and the length of the arc PQ is also r. As the arc is the same length as the radius of the circle, the angle θ is 1 radian. This is often abbreviated to 1^c or 1 rad.

The angle made at the centre of the circle by an arc of length r is 1^c. The circumference of the circle is $2\pi r$ which is 2π times the length of this arc. So the angle of one revolution is $2\pi \times 1^c = 2\pi^c$ (two pi radians).

One full turn is 360 degrees. One full turn in radians is 2π radians. So:

$$360° = 2\pi^c$$

Halving gives: $180° = \pi^c$

and halving again: $90° = \frac{\pi^c}{2}$

Example 1

Convert 60° to radians.

$$\begin{aligned} 360° &= 2\pi^c \\ 1° &= \frac{2\pi^c}{360} \\ 60° &= \frac{2\pi^c}{360} \times 60 \\ &= \frac{\pi^c}{3} \end{aligned}$$

Example 2

Convert $\frac{5\pi^c}{6}$ to degrees.

$$\begin{aligned} 2\pi^c &= 360° \\ 1^c &= \frac{360°}{2\pi} \\ \frac{5\pi^c}{6} &= \frac{360}{2\pi} \times \frac{5\pi}{6} \text{ degrees} \\ &= 150° \end{aligned}$$

2.2 Length of an arc of a circle

Suppose that the arc PQ of a circle subtends an angle θ radians at the centre of the circle. As the complete angle at the centre of the circle is of size 2π radians:

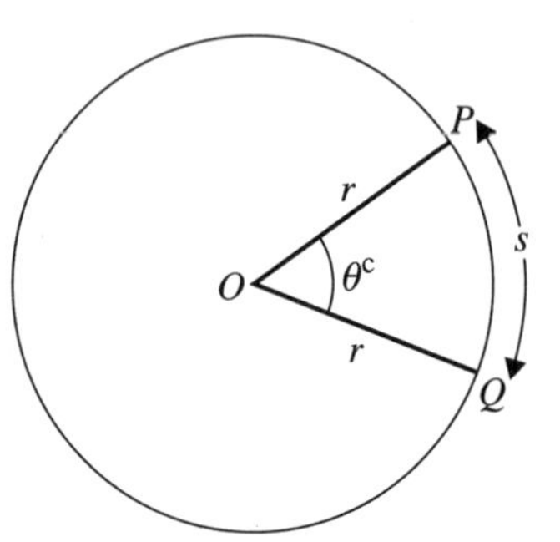

$$\frac{\text{length of arc } PQ}{\text{circumference of circle}} = \frac{\theta}{2\pi}$$

The circumference of the circle is $2\pi r$. If we say the length of the arc PQ is s then:

$$\frac{s}{2\pi r} = \frac{\theta}{2\pi}$$

So: $s = \frac{\theta}{2\pi} \times 2\pi r$

and: $s = r\theta$

- **In general the length s of an arc of a circle is: $s = r\theta$, where r is the radius and θ is the angle in radians subtended by the arc at the centre of the circle.**

Example 3

Find the length of the minor arc of a circle of radius 8 cm, given that the arc subtends an angle of 2 radians at the centre of the circle.

$s = r\theta = 8 \times 2\,\text{cm} = 16\,\text{cm}$

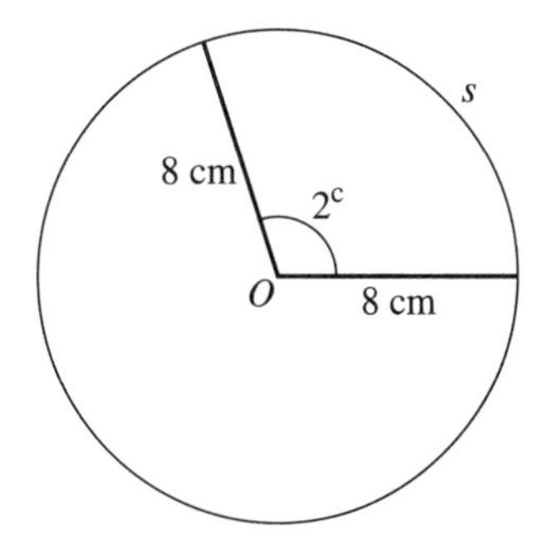

Example 4
An arc of a circle of length 5 cm subtends an angle of $1\frac{1}{2}$ radians at the centre of the circle. Calculate the radius of the circle.

$$s = r\theta$$
$$5 = r \times 1\tfrac{1}{2}$$
$$r = \frac{5}{1\frac{1}{2}} = 3\tfrac{1}{3}$$

The radius is $3\frac{1}{3}$ cm.

2.3 Area of a sector of a circle

You can find the area of the sector OPQ in a similar way to that in which you found the length of the arc PQ.

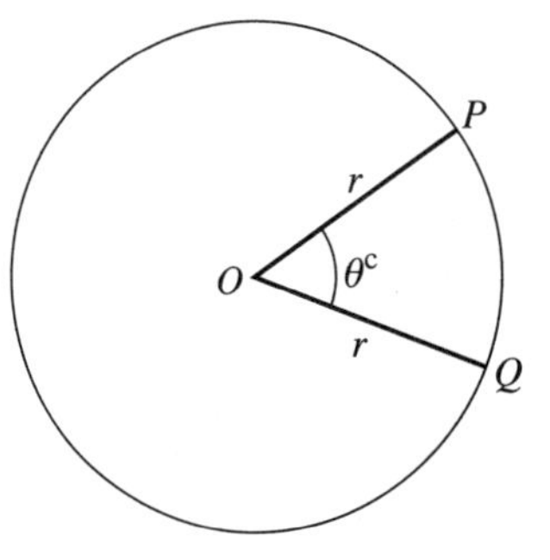

$$\frac{\text{area of sector } OPQ}{\text{area of circle}} = \frac{\theta}{2\pi}$$

The area of the circle is πr^2 so if we call the area of the sector A, then:

$$\frac{A}{\pi r^2} = \frac{\theta}{2\pi}$$
$$A = \frac{\theta}{2\pi} \times \pi r^2$$
$$A = \tfrac{1}{2}r^2\theta$$

- **In general, the area A of a sector of a circle is:**

$$A = \tfrac{1}{2}r^2\theta$$

where r is the radius and θ is the angle in radians subtended by the arc at the centre of the circle.

Example 5
Find the area of the minor sector OPQ of a circle, given that the arc PQ subtends an angle 1.2 radians at the centre of the circle and that the radius of the circle is 5 cm.

$$A = \tfrac{1}{2}r^2\theta$$
$$= \tfrac{1}{2} \times 5^2 \times 1.2\,\text{cm}^2$$
$$= 15\,\text{cm}^2$$

Example 6

In the diagram the area of the minor sector OPQ is $18\,\text{cm}^2$. The angle POQ is of size 0.64 radians. Calculate the radius of the circle.

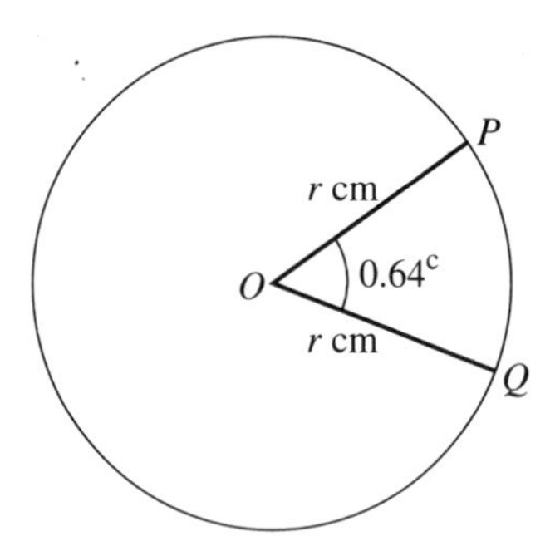

$$A = \tfrac{1}{2}r^2\theta$$
$$18 = \tfrac{1}{2} \times r^2 \times 0.64$$
$$r^2 = \frac{18}{0.32} = 56.25$$
$$r = 7.5$$

The radius of the circle is 7.5 cm.

Exercise 2A

1 Convert the following to degrees:

(a) π^c (b) $\frac{\pi^c}{2}$ (c) $\frac{\pi^c}{6}$ (d) $\frac{\pi^c}{4}$ (e) $\frac{3\pi^c}{2}$

(f) $\frac{3\pi^c}{4}$ (g) $4\pi^c$ (h) $3\pi^c$ (i) $\frac{5\pi^c}{4}$ (j) $\frac{5\pi^c}{6}$

2 Convert the following to radians, leaving your answer in terms of π:

(a) $22\frac{1}{2}^\circ$ (b) 15° (c) 180° (d) 210° (e) 120°
(f) 135° (g) 225° (h) 330° (i) 300° (j) 315°

3 Convert the following to degrees, giving your answer to the nearest degree:

(a) 1^c (b) 2.4^c (c) 5^c (d) 3.5^c (e) 0.4^c
(f) 1.7^c (g) 4.3^c (h) 5.5^c (i) 4^c (j) 6^c

4 Convert the following to radians, giving your answer to two decimal places:

(a) 20° (b) 50° (c) 70° (d) 130° (e) 170°
(f) 230° (g) 250° (h) 85° (i) 38° (j) 152°

5 Find the length of the arc PQ and the area of the corresponding sector in each case:

(a) $OP = 3$ cm, $\theta = 1^c$ (b) $OP = 7$ cm, $\theta = 2^c$
(c) $OP = 8$ cm, $\theta = 1.5^c$ (d) $OP = 9.5$ cm, $\theta = 3^c$
(e) $OP = 5.5$ cm, $\theta = 0.5^c$ (f) $OP = 9$ cm, $\theta = 2.8^c$
(g) $OP = 11$ cm, $\theta = 4.5^c$ (h) $OP = 9.5$ cm, $\theta = 5^c$
(i) $OP = 11.2$ cm, $\theta = 4.7^c$ (j) $OP = 8.3$ cm, $\theta = 6.1^c$

6 The arc PQ of a circle has length 15 cm and the radius of the circle is 6 cm. Find the angle, in radians, subtended by the arc at the centre of the circle.

7 An arc is of length 9 cm and it subtends an angle of 2.5 radians at the centre of a circle. Calculate the radius of the circle.

8 Given that the area of the minor sector OPQ in the diagram for question 5 is $10\,\text{cm}^2$ and that $OP = 4\,\text{cm}$, calculate θ in radians.

9 Given that the area of the minor sector OPQ is $50\,\text{cm}^2$ and that $\theta = 1.4$ radians, calculate the radius of the circle.

10 Calculate the area of the shaded region in the diagram, given that:

(a) $OC = 1\,\text{cm}$, $CA = 1\,\text{cm}$, $\theta = 2^c$
(b) $OC = 1\,\text{cm}$, $CA = 0.5\,\text{cm}$, $\theta = 1.5^c$
(c) $OC = 1.5\,\text{cm}$, $CA = 0.5\,\text{cm}$, $\theta = 2.5^c$
(d) $OC = 2\,\text{cm}$, $CA = 2.5\,\text{cm}$, $\theta = 0.7^c$
(e) $OC = 2.7\,\text{cm}$, $CA = 1.2\,\text{cm}$, $\theta = 1.6^c$

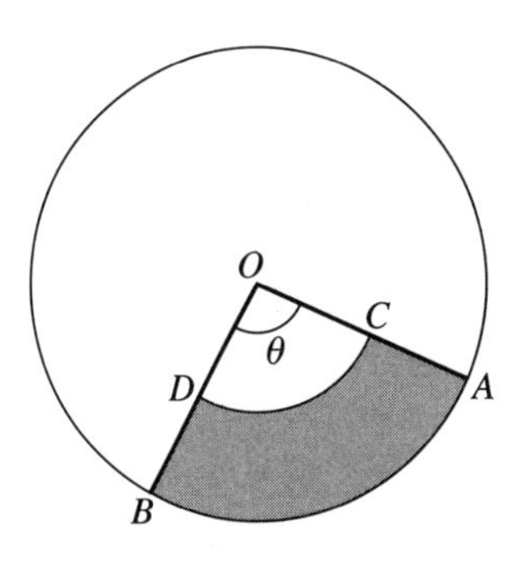

2.4 The three basic trigonometric functions for *any* angle

The theory of trigonometry which applies to acute angles can be extended to angles of any size. Think of an arm, OA, that is fixed at O and turns in an anticlockwise direction, starting along the positive x-axis. As it turns, it makes an angle with the positive x-axis.

The quadrant between the positive x-axis and the positive y-axis is called the **first quadrant**. In this quadrant any angle θ is acute. So in this diagram θ is an angle such that $0 < \theta < 90°$

$\left(\text{or } 0 < \theta < \frac{\pi^c}{2}\right)$.

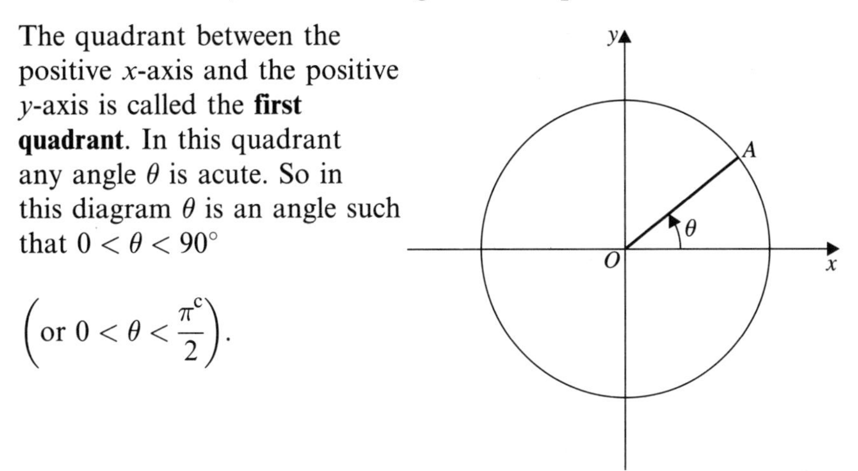

The three trigonometric ratios for positive, acute angles are defined on p. 222.

The angle θ is in the **second quadrant** when the arm OA lies between the negative x-axis and the positive y-axis. Here the angle is obtuse – it lies between 90° and 180°

$\left(\text{or } \dfrac{\pi^c}{2} \text{ and } \pi^c\right)$.

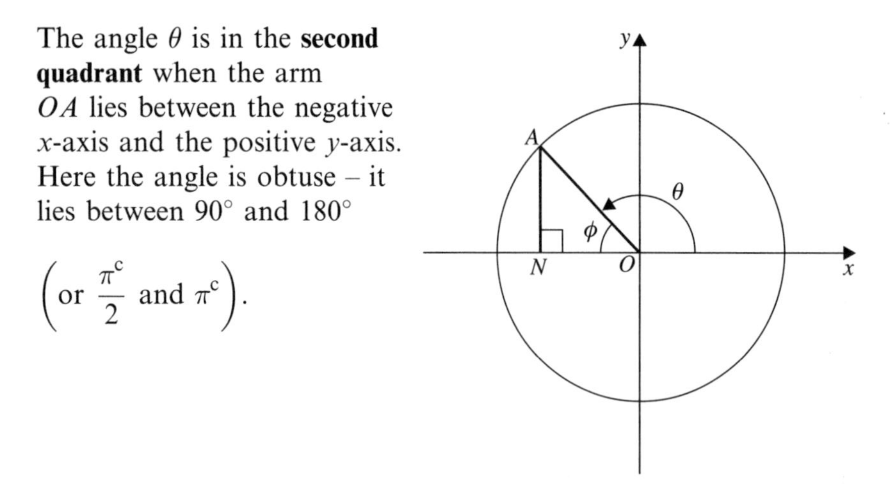

To find the trigonometric ratios of angles in the second quadrant drop a perpendicular from A to the x-axis at N and consider the acute angle ϕ which OA makes with the x-axis. In the triangle ANO, the length of the arm AO is defined to be a positive quantity. AN is upwards in the positive y direction so its length is a positive quantity. ON is in the negative x direction and so is negative. We define $\sin\theta$, $\cos\theta$ and $\tan\theta$ to be equal in size to $\sin\phi$, $\cos\phi$ and $\tan\phi$ but with the appropriate sign.

Thus
$$\sin\phi = \frac{AN}{AO} = \frac{+\text{ive}}{+\text{ive}} = +\text{ive}$$

$$\cos\phi = \frac{NO}{AO} = \frac{-\text{ive}}{+\text{ive}} = -\text{ive}$$

$$\tan\phi = \frac{AN}{NO} = \frac{+\text{ive}}{-\text{ive}} = -\text{ive}$$

So in the second quadrant:

$$\sin\theta = +\sin\phi$$
$$\cos\theta = -\cos\phi$$
$$\tan\theta = -\tan\phi$$

When the position of the arm OA is such that $180° < \theta < 270°$, it is said to be in the **third quadrant**. As before, to find the trigonometric ratios in the third quadrant, drop a perpendicular from A to N on the x-axis and consider the acute angle ϕ lying in the right-angled triangle AON.

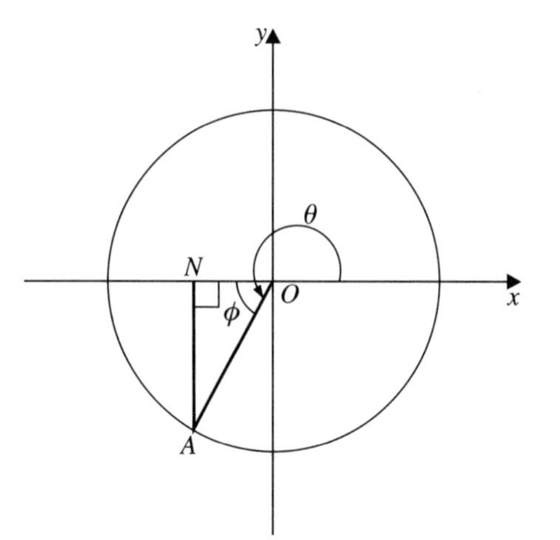

AO is again defined positive.

ON is negative.

AN is negative.

So in the third quadrant:

$$\sin\phi = \frac{AN}{AO} = \frac{-\text{ive}}{+\text{ive}} = -\text{ive}$$

$$\cos\phi = \frac{NO}{AO} = \frac{-\text{ive}}{+\text{ive}} = -\text{ive}$$

$$\tan\phi = \frac{AN}{NO} = \frac{-\text{ive}}{-\text{ive}} = +\text{ive}$$

Again, we define $\sin\theta$, $\cos\theta$ and $\tan\theta$ in the third quadrant to be $\sin\phi$, $\cos\phi$ and $\tan\phi$ with the appropriate signs.

So in the third quadrant:

$$\sin\theta = -\sin\phi$$
$$\cos\theta = -\cos\phi$$
$$\tan\theta = +\tan\phi$$

In the **fourth quadrant**, when $270° < \theta < 360°$, the perpendicular AN is again dropped to the x-axis so that the acute angle ϕ can be considered in the triangle ANO.

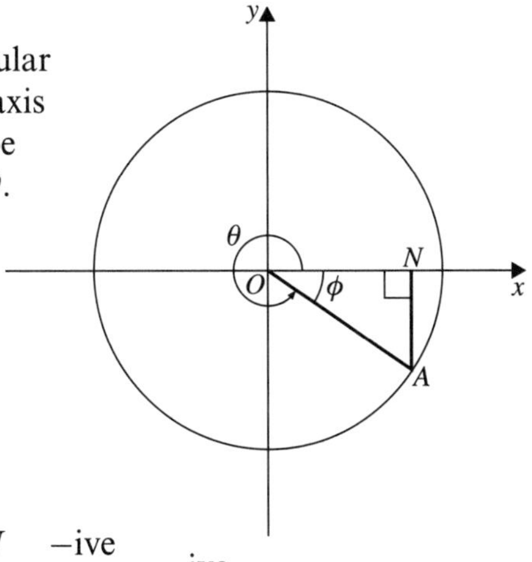

In this quadrant:

AO is positive.

ON is positive.

AN is negative.

In the fourth quadrant:

$$\sin\phi = \frac{AN}{AO} = \frac{-\text{ive}}{+\text{ive}} = -\text{ive}$$

$$\cos\phi = \frac{NO}{AO} = \frac{+\text{ive}}{+\text{ive}} = +\text{ive}$$

$$\tan\phi = \frac{AN}{NO} = \frac{-\text{ive}}{+\text{ive}} = -\text{ive}$$

$\sin\theta$, $\cos\theta$ and $\tan\theta$ are defined as before to be $\sin\phi$, $\cos\phi$ and $\tan\phi$ with the appropriate signs.

That is, in the fourth quadrant:

$$\sin\theta = -\sin\phi$$

$$\cos\theta = +\cos\phi$$

$$\tan\theta = -\tan\phi$$

Angles can lie outside the range 0–360° but they always lie in one of the four quadrants and so the same results apply.

The angle 410° lies in the first quadrant.

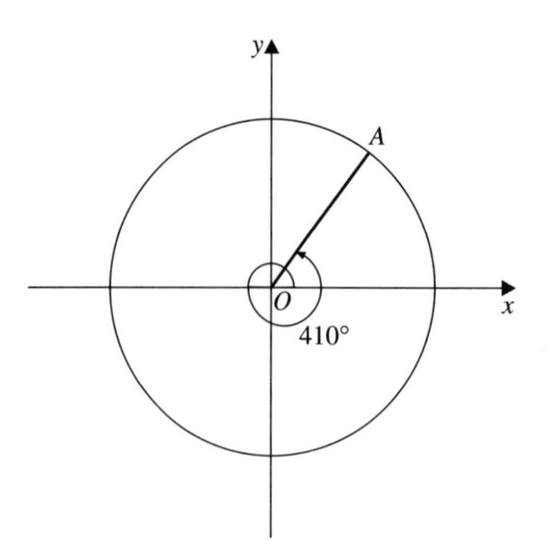

If the arm OA moves in a clockwise direction, then the angle through which it turns is defined to be negative.

The angle $-140°$ lies in the third quadrant.

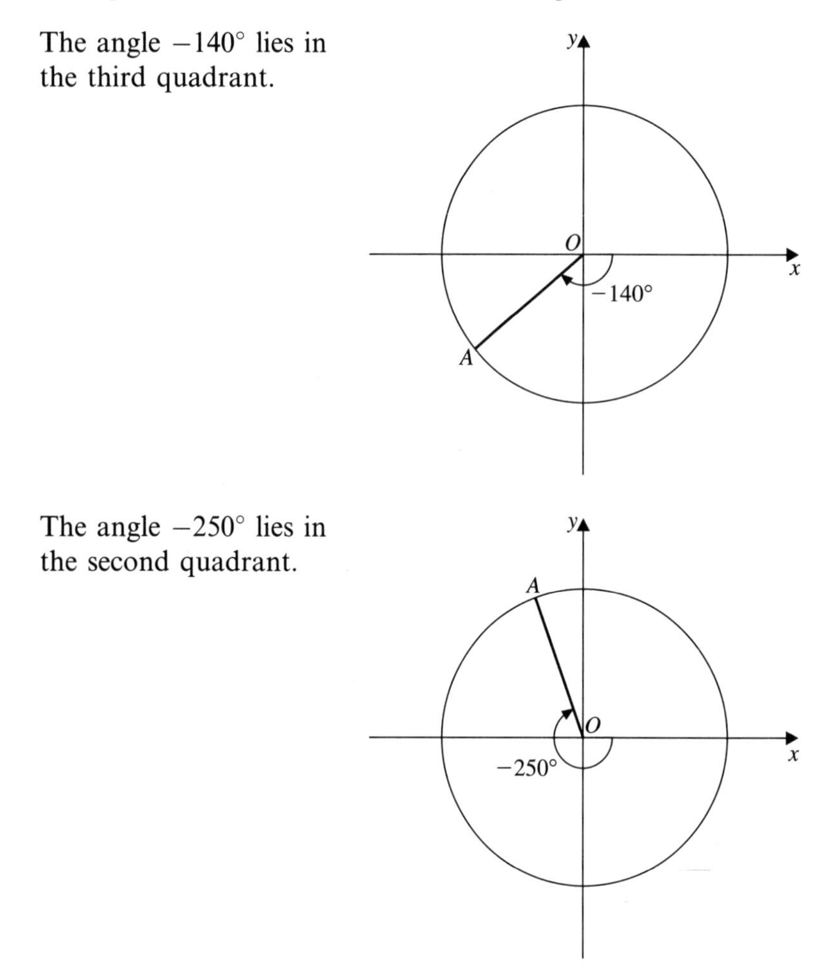

The angle $-250°$ lies in the second quadrant.

Remember, always drop a perpendicular from A to N on the x-axis (*never* to the y-axis), consider the acute angle in the triangle ANO and give it an appropriate sign.

- **The easiest way to remember which trigonometric ratios are positive and which negative in each quadrant is to remember which ratios are positive and remember that the others are negative. In the first quadrant all the ratios are positive, in the second quadrant the sine is positive, in the third quadrant the tangent is positive and in the fourth quadrant cosine is positive. These results can be shown diagrammatically.**

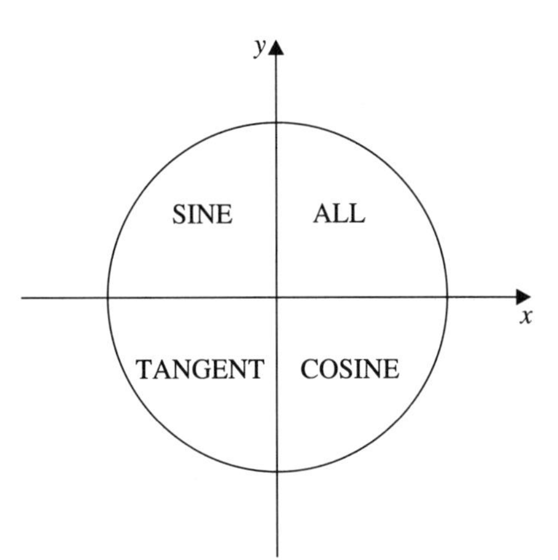

A good way to remember these is by the mnemonic 'positively All Silver Tea Cups'.

Some special results

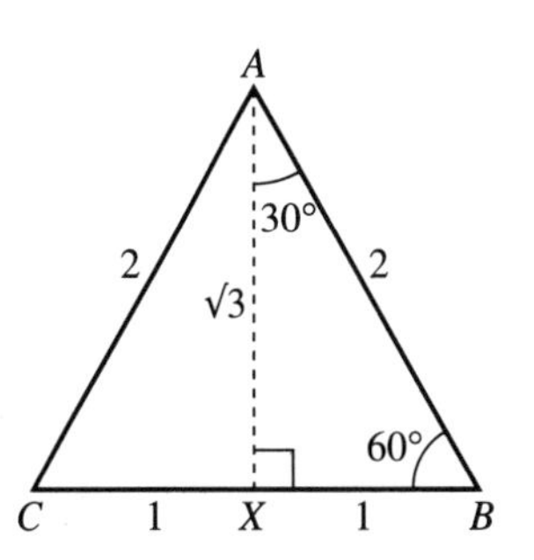

Consider an equilateral triangle ABC of side 2.

Obviously, each of its angles is of size 60°.

If you drop a perpendicular from A to X, the mid-point of BC, then $\angle XAB = 30°$ and $CX = XB = 1$.

Use Pythagoras' theorem in the right-angled triangle AXB:

$$AX^2 = 2^2 - 1^2 = 4 - 1 = 3$$

So $$AX = \sqrt{3} \text{ units}$$

Thus in the triangle AXB:

$$\sin 30° = \tfrac{1}{2}$$

$$\cos 30° = \frac{\sqrt{3}}{2}$$

and $$\tan 30° = \frac{1}{\sqrt{3}} \text{ or } \frac{\sqrt{3}}{3}$$

Also, $$\sin 60° = \frac{\sqrt{3}}{2}$$

$$\cos 60° = \tfrac{1}{2}$$

and $$\tan 60° = \sqrt{3}$$

Now consider an isosceles, right-angled triangle PQR in which $PQ = QR = 1$. As the triangle is isosceles $\angle QPR = \angle QRP = 45°$. Pythagoras' theorem gives:

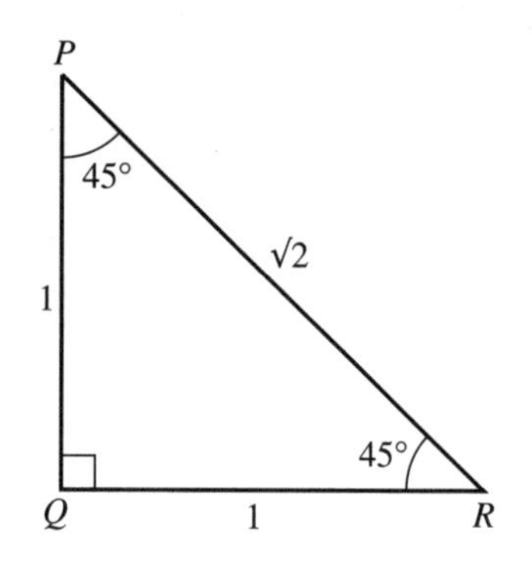

$$PR^2 = 1^2 + 1^2 = 2$$

$$PR = \sqrt{2}$$

So: $$\sin 45° = \cos 45° = \frac{1}{\sqrt{2}} \text{ or } \frac{\sqrt{2}}{2}$$

and $$\tan 45° = \frac{1}{1} = 1$$

- **The angles 30°, 45° and 60° occur frequently in trigonometry. Memorise their exact trigonometric ratios.**

	30°	**45°**	**60°**
sin	$\frac{1}{2}$	$\frac{1}{\sqrt{2}}$	$\frac{\sqrt{3}}{2}$
cos	$\frac{\sqrt{3}}{2}$	$\frac{1}{\sqrt{2}}$	$\frac{1}{2}$
tan	$\frac{1}{\sqrt{3}}$	**1**	$\sqrt{3}$

Example 7

Write $\cos 127°$ as a trigonometric ratio of an acute angle.

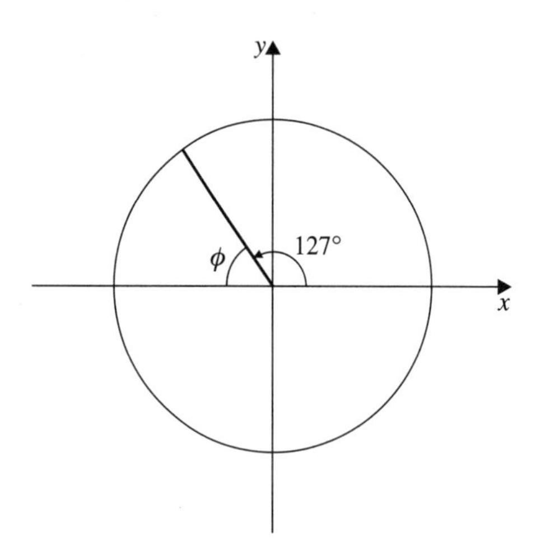

127° lies in the second quadrant. So consider

$$\phi = 180° - 127° = 53°$$

In the second quadrant cosine is negative, so:

$$\cos 127° = -\cos 53°$$

Example 8

Write $\sin(-120°)$ as the trigonometric ratio of an acute angle.

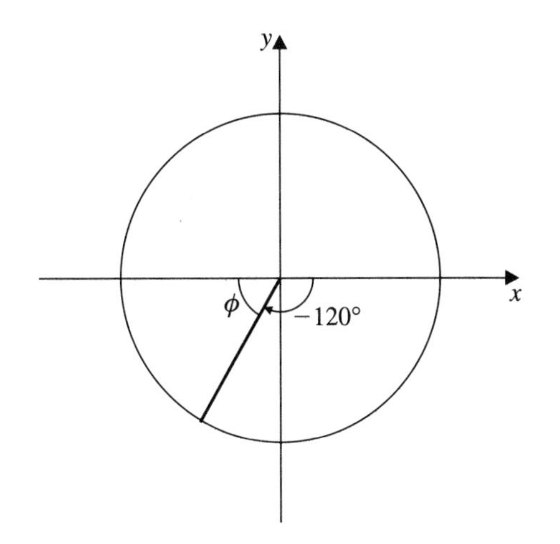

$-120°$ lies in the third quadrant, so consider

$$\phi = 180° - 120° = 60°$$

In the third quadrant sine is negative so

$$\sin(-120°) = -\sin 60°$$

Example 9

Write $\tan 317°$ as the trigonometric ratio of an acute angle.

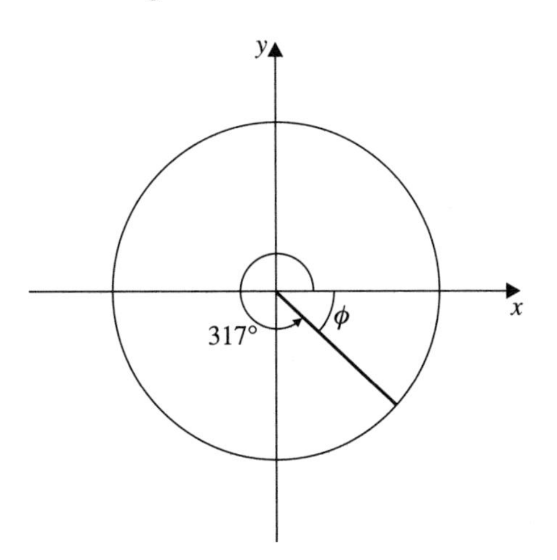

$317°$ lies in the fourth quadrant. So consider

$$\phi = 360° - 317° = 43°$$

In the fourth quadrant, tangent is negative, so:

$$\tan 317° = -\tan 43°$$

Example 10
Write $\sin 380°$ as the trigonometric ratio of an acute angle.

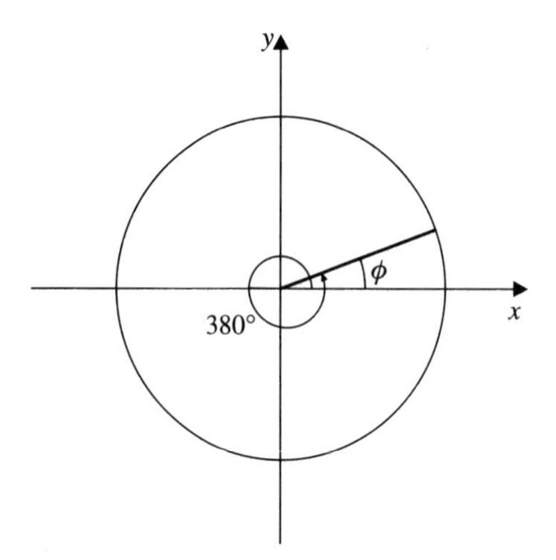

380° lies in the first quadrant. So consider

$$\phi = 380° - 360° = 20°$$

Sine is positive in the first quadrant, so

$$\sin 380° = +\sin 20°$$

Example 11
Find $\sin 315°$, leaving your answer in terms of surds.

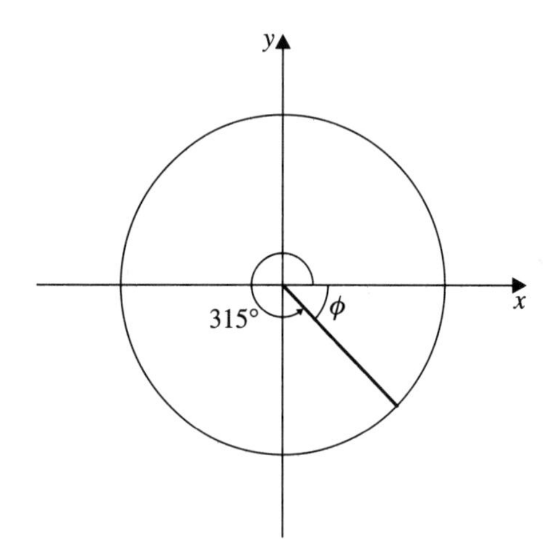

315° lies in the fourth quadrant, so

$$\phi = 360° - 315° = 45°$$

In the fourth quadrant sine is negative, so

$$\sin 315° = -\sin 45° = -\frac{1}{\sqrt{2}}$$

2.5 Drawing the graphs of sin *x*, cos *x* and tan *x*

Section 2.4 shows how to find the sine, cosine or tangent of any angle. This section shows you how to draw the graphs of these functions.

Graphing sin *x*

Here is a table of values for the function $y = \sin x$ when $0 \leqslant x \leqslant 360°$. Each value of y is given to two decimal places:

x	0	30°	45°	60°	90°	120°	135°	150°	180°	210°	225°	240°	270°	300°	315°	330°	360°
y	0	0.5	0.71	0.87	1	0.87	0.71	0.5	0	−0.5	−0.71	−0.87	−1	−0.87	−0.71	−0.5	0

If you plot these figures on a graph, it looks like this:

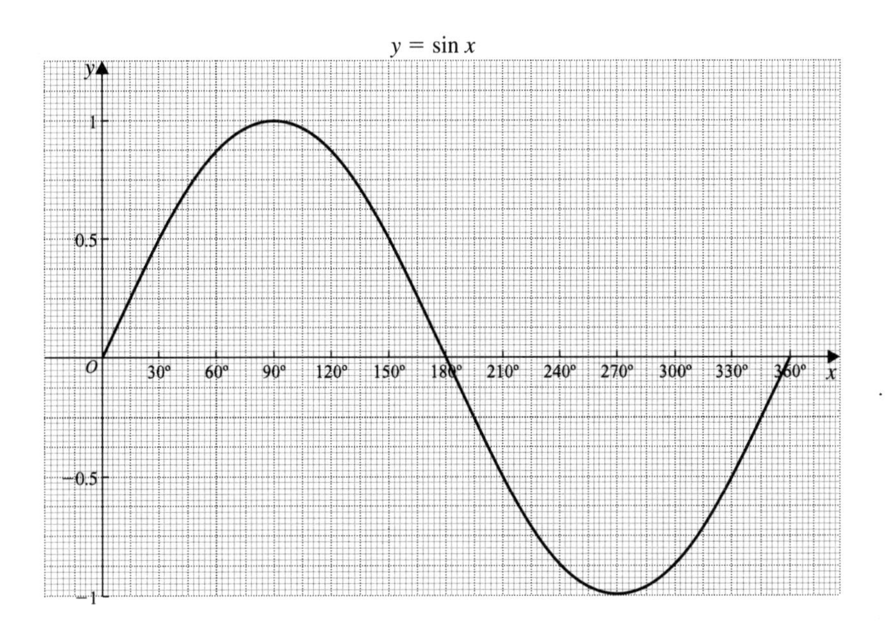

The curve will repeat itself again and again for values of x above 360° and below 0. The function has a maximum value of $+1$ and a minimum value of -1. The maximum value occurs at 90°, 450°, 810°, and so on; in other words at $90° \pm 360n°$, where n is an integer $\left(\text{or at } \frac{\pi^c}{2} \pm 2n\pi^c\right)$. The minimum value occurs at 270°, 630°, 990°, and so on; in other words at $270° \pm 360n°$ $\left(\text{or at } \frac{3\pi^c}{2} \pm 2n\pi^c\right)$.

The other main feature of the curve is that it cuts the x-axis at 0, 180°, 360°, 540°, 720°, and so on; in other words every $180n°$ (or $n\pi^c$).

As the curve starts at O and continues to 360° before repeating itself, we say that the sine curve has a **period** of 360° (or $2\pi^c$).

Graphing cos x

Here is a table of values for the function $y = \cos x$ when $0 \leqslant x \leqslant 360°$. Again each value of y is given to 2 decimal places:

x	0	30°	45°	60°	90°	120°	135°	150°	180°	210°	225°	240°	270°	300°	315°	330°	360°
y	1	0.87	0.71	0.5	0	−0.5	−0.71	−0.87	−1	−0.87	−0.71	−0.5	0	0.5	0.71	0.87	1

The curve looks like this:

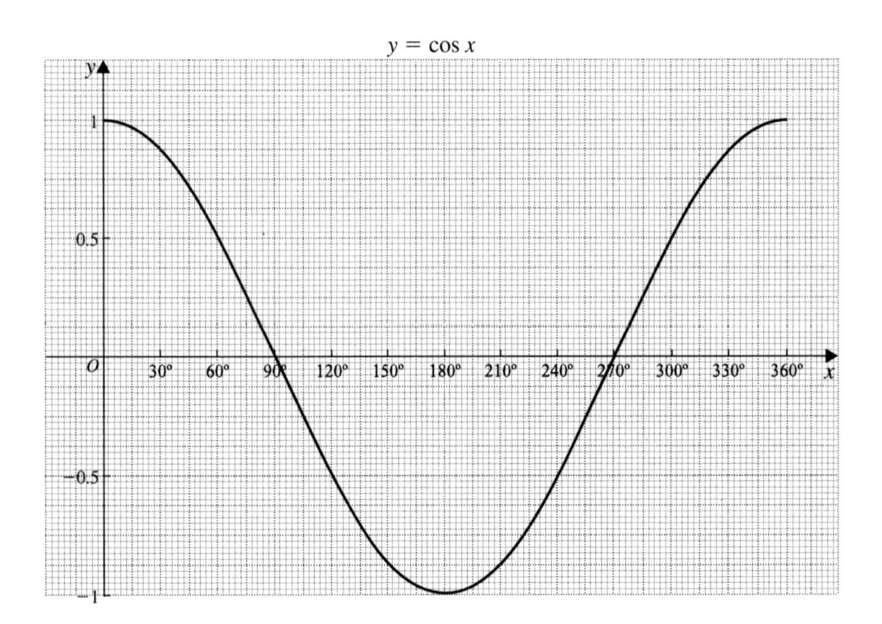

Again the main feature of the curve is that it is periodic (or cyclic). It has a period of 360°. Its maximum is again 1 and its minimum −1. But this time the maximum value occurs at 0, 360°, 720°, 1080°, and so on; in other words at $360n°$ (or $2n\pi^c$). The minimum value occurs at 180°, 540°, 900°, and so on; in other words at $180° \pm 360n°$ (or at $\pi^c \pm 2n\pi^c$). The curve cuts the x-axis at 90°, 270°, 450°, 630°, and so on; in other words at $\pm(2n+1)90°$ or $\pm(2n+1)\dfrac{\pi^c}{2}$.

Graphing tan x

Here is a table of values for $y = \tan x$ when $0 \leqslant x \leqslant 360°$:

x	0	30°	45°	60°	90°	120°	135°	150°	180°	210°	225°	240°	270°	300°	315°	330°	360°
y	0	0.58	1	1.73	$\pm\infty$	−1.73	−1	−0.58	0	0.58	1	1.73	$\pm\infty$	−1.73	−1	−0.58	0

The symbol ∞ means **infinity**.

The curve looks like this:

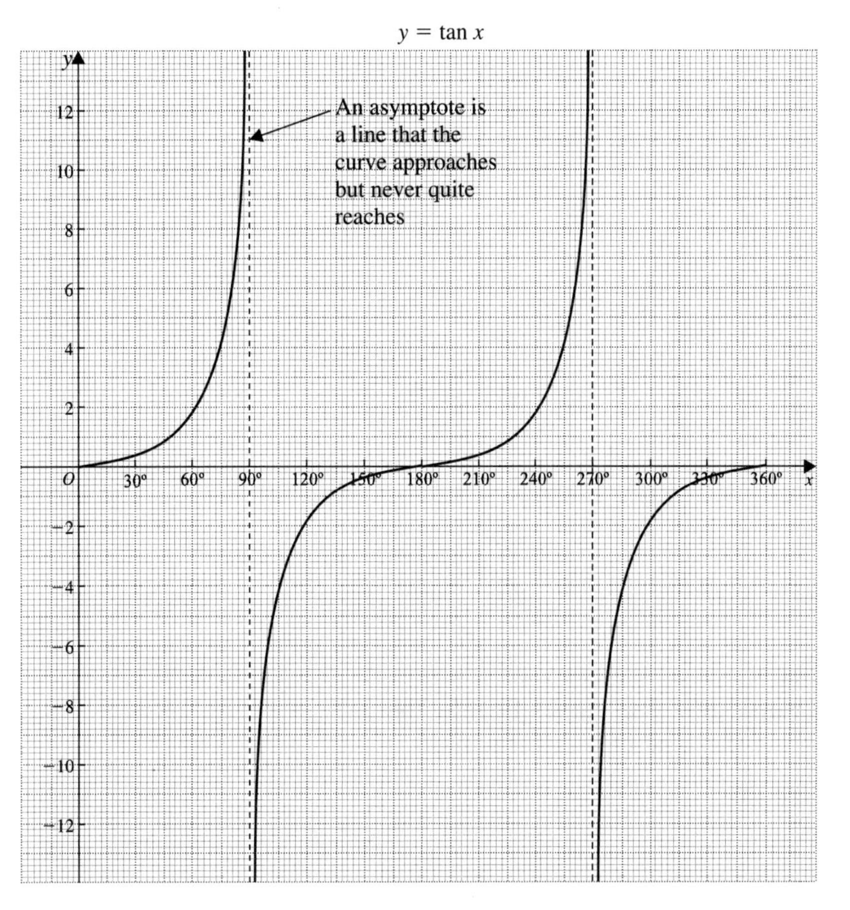

Again the curve is periodic but this time the period is 180° (π^c). The curve cuts the x-axis at 0, 180°, 360°, 540°, 720°, and so on; in other words at $180n°$ (or $n\pi^c$). It has **asymptotes** (lines that the curve approaches but never actually reaches) at 90°, 270°, 450°, and so on; in other words at $\pm(2n+1)90°$ $\left(\text{or } \pm(2n+1)\frac{\pi^c}{2}\right)$.

The graph of $y = \tan x$ has no maximum or minimum.

You must memorise the main features of the curves $y = \sin x$, $y = \cos x$ and $y = \tan x$ – their shape, their maximum and minimum values and where they occur, the points at which they cut the x-axis, and the position of the asymptotes. For an advanced course in mathematics you must be able to sketch these three curves from memory.

Graphing sin nx, cos nx and tan nx

What happens to functions such as $y = \sin x$ when x is multiplied by a number to give $y = \sin 2x$, $y = \sin 3x$ and so on? Here is the table of values of $y = \sin 2x$:

x	0	30°	45°	60°	90°	120°	135°	150°	180°	210°	225°	240°	270°	300°	315°	330°	360°
y	0	0.87	1	0.87	0	−0.87	−1	−0.87	0	0.87	1	0.87	0	−0.87	−1	−0.87	0

The curve looks like this:

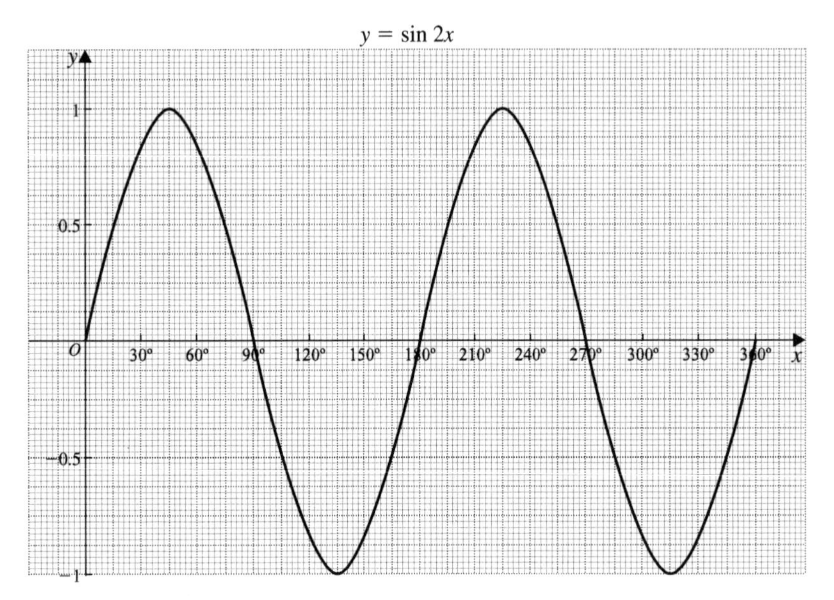

The curve of $y = \sin 2x$ is the same shape as the curve of $y = \sin x$ but the period is 180°. In other words *two* sine curves fit into the range 0–360°. This is generally true of this type of transformation.

The sketch of $y = \cos \frac{1}{2}x$ looks like this:

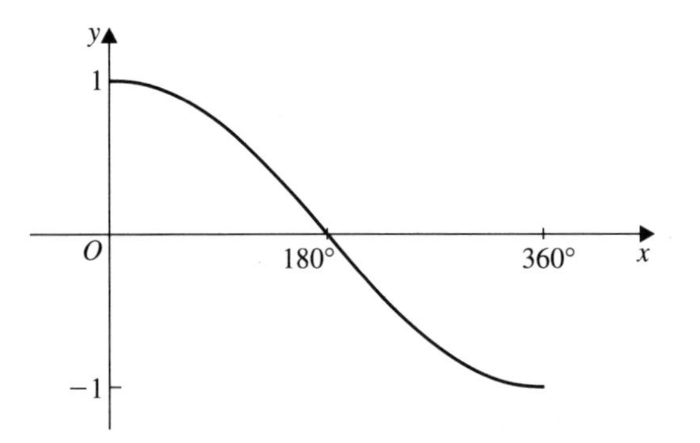

The curve of $y = \cos \frac{1}{2}x$ is the same shape as the curve of $y = \cos x$ but only *half* the graph appears in the range 0–360°.

- **In general the curve of $y = \sin nx$, $y = \cos nx$ or $y = \tan nx$ where n is a positive real number ($n \in \mathbb{R}^+$) is the same shape as the curve of $y = \sin x$, $y = \cos x$ or $y = \tan x$, but n of each curve fit into the range 0–360°.**

Example 12
Sketch the graph of $y = \tan 3x$ in the range $0 \leqslant x \leqslant 360°$.

$y = \tan 3x$ tells us that three curves with the same shape as $y = \tan x$ fit into the range $0 \leqslant x \leqslant 360°$. The curve looks like this:

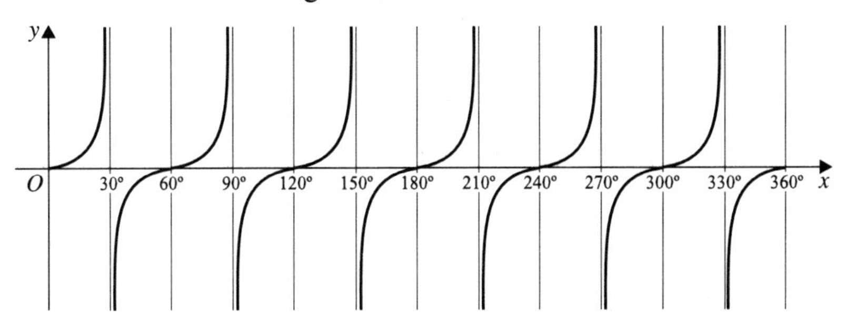

Example 13
Sketch the curve of $y = \cos \frac{1}{3}x$ in the range $0 \leqslant x \leqslant 360°$.

The shape of $y = \cos \frac{1}{3}x$ is the same as that of $y = \cos x$ but only one-third of the curve fits in the range $0 \leqslant x \leqslant 360°$. So the sketch of $y = \cos \frac{1}{3}x$ looks like this:

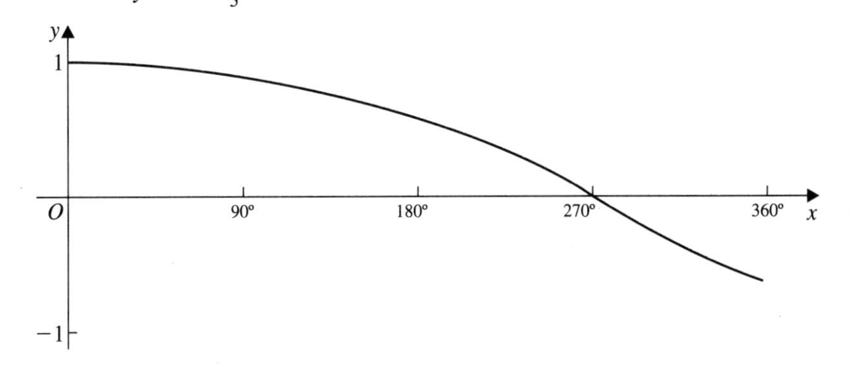

Graphing −sin x, −cos x and −tan x

Transformations such as $y = -\sin x$ are easy to deal with as for any function the curve of $y = -f(x)$ is a reflection of the curve of $y = f(x)$ in the x-axis.

Here is the curve of $y = -\sin x$ for $0 \leqslant x \leqslant 360°$:

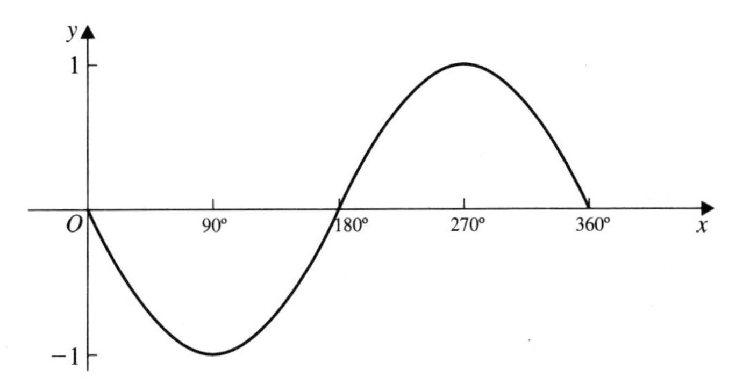

Graphing sin(−x), cos(−x) and tan(−x)

It is harder to graph the transformation that occurs when the minus sign is *inside* the function: for example, $y = \sin(-x)$. From the work done on page 54 on the trigonometric ratios of negative angles:

$$\sin(-x) = -\sin x$$

The graph of $y = \sin(-x)$ is the same as that shown above for $y = -\sin x$.

$$\cos(-x) = +\cos x$$
$$\tan(-x) = -\tan x$$

You can also check these from the graphs of the functions.

To draw the curve of $y = \cos(-x)$ rewrite the function as $y = \cos x$. The sketch of the curve is the same as that of $y = \cos x$ on page 60.

Example 14

Sketch the curve of $y = \tan(-2x)$.

$y = \tan(-2x)$ can be written $y = -\tan 2x$.

The graph of $y = \tan 2x$ has the same shape as the graph of $y = \tan x$, but two of these curves fit into $0 \leqslant x \leqslant 360°$. The graph of $y = -\tan 2x$ is a reflection of the graph of $y = \tan 2x$ in the x-axis. Here is the sketch:

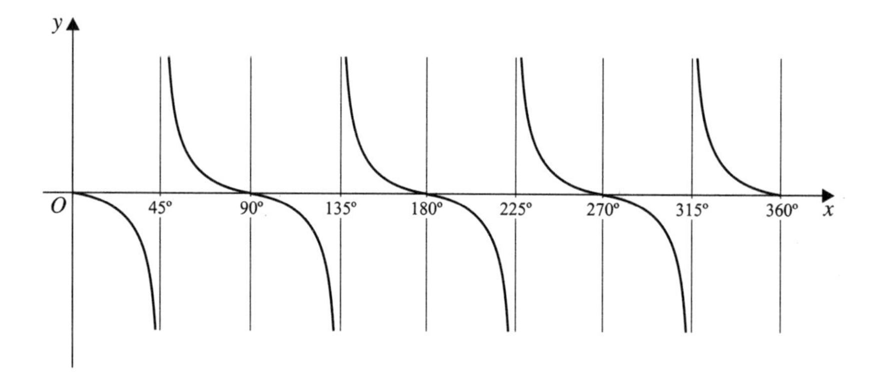

Graphing *n* sin *x*, *n* cos *x* and *n* tan *x*

The curve of $y = 3\text{f}(x)$ represents a stretch of the curve $y = \text{f}(x)$ from the x-axis of scale factor 3. The transformation $y = 3\sin x$ is a stretch of the curve $y = \sin x$ from the x-axis of scale factor 3. The transformation $y = 5\tan x$ is a stretch of the curve $y = \tan x$ from the x-axis of scale factor 5. For example, here is the curve of $y = 3\cos x$:

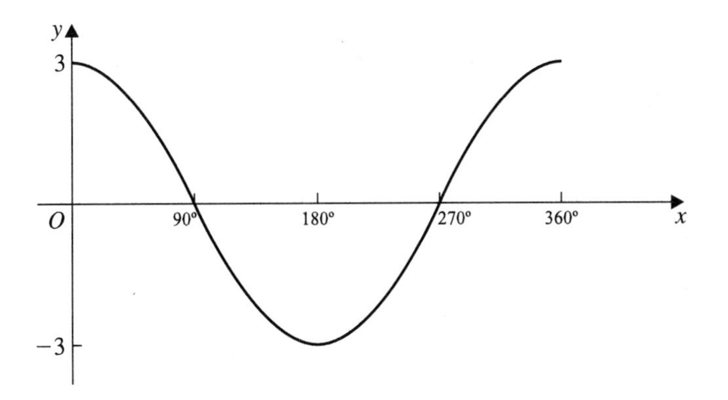

Graphing sin(x + n), cos(x + n) and tan(x + n)

The final transformation we shall deal with is one of the form $y = \mathrm{f}(x + n)$ where n is a constant. Here is a table of values for the function $y = \sin(x + 30°)$:

x	0	30°	45°	60°	90°	120°	135°	150°	180°	210°	225°	240°	270°	300°	315°	330°	360°
y	0.5	0.87	0.97	1	0.87	0.5	0.26	0	−0.5	−0.87	−0.97	−1	−0.87	−0.5	−0.26	0	0.5

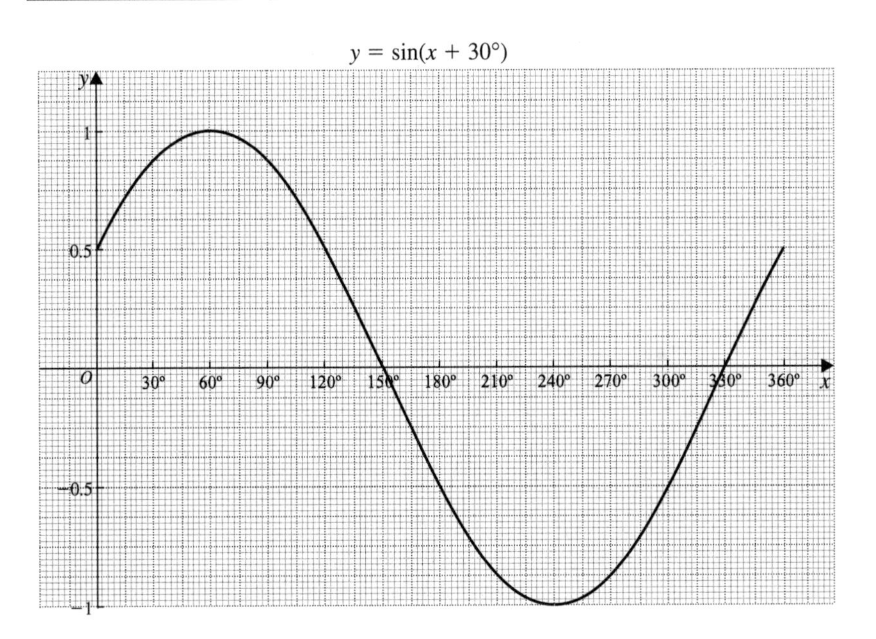

The curve is the same shape as that of $y = \sin x$ but it has been shifted 30° *to the left*.

- **Generally, the graph of $y = \mathrm{f}(x + n)$ is the same as that of $y = \mathrm{f}(x)$ but translated n to the left. Similarly, the graph of $y = \mathrm{f}(x - n)$ is the same as that of $y = \mathrm{f}(x)$ but translated n to the right.**

Remember that when a *positive* number is added the graph is translated that amount in the *negative* x direction. When a *negative* number is added the graph is translated that amount in the *positive* x direction.

Example 15

Sketch the curve of $y = 1 + 2\sin(x - 90°)$ for $0 \leqslant x \leqslant 360°$.

The '$-90°$' means that the sine graph is translated 90° to the right:

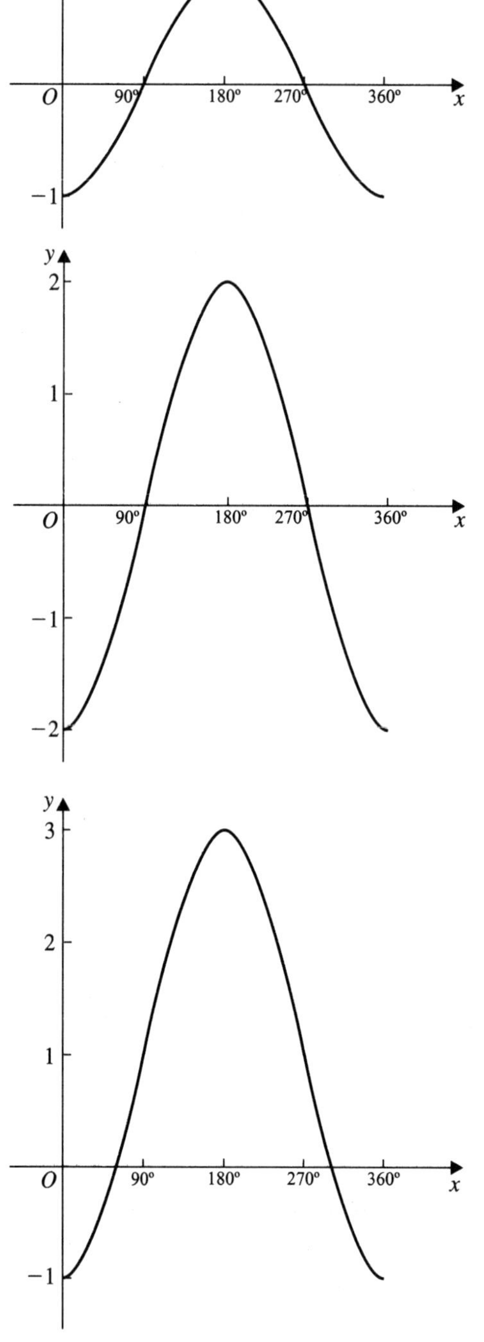

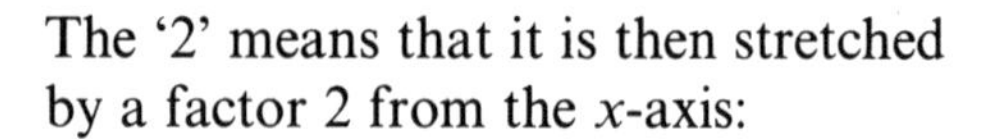

The '2' means that it is then stretched by a factor 2 from the x-axis:

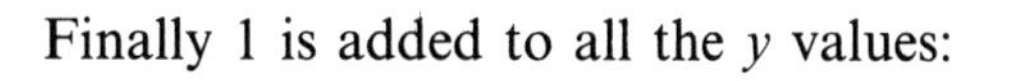

Finally 1 is added to all the y values:

Exercise 2B

1 Write each of the following as trigonometric ratios of positive acute angles:

(a) $\sin 260°$ (b) $\cos 140°$ (c) $\tan 185°$
(d) $\tan 355°$ (e) $\cos 137°$ (f) $\sin 414°$
(g) $\sin(-194°)$ (h) $\cos(-336°)$ (i) $\tan 396°$
(j) $\tan 148°$ (k) $\sin(-443)°$ (l) $\cos 248°$
(m) $\sin 331°$ (n) $\cos 293°$ (o) $\tan(-96°)$
(p) $\tan(216°)$ (q) $\sin 339°$ (r) $\cos 481°$
(s) $\tan 127°$ (t) $\sin(-198°)$

2 Write down the values of the following leaving your answers in terms of surds where appropriate:

(a) $\sin 120°$ (b) $\cos 150°$ (c) $\tan 225°$
(d) $\cos 300°$ (e) $\sin(-30°)$ (f) $\cos(-120°)$
(g) $\sin 240°$ (h) $\sin 420°$ (i) $\cos 315°$
(j) $\sin(-135°)$ (k) $\tan(-150°)$ (l) $\tan 150°$
(m) $\sin 330°$ (n) $\cos 210°$ (o) $\tan 240°$
(p) $\tan 390°$ (q) $\cos(-300°)$ (r) $\cos(-420°)$
(s) $\sin 405°$ (t) $\tan 150°$

3 Plot on the same axes the graphs of $y = \sin x$ and $y = 0.5 + \cos x$ for $0 \leqslant x \leqslant 360°$.
Use your graph to find approximate solutions of:

(a) $\sin x = 0.5 + \cos x$ (b) $0.5 + \cos x = 1.1$
(c) $\cos x = 0.3$

4 Given that:

$$f(x) = \tan x° + \frac{2}{\tan x°}$$

copy and complete the following table at the top of a piece of graph paper.

x	35	40	45	50	55	60	65	70
f(x)	3.56	3.22	3			2.89	3.08	3.48

Draw an accurate graph of $y = f(x)$ for $35 \leqslant x \leqslant 70$.
Use your graph to estimate the set of values of x, within the above range, for which

$$f(x) \leqslant 3.3$$

[E]

5 (a) Copy and complete the following table for which

$$f(x) \equiv \cos^2 x^\circ - \sin x^\circ, \ 0 \leqslant x \leqslant 180$$

giving your answers to 2 decimal places where appropriate.

x	0	30	60	90	120	150	180
f(x)	1	0.25					

(b) On graph paper and using a scale of 2 cm to represent 30° on the x-axis and 4 cm to represent 1 unit on the y-axis, draw the graph of $y = f(x)$.

(c) From your graph estimate the solutions of the equation

$$\cos^2 x^\circ - \sin x^\circ = 0, \ 0 \leqslant x \leqslant 180.$$ [E]

6 Sketch the graphs of the following for $0 \leqslant x \leqslant 360°$:

(a) $y = -\cos x$ (b) $y = 2\sin x$
(c) $y = \tan(-x)$ (d) $y = \tan(x + 30°)$
(e) $y = -3\sin x$ (f) $y = 1 + \tan x$
(g) $y = 1 - 2\cos x$ (h) $y = 1 + \cos(-2x)$
(i) $y = 2\cos(-2x)$ (j) $y = 2\sin(x + 60°)$

7 Sketch the graphs of the following for $-180° \leqslant x \leqslant 180°$:

(a) $y = \sin(x - 30°)$ (b) $y = \cos(x + 30°)$
(c) $y = \tan(x + 60°)$ (d) $y = 1 + 2\cos x$
(e) $y = \cos(60° - x)$ (f) $y = \sin(30° - x)$
(g) $y = 2\tan(90° - x)$ (h) $y = 1 + 3\cos(90° - x)$
(i) $y = 1 - 25\sin\frac{1}{3}x$ (j) $y = 1 - \frac{1}{2}\sin 2x$

8 Using the same scales and axes, sketch, for $-\pi \leqslant x \leqslant \pi$, the graphs of:

(a) $y = \sin x$ (b) $y = \dfrac{2x}{\pi}$

where x is measured in radians.
On your sketch mark the two points P and Q other than the origin where your graphs intersect.
Give the coordinates of P and Q. [E]

9 Sketch the graph of $y = \tan x$ for $0 \leqslant x \leqslant 3\pi$ and use your graph to find approximately the solutions, in the range $0 \leqslant x \leqslant 3\pi$, of the equations

$$\tan x = \sqrt{3} \text{ and } \tan x = -\sqrt{3}$$

10 Draw the graph of $y = 3\sin x - 5\cos x$, for values of x from $-100°$ to $200°$.
Use your graph to estimate the coordinates of the minimum point and the maximum point on the graph.

2.6 The identities $\tan\theta \equiv \dfrac{\sin\theta}{\cos\theta}$ and $\cos^2\theta + \sin^2\theta \equiv 1$

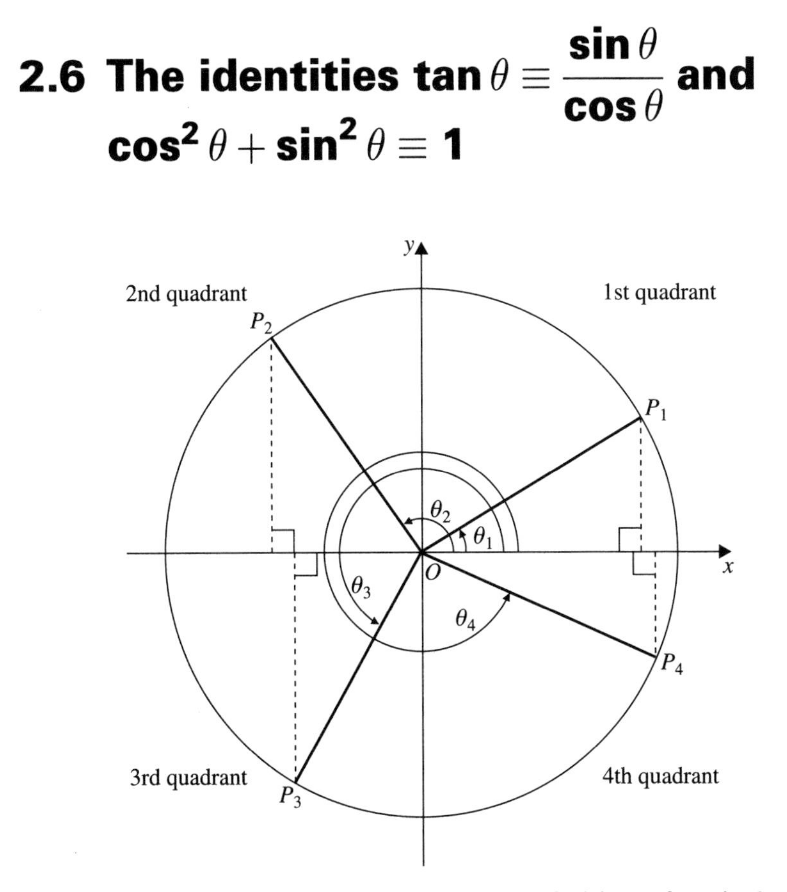

You have seen in section 2.4 how the definitions for $\sin\theta$, $\cos\theta$ and $\tan\theta$ are extended to include any size of angle. In the diagram, the circle has radius r and the points P_1, P_2, P_3 and P_4 have coordinates (x_1, y_1), (x_2, y_2), (x_3, y_3) and (x_4, y_4).

For any point $P(x, y)$ on the circle you have:

$$\sin\theta = \frac{y}{r}, \quad \cos\theta = \frac{x}{r} \quad \text{and} \quad \tan\theta = \frac{y}{x}$$

where θ is the angle between the positive x-axis and OP measured in an anticlockwise sense.

These results apply in particular for θ_1, θ_2, θ_3 and θ_4 corresponding to the points P_1, P_2, P_3 and P_4.

So you have, in general,

$$\tan\theta = \frac{y}{x}$$

Also: $$\frac{\sin\theta}{\cos\theta} = \frac{\frac{y}{r}}{\frac{x}{r}} = \frac{y}{r}\times\frac{r}{x} = \frac{y}{x}$$

■ **So** $$\tan\theta \equiv \frac{\sin\theta}{\cos\theta} \text{ for all values of } \theta$$

Also $x^2 + y^2 = r^2$ for any point P on the circle.

Divide by r^2: $$\frac{x^2}{r^2} + \frac{y^2}{r^2} = 1$$

That is: $$\left(\frac{x}{r}\right)^2 + \left(\frac{y}{r}\right)^2 = 1$$

$$(\cos\theta)^2 + (\sin\theta)^2 = 1$$

Using the notation $\cos^2\theta \equiv (\cos\theta)^2$ and $\sin^2\theta \equiv (\sin\theta)^2$ gives you:

■ $\cos^2\theta + \sin^2\theta \equiv 1$**, for all values of** θ

Example 16

Given that the angle α is obtuse and that $\cos\alpha = -\frac{4}{5}$, find the values of (a) $\sin\alpha$ (b) $\tan\alpha$.

(a) As α is obtuse, it is in the second quadrant. Using the identity $\cos^2\theta + \sin^2\theta \equiv 1$, you have

$$\left(-\tfrac{4}{5}\right)^2 + \sin^2\alpha = 1$$

$$\sin^2\alpha = 1 - \tfrac{16}{25} = \tfrac{9}{25}$$

$\sin\alpha > 0$ in second quadrant so $\sin\alpha = \frac{3}{5}$.

(b) Now $\tan\alpha = \dfrac{\sin\alpha}{\cos\alpha} = \dfrac{\frac{3}{5}}{-\frac{4}{5}} = -\frac{3}{4}$ and the sign is confirmed as negative because the angle is in the second quadrant.

Example 17

Given that $-90° < \beta < 0$ and $\tan\beta = -\frac{2}{3}$ find the exact value of $\cos\beta$.

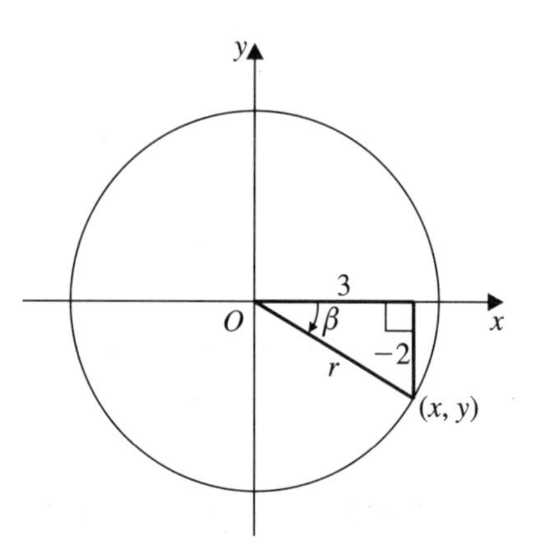

The angle β is in the fourth quadrant, as shown.
Let r be the radius of the circle.
Using Pythagoras, $r = \sqrt{[(-2)^2 + 3^2]} = \sqrt{13}$

$$\cos\beta = \frac{x}{r} = \frac{3}{\sqrt{13}}$$

The word 'exact' implies that a surd answer is required here.

2.7 Solving simple equations in a given interval

Equations in trigonometry can have many solutions.

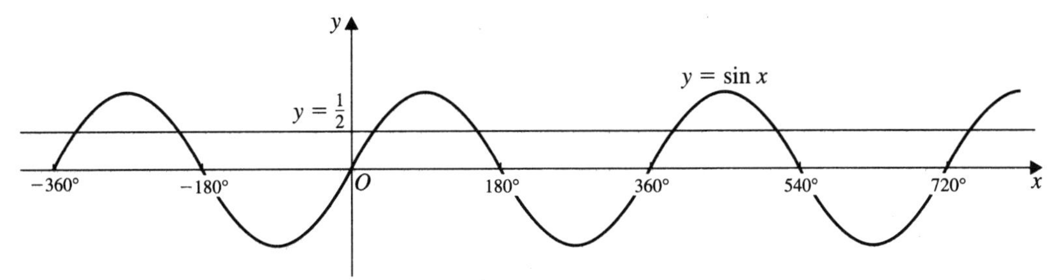

For example, an equation $\sin x = \frac{1}{2}$ has solutions at all points where the curve with equation $y = \sin x$ meets the line $y = \frac{1}{2}$, and seven of these are shown. So it is usual to limit the set of possible values of x.

Example 18
Find the values of x in the interval $0 \leqslant x \leqslant 360°$ for which

(a) $\sin x = \frac{1}{2}$ (b) $\cos x = \frac{1}{\sqrt{2}}$ (c) $\tan x = -\frac{1}{\sqrt{3}}$.

(a) From the above graph $\sin x = \frac{1}{2}$ has two solutions between 0 and 360°:

$$x = 30° \text{ or } 180° - 30° = 150°$$

So solutions are 30°, 150°.

(b) $\cos x = \frac{1}{\sqrt{2}}$ produces an acute angle solution of 45°.

As the cosine is positive also in the fourth quadrant, $\cos(360° - 45°) = \cos 315° = \frac{1}{\sqrt{2}}$ too. Answers are 45°, 315°.

(c) $\tan x = \frac{1}{\sqrt{3}}$ produces an acute angle solution of 30°. The tangent is negative in the second and fourth quadrants.

The solutions of $\tan x = -\frac{1}{\sqrt{3}}$ are therefore

$$180° - 30° \text{ and } 360° - 30°$$

That is, 150° and 330° are the solutions.

Example19
Solve the equation $\cos x = 0.3941$ for $0 < x \leqslant 360°$

$$\cos x = 0.3941$$

So: $x = 66.8°$ from a calculator

But cosine is positive in the first and fourth quadrants. So a second solution lies in the fourth quadrant:

$$x = 360° - 66.8° = 293.2°$$

So the solutions are $x = 66.8°$ and $293.2°$.

Example 20

Solve the equation:

$$\sin 2x = \tfrac{1}{2} \text{ for } 0 < x \leqslant 2\pi$$

$$\sin 2x = \tfrac{1}{2}$$

So: $$2x = 30° = \frac{\pi}{6}$$

But sine is positive in the first *and* second quadrants. So a second solution is:

$$2x = \pi - \frac{\pi}{6} = \frac{5\pi}{6}$$

Another solution in the first quadrant is:

$$2x = 2\pi + \frac{\pi}{6} = \frac{13\pi}{6}$$

and yet another in the second quadrant is:

$$2x = 2\pi + \frac{5\pi}{6} = \frac{17\pi}{6}$$

Dividing by 2 gives: $x = \frac{\pi}{12}, \frac{5\pi}{12}, \frac{13\pi}{12}$ and $\frac{17\pi}{12}$.

Notice that the equation $\sin x = \frac{1}{2}$ has two solutions, namely $\frac{\pi}{6}$ and $\frac{5\pi}{6}$ in the interval $(0, 2\pi)$.

Notice too that $\sin 2x = \frac{1}{2}$ has four solutions in the same interval. Similarly, $\sin 3x = \frac{1}{2}$ has six solutions: you should check this statement for yourself.

Example 21

Solve the equation:

$$3 \sin x = 5 \cos x \text{ for } 0 \leqslant x \leqslant 2\pi$$

Divide each side by $\cos x$:

$$3\frac{\sin x}{\cos x} = 5$$

That is: $$3 \tan x = 5$$

So: $$\tan x = \tfrac{5}{3}$$

$$x = 1.03 \text{ (2 d.p.)}$$

Tan x is also positive in the third quadrant and in this case

$$x = \pi + 1.03$$

$$= 4.17 \text{ (2 d.p.)}$$

Example 22

Find the values of x in the interval $-\pi \leqslant x \leqslant \pi$ for which

$$4\cos^2 x + \cos x + 1 = 2\sin^2 x$$

You know that $\cos^2 x + \sin^2 x \equiv 1$ for all values of x.

So: $$\sin^2 x \equiv 1 - \cos^2 x$$

The equation given is $4\cos^2 x + \cos x + 1 = 2\sin^2 x$, that is:

$$4\cos^2 x + \cos x + 1 = 2(1 - \cos^2 x)$$

So: $$4\cos^2 x + \cos x + 1 = 2 - 2\cos^2 x$$

and: $$6\cos^2 x + \cos x - 1 = 0$$

This is a quadratic in $\cos x$ which factorises into

$$(3\cos x - 1)(2\cos x + 1) = 0$$

So $$\cos x = \tfrac{1}{3} \text{ or } \cos x = -\tfrac{1}{2}$$

For $-\pi \leqslant x \leqslant \pi$, you know that $\cos x$ has positive and negative values like this:

x	$-\pi \leqslant x < -\frac{\pi}{2}$	$-\frac{\pi}{2} < x < 0$	$0 \leqslant x < \frac{\pi}{2}$	$\frac{\pi}{2} < x \leqslant \pi$
$\cos x$ (signs)	$-$	$+$	$+$	$-$

$\cos x = \frac{1}{3} \Rightarrow x = 1.23$ or $x = -1.23$

$\cos x = -\frac{1}{2} \Rightarrow x = 2.09$ or $x = -2.09$

Exercise 2C

1 Plot the graph of $y = 1 + \cos 2x$ for $0 \leqslant x \leqslant 360°$. From your graph find approximate solutions to:

(a) $1 + \cos 2x = 1.7$ (b) $1 + \cos 2x = 0.3$

(c) $\cos 2x = 0.3$.

2 Plot the graph of $y = \cos \frac{1}{2}x$ for $0 \leqslant x \leqslant 360°$. Hence find approximate solutions to:

(a) $\cos \frac{1}{2}x = 0.7$ (b) $\cos \frac{1}{2}x = -0.3$.

Solve the following equations, giving all solutions in the interval $0 < x \leqslant 360°$.

3 $\sin x = 0.332$

4 $\tan x = -1.102$

5 $\cos x = -0.7486$

6 $\tan x = 0.3842$

7 $\cos(x + 20°) = 0.7615$

8 $\tan(x - 33°) = 0.9451$

9 $\sin 2x = 0.4751$

10 $\cos 3x = 0.5196$

11 $\sin(2x - 20°) = 0.6348$

12 $\sin(70° - x) = 0.3313$

13 $\cos\left(\frac{1}{2}x - 39.6°\right) = 0.9144$

14 $\tan(3x - 40°) = 0.61$

15 $\sin x - (3 \sin x - 1) = 0$

16 $\cos x(4 \sin x - 3) = 0$

17 $8 \sin^2 x - 6 \sin x + 1 = 0$

18 $\tan^2 x + 3 \tan x + 2 = 0$

19 $\sin x - 1 = \cos^2 x$

20 $6 \cos x - 5 \sin x = 0$

21 $\cos x + \sin x = 3 \sin x$

22 $4 \sin x \cos x - 2 \sin x - 2 \cos x + 1 = 0$

23 $(3 \sin x - 1)(2 \tan x + 3) = 0$

24 $\tan 3x = 3$

25 $9 \sin^2 x - 6 \sin x + \cos^2 x = 0$

26 Find the values of x in the interval $0 < x < \pi$ which satisfy the equation

$$\cos 2x = -0.4$$

giving your answers in radians to 2 decimal places. [E]

27 Given that $0 < \theta < 180$, find θ when:

(a) $\tan \theta° = -1$

(b) $\sin 2\theta° = -\frac{1}{2}$ [E]

Find for $0 \leqslant x \leqslant 2\pi$ those values of x for which:

28 $\sin^2 x = \frac{1}{4}$

29 $\cos^2 x = \frac{1}{4}$

30 $\tan^2 x - 2 \tan x = 0$

31 $3 \sin^2 x - 2 \sin x - 1 = 0$

32 $3 \cos^2 x - 2 \cos x = 0$

33 $\sin^2 x + 3 \cos^2 x = 2$

34 $3 \sin x + 3 = \cos^2 x$

35 $3 \sin^2 x - 2 \cos x + 1 = 0$

SUMMARY OF KEY POINTS

1. 1 radian is the angle subtended at the centre of a circle by an arc with length that is equal to the radius of the circle.
2. The length of an arc of a circle, radius r, which subtends an angle θ^c at the centre of the circle is $r\theta$.
3. The area of a sector of a circle, radius r, bounded by an arc which subtends an angle θ^c at the centre of the circle, is $\frac{1}{2}r^2\theta$.
4. In the second quadrant only $\sin\theta$ is positive.
5. In the third quadrant only $\tan\theta$ is positive.
6. In the fourth quadrant only $\cos\theta$ is positive.
7. $\sin 30^\circ = \frac{1}{2}$ $\cos 30^\circ = \frac{\sqrt{3}}{2}$ $\tan 30^\circ = \frac{1}{\sqrt{3}}$
8. $\sin 60^\circ = \frac{\sqrt{3}}{2}$ $\cos 60^\circ = \frac{1}{2}$ $\tan 60^\circ = \sqrt{3}$
9. $\sin 45^\circ = \frac{1}{\sqrt{2}}$ $\cos 45^\circ = \frac{1}{\sqrt{2}}$ $\tan 45^\circ = 1$
10. The curve of $y = \sin nx$, $y = \cos nx$ or $y = \tan nx$ $(n \in \mathbb{R}^+)$ is the same shape as the curve of $y = \sin x$, $y = \cos x$ or $y = \tan x$, but n of each curve fit into the range $0 \leqslant x \leqslant 360^\circ$.
11. The graph of $y = -\sin x$ is a reflection of $y = \sin x$ in the x-axis. The same is true for the cosine and tangent functions.
12. To sketch the graphs of $y = \sin(-x)$, $y = \cos(-x)$ and $y = \tan(-x)$ we use the fact that $\sin(-x) = -\sin x$, $\cos(-x) = +\cos x$ and $\tan(-x) = -\tan x$.
13. The graph of $y = a\sin x$ $(a \in \mathbb{R}^+)$ is a stretch of the graph of $y = \sin x$, scale factor a, from the x-axis. The same is true for the cosine and tangent fucntions.
14. The graph of $y = \sin(x + n)$, where x and n are measured in degrees, is the same shape as the graph of $y = \sin x$ but is translated n° to the left. The same is true for the cosine and tangent functions.
15. $\cos^2\theta + \sin^2\theta \equiv 1$
16. $\tan\theta \equiv \dfrac{\sin\theta}{\cos\theta}$

Coordinate geometry in the *xy*-plane

3

3.1 The equation of a straight line in the form $y - y_1 = m(x - x_1)$

Suppose you are asked to find the equation of the straight line with gradient 3 which passes through the point (2, 11). Take A to be the point (2, 11) and let $P(x, y)$ be *any* other point on the line. Here is a diagram of the situation:

$P(x, y)$ means the point P with coordinates (x, y).

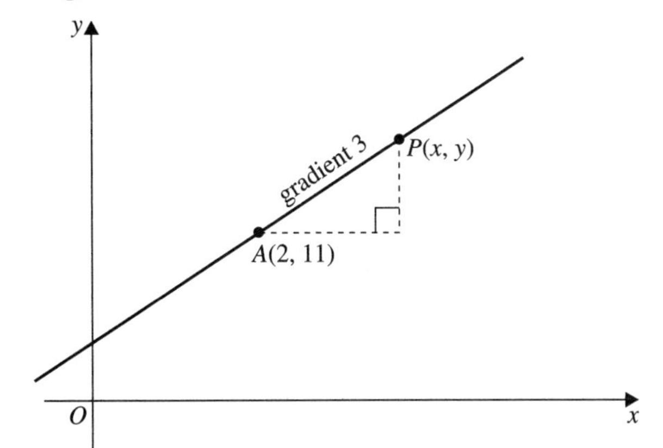

You know that the gradient of the straight line joining the points (x_1, y_1) and (x_2, y_2) is $\frac{y_2 - y_1}{x_2 - x_1}$. So the gradient of AP is $\frac{y - 11}{x - 2}$. But you are told that the gradient of this line is 3. So:

$$\frac{y - 11}{x - 2} = 3$$

i.e.

$$y - 11 = 3(x - 2)$$

$$y - 11 = 3x - 6$$

$$y = 3x + 5$$

This is the equation of the required straight line.

General method of finding the equation of a straight line from one point and the gradient

Let's try to generalise this method. Suppose you wish to find the equation of the straight line which passes through the point with coordinates (x_1, y_1) and which has a gradient of m. Again, let the point with given coordinates be $A(x_1, y_1)$. Take *any* other point on the line to be P and let it have coordinates (x, y). Here is a diagram of the situation:

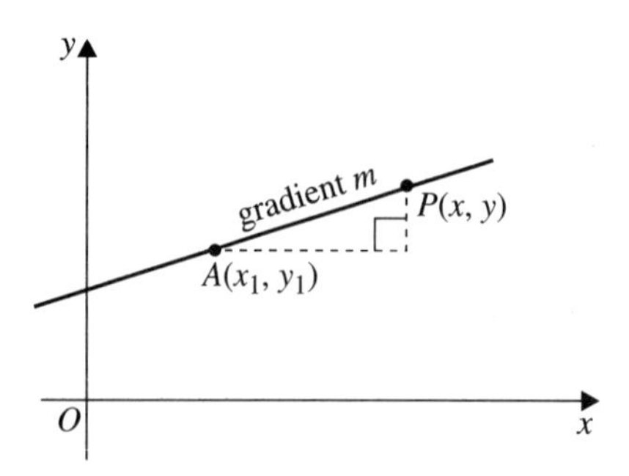

The gradient of AP is:

$$\frac{y - y_1}{x - x_1}$$

You know that the gradient of the line is m, so you have:

$$\frac{y - y_1}{x - x_1} = m$$

or $$y - y_1 = m(x - x_1)$$

- **The equation of a line with gradient m, passing through the point with coordinates (x_1, y_1) is**

$$\boldsymbol{y - y_1 = m(x - x_1)}$$

Example 1

Find the equation of the straight line with gradient $-\frac{1}{2}$ which passes through the point (3, 2).

Assume that the point (x, y) is any point on the line other than (3, 2). Then using the equation

$$\frac{y - y_1}{x - x_1} = m$$

gives: $$\frac{y - 2}{x - 3} = -\frac{1}{2}$$

or $$y - 2 = -\tfrac{1}{2}(x - 3)$$

This can be written: $$2y - 4 = -x + 3$$

or $$2y + x - 7 = 0$$

Example 2

Find the equation of the straight line with gradient $\frac{2}{3}$ which passes through the point $(-1, -4)$.

If the point (x, y) is any point on the line other than $(-1, -4)$ then:

$$\frac{y - y_1}{x - x_1} = m$$

gives:
$$\frac{y - (-4)}{x - (-1)} = \tfrac{2}{3}$$

$$\frac{y + 4}{x + 1} = \tfrac{2}{3}$$

or:
$$y + 4 = \tfrac{2}{3}(x + 1)$$

This can be written:

$$3y + 12 = 2x + 2$$

or:
$$3y - 2x + 10 = 0$$

3.2 The equation of a straight line in the form $\dfrac{y - y_1}{y_2 - y_1} = \dfrac{x - x_1}{x_2 - x_1}$

Now consider the situation where the gradient of the line is not given: instead, *two* points are given, each of which lies on the line. For example, try to find the equation of line which passes through the points (2, 1) and (5, 7).

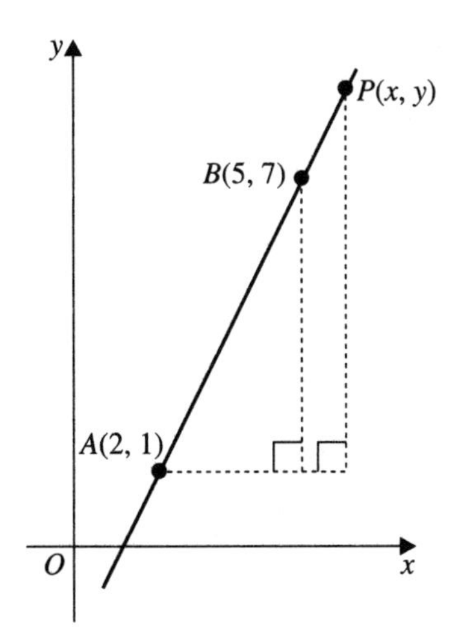

Choose $P(x, y)$ to be any point on the line other than (2, 1) or (5, 7) and use the formula for the gradient of the line joining the two points (x_1, y_1) and (x_2, y_2), which is $\dfrac{y_2 - y_1}{x_2 - x_1}$.

The gradient of the line AP is $\dfrac{y - 1}{x - 2}$ and the gradient of the line AB is $\dfrac{7 - 1}{5 - 2}$.

But ABP is just one straight line and so AP and AB must have the same gradient. Thus:

$$\frac{y-1}{x-2} = \frac{7-1}{5-2} = \frac{6}{3} = 2$$

and:

$$y - 1 = 2(x - 2)$$
$$y - 1 = 2x - 4$$
$$y = 2x - 3$$

General method of finding the equation of a straight line from two points on the line

As before, you can generalise this to find a formula that you can use to solve all such problems. Consider the straight line which passes through the points (x_1, y_1) and (x_2, y_2).

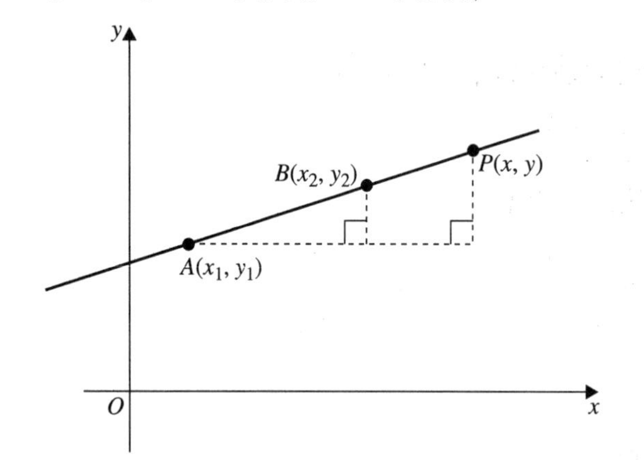

If $P(x, y)$ is any point on the line other than (x_1, y_1) or (x_2, y_2) then the gradient of AP is:

$$\frac{y - y_1}{x - x_1}$$

and the gradient of AB is:

$$\frac{y_2 - y_1}{x_2 - x_1}$$

But as AP is part of the same straight line as AB, it must have the same gradient. So:

$$\frac{y - y_1}{x - x_1} = \frac{y_2 - y_1}{x_2 - x_1}$$

or:

$$\frac{y - y_1}{y_2 - y_1} = \frac{x - x_1}{x_2 - x_1}$$

- **The equation of a line passing through the points with coordinates (x_1, y_1) and (x_2, y_2) is**

$$\frac{y - y_1}{y_2 - y_1} = \frac{x - x_1}{x_2 - x_1}$$

Example 3

Find the equation of the straight line that passes through the points (2, –1) and (3, 7).

If (x, y) is any point on the line other than (2, –1) or (3, 7) then:

$$\frac{y-(-1)}{7-(-1)} = \frac{x-2}{3-2}$$

$$\frac{y+1}{8} = \frac{x-2}{1}$$

$$y+1 = 8(x-2)$$

or: $$y+1 = 8x-16$$

so: $$y = 8x-17$$

Example 4

Find the equation of the straight line which passes through the points (–4, – 2) and (3, 6).

If (x, y) is any point on the line other than (–4, – 2) or (3, 6) then:

$$\frac{y-(-2)}{6-(-2)} = \frac{x-(-4)}{3-(-4)}$$

$$\frac{y+2}{6+2} = \frac{x+4}{3+4}$$

So: $$7(y+2) = 8(x+4)$$

i.e. $$7y+14 = 8x+32$$

or $$7y = 8x+18$$

You could use a slightly different approach in examples 3 and 4 by finding the gradient of the line and then using the form $y - y_1 = m(x - x_1)$ described on p. 78. The equation found is then checked by using the point (x_2, y_2).

Example 5

Find the equation of the straight line which passes through the points (2, –1) and (3, 7).

$$m = \frac{7-(-1)}{3-2} = 8$$

The equation of the line is

$$y-(-1) = 8(x-2)$$

$$y+1 = 8(x-2)$$

Checking with (3, 7) gives:

$$\text{LHS} = 7 + 1 = 8$$
$$\text{RHS} = 8(3 - 2) = 8$$

showing that the equation is correct. You could rearrange the equation into the more usual form:

$$y + 1 = 8(x - 2)$$
$$y + 1 = 8x - 16$$
$$y = 8x - 16 - 1$$

So: $$y = 8x - 17$$

Example 6

Find the equation of the straight line which passes through the points (−4, − 2) and (3, 6).

$$m = \frac{6 - (-2)}{3 - (-4)} = \frac{6 + 2}{3 + 4} = \tfrac{8}{7}$$

The equation of the line is:

$$y - (-2) = \tfrac{8}{7}\,[x - (-4)]$$
$$y + 2 = \tfrac{8}{7}\,(x + 4)$$

Checking with (3, 6) gives:

$$\text{LHS} = 6 + 2 = 8$$
$$\text{RHS} = \tfrac{8}{7}\,(3 + 4) = \tfrac{8}{7} \times 7 = 8$$

Once again, you can rearrange the equation into the more usual form:

$$y + 2 = \tfrac{8}{7}\,(x + 4)$$
$$7(y + 2) = 8(x + 4)$$
$$7y + 14 = 8x + 32$$
$$7y = 8x + 32 - 14$$

So: $$7y = 8x + 18$$

3.3 The general equation of a straight line

Any straight line equation can be expressed in the form $ax + by + c = 0$, where a, b and c are constants and the point $P(x, y)$ lies on the line. You can deduce a number of facts from this equation.

(i) By putting $y = 0$, you can see that $ax + c = 0$, that is $x = -\frac{c}{a}$.

So the line meets the x-axis at the point $\left(-\frac{c}{a}, 0\right)$.

(ii) By putting $x = 0$, you can see that $by + c = 0$, that is $y = -\frac{c}{b}$.

So the line meets the y-axis at the point $\left(0, -\frac{c}{b}\right)$.

(iii) If you rearrange the equation $ax + by + c = 0$ into the form $y = -\frac{a}{b}x - \frac{c}{b}$, you can compare it with the equation $y = mx + c'$ and deduce that m, the gradient of the line, is $-\frac{a}{b}$, and that c', the y-intercept, is $-\frac{c}{b}$ as shown in (ii) above.

When sketching a line whose equation is given in the form $ax + by + c = 0$, all that you need to do is find the x- and y-intercepts.

Example 7

Sketch the line with equation $3x - 4y - 12 = 0$, showing the x- and y-intercepts. Find the gradient of the line.

Consider the equation $3x - 4y - 12 = 0$.

When $y = 0$,

$$3x - 12 = 0 \Rightarrow x = 4$$

When $x = 0$,

$$-4y - 12 = 0 \Rightarrow y = -3$$

So the points where the line meets the coordinate axes have coordinates (4, 0) and (0, −3), as shown below.

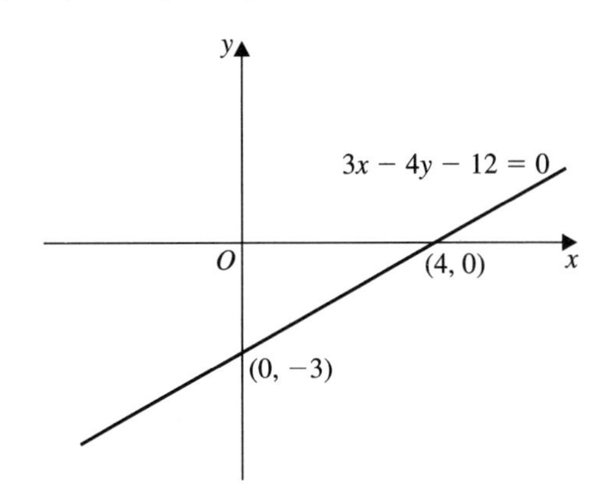

You can find the gradient in either of these ways:

Rearrange the equation $3x - 4y - 12 = 0$ into the $y = mx + c$ form which is

$$y = \tfrac{3}{4}x - 3$$

and then $m = \frac{3}{4}$.

OR

From triangle OAB you can see that

$$\frac{\text{increase in } y}{\text{increase in } x} = \frac{3}{4}$$

This is the gradient of the line by definition.

Exercise 3A

A straight line has gradient m and passes through the point (x_1, y_1). Find, in the form $ax + by + c = 0$, an equation of the line when:

1 $m = 2$; $(x_1, y_1) = (5, 1)$

2 $m = -3$; $(x_1, y_1) = (8, 3)$

3 $m = \frac{1}{2}$; $(x_1, y_1) = \left(-\frac{1}{2}, 5\right)$

4 $m = -\frac{4}{5}$; $(x_1, y_1) = (3, -2)$

5 $m = 3$; $(x_1, y_1) = \left(1, -\frac{1}{2}\right)$

6 $m = -7$; $(x_1, y_1) = (3, 6)$

In questions 7–14, the coordinates of two points on a line are given. Find an equation of the line in the form $ax + by + c = 0$.

7 (5, 3) and (2, 1)

8 (3, 7) and (2, 4)

9 (−2, 3) and (3, −2)

10 (−1, 4) and (2, 3)

11 (−1, −7) and (−2, 6)

12 (4, −2) and (3, −8)

13 (4, −3) and (6, 0)

14 (−1, −5) and (−2, −3)

In questions 15–20, find the coordinates of the points where the line whose equation is given cuts the coordinate axes.

15 $2x + 3y - 6 = 0$

16 $3x - 5y + 15 = 0$

17 $2x + y + 7 = 0$

18 $4x - 3y - 9 = 0$

19 $5x - 11y + 22 = 0$

20 $4y - 6x - 16 = 0$

In questions 21–26, find the gradient of the line whose equation is given.

21 $2x - y - 7 = 0$

22 $x - 3y + 2 = 0$

23 $3y - 4x = 12$

24 $5x + 9y - 12 = 0$

25 $x = 3y - 7$

26 $13y + 17x = 102$

27 The lines with equations $7x + y - 10 = 0$ and $2x + y = 0$ meet at the point A. Find an equation of the line which passes through A and the point B whose coordinates are $(5, -3)$.

28 The lines with equations $5x + 6y = 45$ and $3y - x - 5 = 0$ meet at the point A. Find an equation of the line through A whose gradient is 2.

29 A line whose gradient is $\frac{2}{3}$ passes through the point of intersection of the lines with equations $3x + 2y = 13$ and $2x + 3y = 12$. Show that this line passes through the origin O.

30 The lines AB and AC have gradient 1 and -1 respectively and A is the point $(0, -1)$. The line BC has equation $y = 1$. Find:

(a) equations for the lines AB and AC

(b) the area of $\triangle ABC$

3.4 Parallel and perpendicular lines

Two lines are parallel if their gradients are equal. For example, the lines with equations $y = 3x + 2$ and $y = 3x - 4$ are parallel, because each has a gradient equal to 3.

Example 8

Find an equation of the line parallel to the line with equation $3x - 4y = 7$ and which passes through the point $(2, -3)$.

The line $3x - 4y = 7$ can be written as:

$$4y = 3x - 7$$

that is:

$$y = \tfrac{3}{4}x - \tfrac{7}{4}$$

This line has gradient $\frac{3}{4}$.

So a line parallel to this also has gradient $\frac{3}{4}$. The line with gradient $\frac{3}{4}$ passing through the point $(2, -3)$ has equation.

$$y - (-3) = \tfrac{3}{4}(x - 2)$$

$$y + 3 = \tfrac{3}{4}x - \tfrac{6}{4}$$

$$4y + 12 = 3x - 6$$

That is: $$3x - 4y = 18$$

The line l with equation $y = mx + c$ is shown, where l meets the y-axis at $P(0, c)$ and the x-axis at $Q\left(-\frac{c}{m}, 0\right)$.

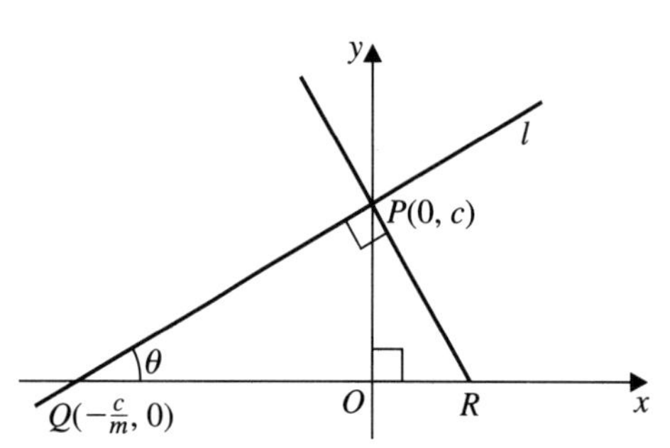

The line PR is perpendicular to l and $\angle PQR = \theta$.

In $\triangle PQR$, $\angle PRQ = 90° - \theta$ (angles in a triangle add up to 180°).

In $\triangle POR$, $\angle OPR = 90° - (90° - \theta) = 90° - 90° + \theta = \theta$

(since again, angles in a triangle add up to 180°).

Now in $\triangle POR$: $\tan\theta = \frac{OR}{OP}$

But $OP = c$

so $\frac{OR}{c} = \tan\theta$

$OR = c\tan\theta$

The gradient of l is m, from $y = mx + c$. The gradient of l can also be found by looking at $\triangle POQ$:

$$\text{gradient of line } l = \frac{OP}{OQ} = \tan\theta$$

So $m = \tan\theta$

and $OR = cm$

Now a line which slopes down from the left to right has a negative gradient (see p. 216).

So gradient of line $PR = -\frac{OP}{OR} = -\frac{c}{cm} = -\frac{1}{m}$

Therefore all lines perpendicular to $y = mx + c$ have gradient $-\frac{1}{m}$, because they must have the same gradient as the line PR since they are parallel to PR.

Now (gradient of QP) × (gradient of PR) $= m \times \left(-\frac{1}{m}\right) = -1$

■ **Two lines are perpendicular if the product of their gradients is −1.**

Example 9

Find an equation of the line passing through the point (4, 5) which is perpendicular to the line with equation $2x - 3y = 7$.

Rearranging $2x - 3y = 7$ into the form

$$y = \tfrac{2}{3}x - \tfrac{7}{3}$$

shows that the gradient of this line is $\tfrac{2}{3}$.

Any line perpendicular to the line $2x - 3y = 7$ has gradient m, where

$$m \times \tfrac{2}{3} = -1$$

That is: $$m = \frac{-1}{\frac{2}{3}} = -\frac{3}{2}$$

So an equation of the required line is

$$y - 5 = -\tfrac{3}{2}(x - 4)$$

$$2y - 10 = -3x + 12$$

That is: $$3x + 2y = 22.$$

Example 10

The points $A(-1, 3)$, $B(3, 0)$ and $C(1, 5)$ are given. Find:

(a) the distance between A and B

(b) an equation of the line through C which is perpendicular to AB.

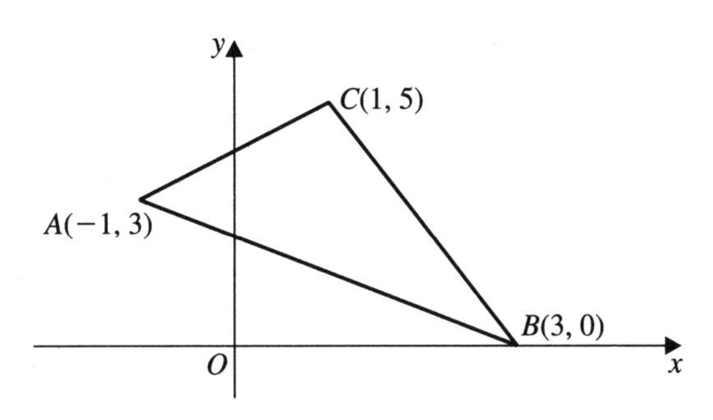

(a) Using $AB^2 = (x_1 - x_2)^2 + (y_1 - y_2)^2$ gives

$$AB^2 = (-1 - 3)^2 + (3 - 0)^2 = 16 + 9 = 25$$

So: $$AB = 5$$

(b) Using $\dfrac{y_2 - y_1}{x_2 - x_1}$,

$$\text{gradient of line } AB = \frac{0 - 3}{3 - (-1)} = \frac{-3}{4} = -\tfrac{3}{4}$$

The gradient of the line through C perpendicular to AB is m, where

$$m \times \left(-\tfrac{3}{4}\right) = -1$$

That is: $$m = \frac{-1}{\left(-\frac{3}{4}\right)} = \frac{4}{3}$$

So an equation of the line is: $$y - 5 = \tfrac{4}{3}(x - 1)$$

$$3y - 15 = 4x - 4$$

That is: $$3y - 4x = 11$$

3.5 The mid-point of the line joining (x_1, y_1) and (x_2, y_2)

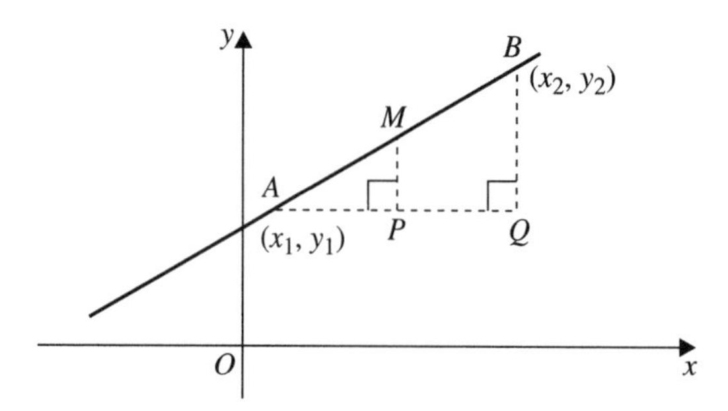

Suppose that M is the mid-point of the line joining $A(x_1, y_1)$ and $B(x_2, y_2)$.

$$AM = MB \text{ so } AP = PQ \text{ (see diagram)}$$

If M has x-coordinate X, then P also has x-coordinate X.

$$AP = X - x_1 \text{ and } PQ = x_2 - X$$

So: $$X - x_1 = x_2 - X$$

$$2X = x_1 + x_2$$

$$X = \tfrac{1}{2}(x_1 + x_2)$$

Similarly, the y-coordinate of M is $\frac{1}{2}(y_1 + y_2)$.

- **The mid-point of the line joining (x_1, y_1) to (x_2, y_2) has coordinates**

$$\left(\tfrac{1}{2}(x_1 + x_2),\ \tfrac{1}{2}(y_1 + y_2)\right)$$

Example 11

The points $A(4, 3)$, $B(6, 1)$ and $C(10, 9)$ are given. The mid-points of AC and BC are M and N respectively.

(a) Find the coordinates of M and N.

(b) Show that MN is parallel to AB.

(a) M is the point $\left(\frac{4+10}{2}, \frac{3+9}{2}\right) \equiv (7, 6)$

The symbol $\equiv$ means **identically equal to**.

N is the point $\left(\frac{6+10}{2}, \frac{1+9}{2}\right) \equiv (8, 5)$

(b) Gradient of $AB = \frac{1-3}{6-4} = \frac{-2}{2} = -1$

Gradient of $MN = \frac{5-6}{8-7} = -1$

Since the gradients of AB and MN are equal the lines are parallel.

Example 12

Find an equation of the perpendicular bisector of the line segment AB where A is $(2, -5)$ and B is $(4, 1)$.

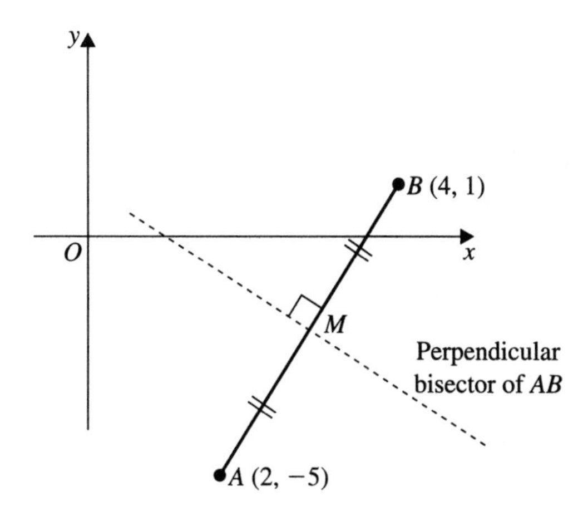

The perpendicular bisector of AB is the line which passes through M, the mid-point of AB, and is at right angles to AB, as shown by the dotted line in the diagram.

M is the point $\left(\frac{2+4}{2}, \frac{-5+1}{2}\right) \equiv (3, -2)$

$$\text{Gradient of } AB = \frac{1-(-5)}{4-2} = \frac{6}{2} = 3$$

So: Gradient of perpendicular bisector $= -\frac{1}{3}$

The equation of the perpendicular bisector of AB is:

$$y - (-2) = -\tfrac{1}{3}(x-3)$$

$$3y + 6 = -x + 3$$

That is: $x + 3y = -3$

Exercise 3B

In questions 1–6 find the coordinates of M, the mid-point of AB, where the coordinates of A and B are:

1 $A(3, 7)$, $B(9, 4)$

2 $A(3, -5)$, $B(6, -3)$

3 $A(-5, -2)$, $B(-7, -6)$

4 $A(-3, 4)$, $B(5, 4)$

5 $A(-3.5, 4.7)$, $B(7.5, 1.3)$

6 $A(2.5, -3.9)$, $B(-8.7, 0.9)$

In questions 7–12, find the equation of the perpendicular bisector of AB for the coordinates given in questions 1–6 respectively.

13 Find an equation of the line which passes through the point $(3, -5)$ and is perpendicular to the line with equation $x + y = 2$.

14 The lines with equations $3x - 5y = 2$ and $4x + ky = 7$ are parallel. Find the value of k.

15 The lines with equations $y = 5x - 6$ and $10x + cy = 8$ are perpendicular. Find the value of c.

16 Find an equation of the straight line:

(a) parallel to the line $y = 4x - 5$, passing through $(2, 3)$

(b) parallel to the line $y = 4 - 6x$, passing through $(-1, 3)$

(c) parallel to the line $2y + 3x = 7$, passing through $(2, -5)$.

17 Find an equation of the straight line:

(a) perpendicular to the line $y = 3x + 5$, passing through $(1, 7)$

(b) perpendicular to the line $y = 2 - 5x$, passing through $(3, -5)$

(c) perpendicular to the line $3y + 2x = 7$, passing through $(1, -\frac{1}{2})$.

18 Show that $\triangle ABC$, where A is $(0, 2)$, B is $(8, 6)$ and C is $(2, 8)$, contains a right angle.

19 $\triangle DEF$, where D is $(-2, 0)$, E is $(\frac{1}{2}, y)$ and F is $(3\frac{1}{2}, -3\frac{1}{2})$ is right-angled at E. Determine the value of y.

20 The quadrilateral $ABCD$, where A is $(4, 5)$ and C is $(3, -2)$, is a square. Find the coordinates of B and D and the area of the square.

21 A, B, and C are the points $(1, 0)$, $(4, -1)$, and $(6, 5)$ respectively. Prove that the angle ABC is a right angle.
If A, B, and C are three vertices of the rectangle $ABCD$, calculate

(a) the coordinates of the point of intersection of the diagonals of the rectangle

(b) the coordinates of D

(c) the area of the rectangle. [E]

22 The line l passes through the point (1, 4) and is perpendicular to the line with equation $x - 2y - 7 = 0$.

(a) Find an equation for l.

(b) Find the coordinates of the point where the lines intersect.

SUMMARY OF KEY POINTS

1 The equation of a line with gradient m, passing through the point with coordinates (x_1, y_1) is

$$y - y_1 = m(x - x_1)$$

2 The equation of a line passing through the points with coordinates (x_1, y_1) and (x_2, y_2) is

$$\frac{y - y_1}{y_2 - y_1} = \frac{x - x_1}{x_2 - x_1}$$

3 The line with equation $ax + by + c = 0$ meets the x-axis at the point $\left(-\frac{c}{a}, 0\right)$, meets the y-axis at the point $\left(0, -\frac{c}{b}\right)$ and has gradient $-\frac{a}{b}$.

4 If two lines are parallel, their gradients are equal.

5 If the lines $y = m_1x + c_1$ and $y = m_2x + c_2$ are perpendicular then $m_1m_2 = -1$.

6 The mid-point of the line joining (x_1, y_1) to (x_2, y_2) has coordinates $\left(\frac{x_1 + x_2}{2}, \frac{y_1 + y_2}{2}\right)$.

Review exercise 1

1 Solve the simultaneous equations

$$2x^2 + 3xy + y^2 = 0$$
$$3x + y = 2$$

[E]

2 Without using a calculator, solve the equation

$$3^{(2x-1)} = 9^{-x}$$

[E]

3 Solve, for $0 \leqslant \theta < 360$, the equation

$$3\cos^2\theta° - 2\sin\theta° = 2$$

giving your solutions to one decimal place where appropriate.

[E]

4 The line with equation $y = 3x + 1$ cuts the curve with equation $y = x^2 + 4x - 5$ at the points A and B. Calculate the length AB. [E]

5 (a) Solve the equation $2x^2 - 7x - 3 = 0$, giving your answers correct to 2 decimal places.

(b) Solve the equation

$$\frac{(3x - 8)}{2} = \frac{(6x + 1)}{5}$$

[E]

6 (a) Sketch the curve with equation

$$y = 3 + \cos x°$$

in the interval $-360 \leqslant x \leqslant 360$.

(b) On the same axes, sketch the curve with equation

$$y = 3\sin x°$$

in the same interval.

(c) On your diagram, shade the regions for which simultaneously

$y < 3\sin x°$ and $y > (3 + \cos x°)$, for $-360 \leqslant x \leqslant 360$. [E]

7 Find the coordinates of the point where the line through $(-3, 13)$ and $(6, 10)$ cuts the line through $(1, 5)$ with gradient 3.

8 Given that $(x + 4)$ is a factor of the expression $3x^3 + x^2 + px + 24$, find the value of the constant p. Hence factorise the expression completely. [E]

9 Solve the following inequalities.

(a) $\frac{1}{2}x + 2 \geqslant x - 5$

(b) $x^2 + 4 < 20$ [E]

10 The curve $y = p \sin x + q \cos x$ passes through the points $(0, 3)$ and $\left(\frac{\pi}{4}, 0\right)$. Find the values of p and q. [E]

11 Calculate the coordinates of the point of intersection of the line joining $(5, -4)$ to $(-3, 4)$ with the line joining $(-13, 4)$ to $(-5, 1)$.

12 The line with equation $y = x + 2$ intersects the curve with equation $x^2 + y^2 = 12 - 2x$ at the points A and B. The coordinates of A are both positive. Calculate the coordinates of (a) the point A, (b) the point B. (c) Hence calculate the length AB. [E]

13 (a) Factorise completely $3x^2 - 3$ and hence simplify

$$\frac{x^2 - 5x + 4}{3x^2 - 3}$$

(b) Solve the equation $3x^2 - 6x - 10 = 0$, giving your answers to 2 decimal places. [E]

14 The diagram represents a circle of radius 10 cm and centre C. Points A and B are taken on the circumference of the circle so that $\angle ACB = 2$ radians. The shaded region R is bounded by the radii CA and CB and the major arc ADB, as shown. Calculate:

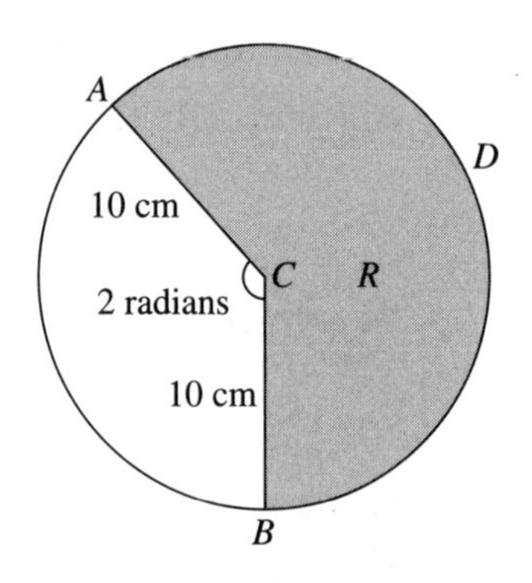

(a) the perimeter of R

(b) the area of R

(c) the area in cm^2 of $\triangle CAB$, giving your answer to 1 decimal place. [E]

15 Find the value of k given that $x^3 + kx^2 - 2x + 1$ is divisible by $x + 2$.

16 Find equations for the lines L_1 and L_2 where (a) L_1 passes through $(3, 2)$ and has gradient 3, (b) L_2 has gradient -2 and y-intercept 13. Find the coordinates of the point of intersection of L_1 and L_2.

17 (a) Solve the equation

$$5x^2 - 7x = 3$$

giving your answers to 2 significant figures.

(b) Multiply $(3x^2 - 2x - 4)$ by $(3 - 5x)$ arranging your answer in ascending powers of x. [E]

18 Rationalise the denominator of $\dfrac{1}{\sqrt{5}(\sqrt{19} + \sqrt{3})}$.

19 Without using a calculator, solve the equation

$$4^{2x-1} = 16^{-\frac{1}{2}x}$$ [E]

20 (a) Solve the equation $3x^2 - 2x - 7 = 0$, giving your answers correct to 3 significant figures.

(b) Given that

$(x^2 - 2x + 3)(x^2 + kx - 2) = x^4 - 15x^3 + qx^2 + px - 6$, find the values of k, p and q. [E]

21 Find in degrees to 1 decimal place in the interval $-180° \leqslant x \leqslant 180°$ the values of x for which

(a) $\sin x = 0.7$ (b) $\cos x = 0.7$ (c) $\tan x = 0.7$ [E]

22 Find the equation of the line joining (2, 7) to (5, 3).

23 Find the set of values of x for which

$$x^2 - x - 12 > 0$$ [E]

24 (a) Find the solutions of the equation

$$x^2 - 5x = 5$$

giving your answers to 2 decimal places.

(b) Find the solution set of the inequality

$$x^2 - 5x > 6$$

25 The curve with equation

$$y = 2 + k \sin x$$

passes through the point with coordinates $\left(\frac{\pi}{2}, -2\right)$. Find the value of k and the greatest value of y. [E]

26 The straight line through the points (1, 4) and (−3, − 4) meets the coordinate axes in the points A and B. Find the area of a square having AB as one of its sides. [E]

27 Given that $0 \leqslant x \leqslant \pi$, find the values of x for which

(a) $\sin 3x = 0.5$

(b) $\tan\left(x + \frac{\pi}{2}\right) = 1$. [E]

28 Given that $p = \sqrt{2}$ and $q = \sqrt{3}$ express $\dfrac{\sqrt{18}+\sqrt{12}}{\sqrt{8}-\sqrt{96}}$ in terms of p and q in its simplest form.

29 Given that for all values of x,

$$3x^2 + 12x + 5 \equiv p(x+q)^2 + r$$

(a) find the values of p, q and r.

(b) Hence, or otherwise, find the minimum value of $3x^2 + 12x + 5$.

(c) Solve the equation $3x^2 + 12x + 5 = 0$, giving your answers to one decimal place.

30 Find the two smallest positive values of x where the curves $y = \sin x$ and $y = -\cos x$ intersect.
Give also the y-coordinates of these intersection points.

31 (a) Solve the equation

$$2x^2 - 7x + 4 = 0$$

giving your answers correct to 3 significant figures.

(b) Multiply $(2x^2 - 3x + 1)$ by $(2 - 3x)$, arranging the answer in descending powers of x. [E]

32 Given that $(x - k)$ is a factor of $\text{f}(x) \equiv kx^3 - 3x^2 - 5kx - 9$, and that $k \in \mathbb{R}$ find the possible values of k and factorise $\text{f}(x)$ for each of these values.

33 The straight line l passes through $A(1, 3\sqrt{3})$ and $B(2+\sqrt{3}, 3+4\sqrt{3})$.

(a) Calculate the gradient of l giving your answer as a surd in its simplest form.

(b) Give the equation of l in the form $y = mx + c$, where constants m and c are surds to be given in their simplest form.

(c) Show that l meets the x-axis at the point $C(-2, 0)$.

(d) Calculate the length of AC.

(e) Find the size of the acute angle between the line AC and the x-axis, giving your answer in degrees. [E]

34 The triangle ABC is equilateral with each side of length 6 cm. With centre A and radius 6 cm, a circular arc is drawn joining B to C. Similar arcs are drawn with centres B and C and with radii 6 cm joining C and A to B respectively, as shown in the diagram. The shaded region R is bounded by the 3 arcs AB, BC and CA.
Calculate, giving your answer in cm² to 3 significant figures,

(a) the area of triangle ABC (b) the area of R. [E]

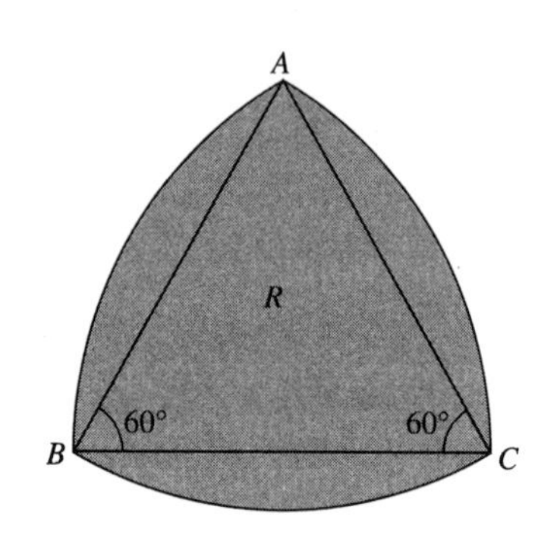

35

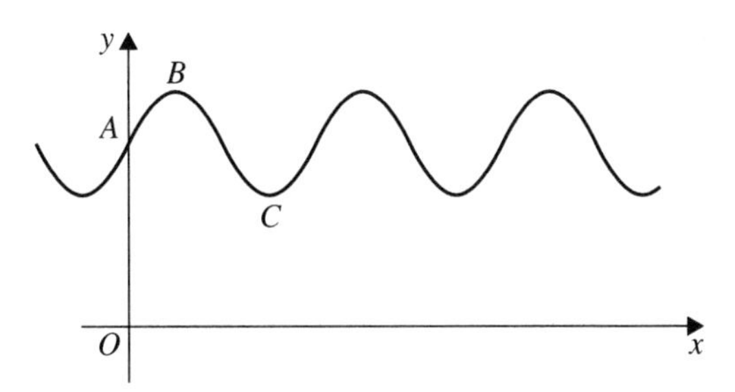

The diagram shows part of the curve with equation $y = 3.1 + \sin 2x$, where x is measured in degrees.

(a) Write down the coordinates of the y-intercept (A), the first maximum for $x > 0$ (B) and the first minimum for $x > 0$ (C), as shown in the diagram.

One solution of the equation $3.1 + \sin 2x = 3.6$ is $x = 15°$.

(b) Find the other solutions of this equation which lie between 0 and 360°. [E]

36 Solve the equation

$$4(2^{x^2}) = 16^x$$ [E]

37 Find the set of values of x for which

$$4x(x - 1) < 3$$ [E]

38 Solve the simultaneous equations

$$x^2 - 2xy + y^2 = 1$$
$$\text{and } 2x - y + 1 = 0$$ [E]

39 (a) Find, as surds, the roots of the equation

$$2(x + 1)(x - 4) - (x - 2)^2 = 0$$

(b) Hence find the set of values of x for which

$$2(x + 1)(x - 4) - (x - 2)^2 > 0$$ [E]

40 For values of x from $-\frac{\pi}{2}$ to $\frac{\pi}{2}$, sketch the curves:

$$y = 0.5 - \sin x$$
$$y = 0.5 + \cos 2x$$

Obtain the coordinates of the point P, at which these curves meet.

41 It is known that $-180° < x < 360°$ and $\tan x = \tan 35°$. Find the possible values of x.

42 (a) Given that $8^y = 4^{2x+1}$, find y in the form $y = px + q$, where p and q are exact fractions.

(b) Solve, giving your answers as exact fractions, the simultaneous equations

$$8^y = 4^{2x+1}$$
$$27^{2y} = 9^{x-3}$$

[E]

43 In the diagram, O is the centre of the circle and AB is a chord. The radius of the circle is 10 cm and $\angle AOB = 150°$. Calculate:

(a) the perimeter of the shaded region

(b) the area of the shaded region.

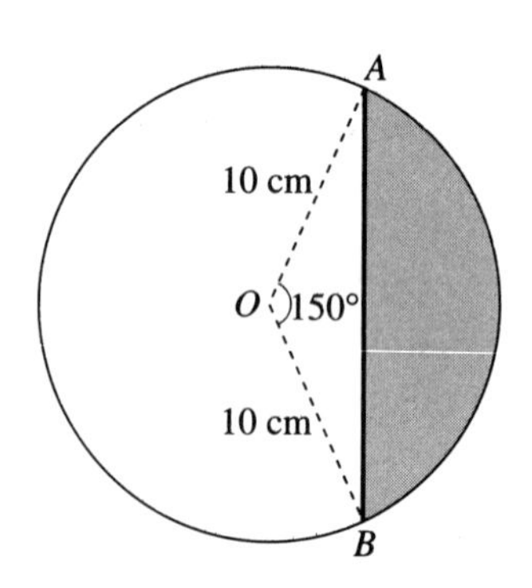

44 Given that

$$(x^2 - 2x + 3)(x^2 + kx - 2) \equiv x^4 - 15x^3 + qx^2 + px - 6$$

find the values of k, p and q. [E]

45 (a) Use algebra to solve $(x - 1)(x + 2) = 18$.

(b) Hence, or otherwise, find the set of values of x for which $(x - 1)(x + 2) > 18$. [E]

46 (a) Find an equation of the line l which passes through the points $A(1, 0)$ and $B(5, 6)$.

The line m with equation $2x + 3y = 15$ meets l at the point C.

(b) Determine the coordinates of C.

The point P lies on m and has x-coordinate -3.

(c) Show, by calculation, that $PA = PB$. [E]

47 Solve, for $0 \leqslant x < 360$,

$$2\sin^2 x° = 1 - \cos x°$$

[E]

48 Given that $27^{3x+1} = 9^y$,

(a) obtain an expression for y in the form $y = ax + b$, where a and b are constants,

(b) solve the equation $27^{3x+1} = 9$, giving your answer as an exact fraction. [E]

49 The line L has equation $3x + 4y - 12 = 0$.

The line M is perpendicular to L and passes through the point $C(6, 6)$.

(a) Find an equation of the line M.

(b) Calculate the perpendicular distance of C from L. [E]

50 Sketch in the same diagram the graphs

$$y = \sin x$$
$$y = \sin 3x \quad \text{for } 0 \leqslant x \leqslant 90^\circ$$

Hence, or otherwise, find the values of x for which
(a) $\sin 3x = 0.5$ (b) $\sin 3x = \sin x$.

51 The polynomial $f(x) \equiv x^3 + px^2 + 2x + q$ is exactly divisible by $(x + 1)$ and by $(x - 2)$. Find the value of p and the value of q and hence factorise $f(x)$.

52 Given that $t^{\frac{1}{3}} = y$, $y \neq 0$,
(a) express $6t^{-\frac{1}{3}}$ in terms of y.
(b) Hence, or otherwise, find the values of t for which $6t^{-\frac{1}{3}} - t^{\frac{1}{3}} = 5$. [E]

53 Solve the simultaneous equations

$$x + 2y = 4$$
$$x^2 + xy - 2y^2 = 4$$
[E]

54 (a) Given that $\tan 75^\circ = 2 + \sqrt{3}$, find in the form $m + n\sqrt{3}$, where m and n are integers, the value of
(i) $\tan 15^\circ$, (ii) $\tan 105^\circ$.
(b) Find, in radians to two decimal places, the values of x in the interval $0 \leqslant x \leqslant 2\pi$ for which $3\sin^2 x + \sin x - 2 = 0$. [E]

55 The vertices of $\triangle ABC$ are the points $A(-1, 5)$, $B(-5, 2)$ and $C(8, -7)$.
(a) Find in the form $px + qy + r = 0$, where p, q and r are integers, an equation of the line passing through B and C.
(b) Show that AB and AC are perpendicular. [E]

56

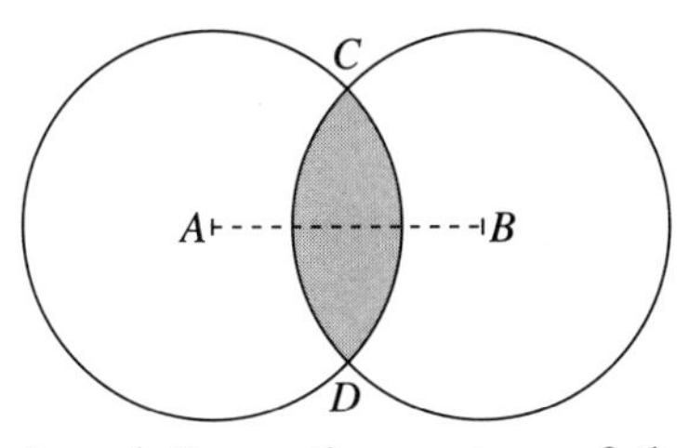

In the diagram, A and B are the centres of the circles, each with radius 10 cm. The circles intersect at C and D. Given that $\angle ACB = 2.33$ radians calculate:
(a) the area of the shaded region
(b) the length of AB to the nearest 0.1 cm.

57 Solve the equation $5(2^x) - 4^x = 4$. [E]

58 Given that $(x+3)$ and $(x-1)$ are both factors of the expression $(x^3 + ax^2 - bx - a)$, calculate the values of the constants a and b. Hence factorise the expression completely.

59 Find the values of a, b and c such that

$$a(x+b)^2 + c \equiv 9x^2 + 72x + 128$$

Hence, or otherwise, find the set of values of x for which

$$9x^2 + 72x + 128 \geqslant 0$$ [E]

60 Find in radians to 2 decimal places in the interval $0 \leqslant x \leqslant 2\pi$ the values of x for which

(a) $\sin x = -0.5$ (b) $\cos x = 0.8$ (c) $\tan x = 3$

61 (a) Find an equation of the straight line passing through the points with coordinates $(-1, 5)$ and $(4, -2)$, giving your answer in the form $ax + by + c = 0$, where a, b and c are integers.
The line crosses the x-axis at the point A and the y-axis at the point B, and O is the origin.
(b) Find the area of $\triangle OAB$.

62 Determine, in degrees, the solutions of each of the equations

(a) $\sin 2x = \frac{1}{2}$ (b) $\sin^2\left(\frac{3x}{2}\right) = \frac{1}{2}$

for which $-180° \leqslant x \leqslant 180°$. [E]

63 Rationalise the denominator of $\dfrac{1}{\sqrt{17} - \sqrt{13}}$

64 Solve the simultaneous equations

$$3x + 2y = 25$$

and

$$xy = 4$$ [E]

65 By substituting $t = x^{\frac{1}{2}}$, or otherwise, find the values of x for which

$$4x + 8 = 33x^{\frac{1}{2}}$$ [E]

66 Find the set of values of x for which
$(x-1)(x-4) < 2(x-4)$. [E]

67 Find the product of $(x-1)$, $(2x+3)$ and $(x+4)$. Express your answer in descending powers of x. [E]

68 (a) Multiply $(2 - 3x - x^2)$ by $(1 - x)$, arranging your answer in ascending powers of x.

(b) Solve the equation $x^2 + 16x - 17 = 0$. [E]

69 Find, giving your answers in terms of π, all values of θ in the interval $0 \leqslant \theta < 2\pi$ for which

(a) $\tan\left(\theta + \frac{\pi}{3}\right) = 1$ (b) $\sin 2\theta = -\frac{\sqrt{3}}{2}$ [E]

70
$$f(x) \equiv 3 + 2\sin(2x + k)^\circ, \; 0 \leqslant x < 360$$

where k is a constant and $0 < k < 360$. The curve with equation $y = f(x)$ passes through the point with coordinates $(15, 3 + \sqrt{3})$.

(a) Show that $k = 30$ is a possible value for k and find the other possible value of k.

Given that $k = 30$,

(b) solve the equation $f(x) = 1$

(c) find the greatest and least values of f

(d) sketch the graph of $y = f(x)$, stating the coordinates of the turning points and the coordinates of the point where the curve meets the y-axis. [E]

71 Rationalise the denominator of $\frac{3\sqrt{11}}{2\sqrt{11} + 5}$.

72 Solve the equation

$$4x^2 + 4x - 7 = 0$$

giving your answers in the form $p \pm q\sqrt{2}$, where p and q are real numbers to be found. [E]

73 (i) Given that $x = 27$, find, without using a calculator, the values of

(a) $x^{\frac{1}{3}}$, (b) $x^{-\frac{2}{3}}$, (c) x^0.

(ii) Given that $2^{2x-1} = \frac{1}{8}$, find the value of x. [E]

74 Solve the simultaneous equations

$$y - 2x = 1$$
$$y^2 = 2x^2 + x$$ [E]

75 Solve, for $-\frac{\pi}{2} < x \leqslant \frac{\pi}{2}$,

(a) $\sin 2x = \cos 2x$, giving your answers in terms of π

(b) $(6\sin x - 1)(\sin x - 2) = -4$, giving your answers to 3 significant figures

(c) $5\sin x - 14 = \frac{2}{\sin x}$, giving your answers to 3 significant figures. [E]

76 Find the set of values of x for which

$$(x - 1)(x - 2) < 6 \qquad \text{[E]}$$

77 Rationalise the denominator of $\dfrac{3 - \sqrt{7}}{\sqrt{7} + 1}$

78 (a) Solve the equation $(2x + 3)^2 = 4$.
(b) Solve the equation $2x^2 + 7x = 11$, giving your answers to one decimal place. [E]

79 The diagram shows a sector OAB of a circle, centre O, of radius 5 cm and a shaded segment of the circle. Given that $\angle AOB = 0.7$ radians, calculate:
(a) the area, in cm^2, of the sector OAB
(b) the area, in cm^2 to 2 significant figures, of the shaded segment. [E]

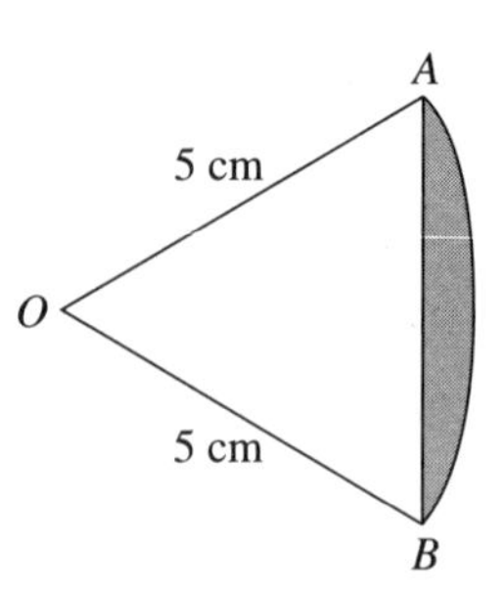

80 The line L passes through the points $A(1, 3)$ and $B(-19, -19)$.
(a) Calculate the distance between A and B.
(b) Find an equation of L in the form $ax + by + c = 0$, where a, b and c are integers. [E]

81 (a) Find the coordinates of the point where the graph of $y = 2\sin(2x + \frac{5}{6}\pi)$ crosses the y-axis.
(b) Find the values of x, where $0 \leqslant x \leqslant 2\pi$, for which $y = \sqrt{2}$. [E]

82 Solve the simultaneous equations

$$2x + 3y = 5$$
$$x^2 + 2xy = 10 + y \qquad \text{[E]}$$

83 List the set of integers n for which

$$2n^2 - 13n + 15 < 0 \qquad \text{[E]}$$

84 Solve, for $-180 \leqslant x \leqslant 180$, to one decimal place where appropriate,
(a) $2\sin 2x° = 3\cos 2x°$
(b) $6\cos^2 x° = 5 - \sin x°$ [E]

85 (a) Given that $(x + 3)$ is a factor of the expression $px^3 + 5x^2 - 16x - 4p$,
(i) find the value of p.

Using the value of p found in (i),

(ii) factorise the expression completely.

(b) Solve the simultaneous equations

$$2x - 3y = 7$$
$$4x^2 - 4xy + 3y^2 = 59$$

[E]

86 Find the set of values of x for which

$$(2x + 1)^2 < 9(4 - x)$$

[E]

87 Rationalise the denominator of $\dfrac{2 - 3\sqrt{5}}{2\sqrt{5} + 1}$

88 Solve the equation $81^x + 81 = 30(9^x)$. [E]

89 (a) Solve the equation $x^2 + x - 9 = 0$, giving your answers correct to 2 decimal places.

(b) Multiply $(x^2 + 2x + 1)$ by $(x - 2)$, arranging your answer in descending powers of x. [E]

90 Solve the simultaneous equations

$$x - 2y = 1$$
$$x^2 - 2xy + 2y^2 = 25$$

[E]

91 (a) Factorise completely $2p^2 - 6q + 6p - 2pq$.

(b) Solve the equation $3x^2 + 3x = 5$, giving your answers to two decimal places. [E]

92 Find the set of values of x for which

$$2x^2 + x - 1 < 0$$

[E]

93 The point P has coordinates (2, 5) and the point Q has coordinates (6, 1).

(a) Write down the coordinates of M, the mid-point of PQ.

(b) Show that the length of PM is $\sqrt{8}$.

(c) Find an equation for the perpendicular bisector of PQ.

The point R with coordinates (x, y) lies on the circle with centre M and diameter PQ.

(d) Using the length of RM, write down an equation in x and y.

Given also that R lies on the perpendicular bisector of PQ,

(e) find the coordinates of the two possible positions of R.

(f) Find the equation of the tangent to the circle at the point with coordinates (6, 5). [E]

94

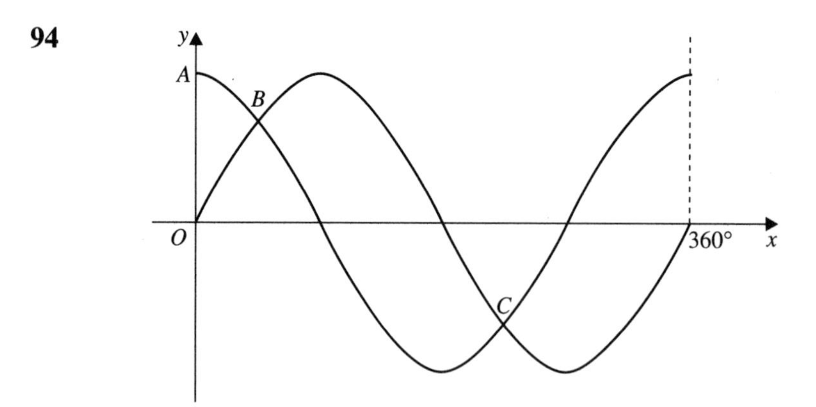

The diagram shows a sketch of $y = \sin x°$ and $y = \cos x°$ for $0 \leqslant x \leqslant 360$.

(a) Sketch a copy of the diagram and identify which graph is $y = \cos x°$.

(b) Find the coordinates of A, B and C.

(c) Sketch the graph of $y = \cos 2x°$ on the same axes.

(d) State how many intersections there are between the curves $y = \sin x°$ and $y = \cos 2x°$ in the interval $0 \leqslant x \leqslant 360$, and hence find the values of x for which $\cos 2x° = \sin x$ in the interval $0 \leqslant x \leqslant 360$.

(e) Name the curve in your diagram which coincides with the curve $y = \sin(x° + 90°)$. [E]

95 Given that $2^{(2x-1)} = \frac{1}{8}$, find the value of x. [E]

96 Given that $(x + 1)$ is a factor of the expression $(2x^3 + ax^2 - 5x - 2)$, find the value of the constant a. Hence factorise the expression completely.

97 Solve the simultaneous equations

$$2y - 3x = 2$$
$$4y^2 - 4xy - 18x^2 = 5$$

[E]

98 The specification for a new rectangular car park states that the length x m is to be 5 m more than the breadth. The perimeter of the car park is to be greater than 32 m.

(a) Form a linear inequality in x.

The area of the car park is to be less than 104 m^2.

(b) Form a quadratic inequality in x.

(c) By solving your inequalities, determine the set of possible values of x. [E]

99 Given that $27^x = 9^{x-1}$, find the value of x. [E]

100

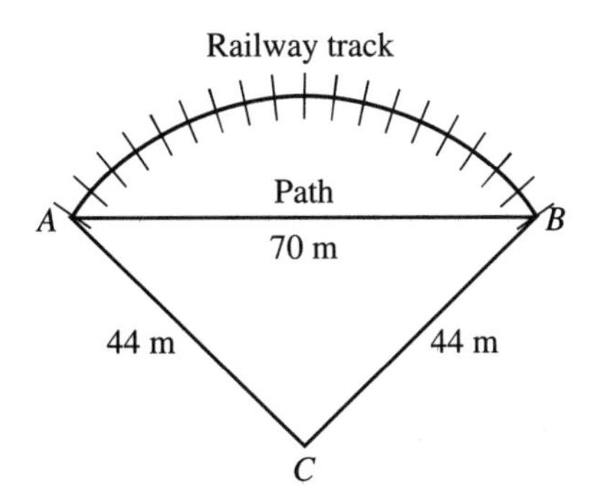

There is a straight path of length 70 m from the point A to the point B. The points are joined also by a railway track in the form of an arc of the circle whose centre is C and whose radius is 44 m, as shown in the diagram.

(a) Show that the size, to 2 decimal places, of $\angle ACB$ is 1.84 radians.

Calculate

(b) the length of the railway track

(c) the shortest distance from C to the path

(d) the area of the region bounded by the railway track and the path. [E]

101 Find the set of values of x for which

$$2x(x-2) < x+3$$ [E]

102 The straight line l_1 passes through the points A and B with coordinates (2, 2) and (6, 0) respectively.

(a) Find an equation of l_1.

The straight line l_2 passes through the point C with coordinates $(-9, 0)$ and has gradient $\frac{1}{4}$.

(b) Find an equation of l_2.

The lines l_1 and l_2 intersect at the point D.

(c) Calculate, to 2 decimal places, the length of AD.

(d) Calculate the area of $\triangle DCB$. [E]

103 Given that $y = 10^x$, show that

(a) $y^2 = 100^x$ (b) $\dfrac{y}{10} = 10^{x-1}$

(c) Using the results from (a) and (b) write the equation

$$100^x - 10\,001(10^{x-1}) + 100 = 0$$

as an equation in y.

(d) By first solving the equation in y, find the values of x which satisfy the given equation in x. [E]

104 (a) Determine the solutions of the equation $\cos(2x - 30)^\circ = 0$ for which $0 \leqslant x \leqslant 360$.

(b)

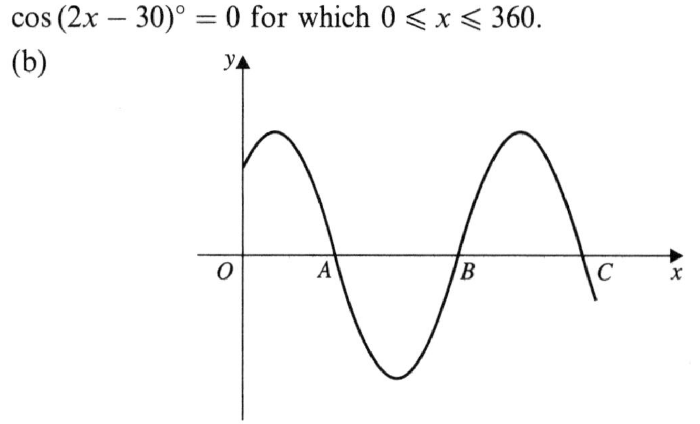

The graph shows part of the curve with equation $y = \cos(px - q)^\circ$, where p and q are positive constants and $q < 180$. The curve cuts the x-axis at points A, B and C, as shown. Given that the coordinates of A and B are (100, 0) and (220, 0) respectively,

(i) write down the coordinates of C,

(ii) find the value of p and the value of q. [E]

105 Solve the equation $5x^2 - 3 = 5x$, giving your answers to 3 significant figures. [E]

106

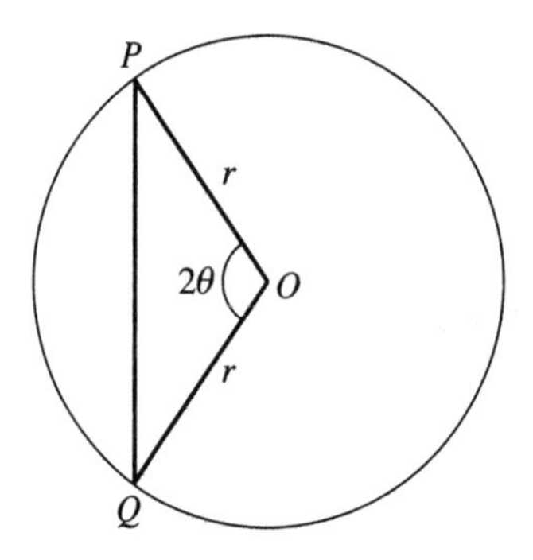

The diagram shows a circle, centre O and radius r. The chord PQ subtends an angle 2θ radians at O, where $0 < \theta < \frac{\pi}{2}$. The length of the minor arc PQ is 1.5 times the length of the chord PQ.

(a) Show that $2\theta - 3\sin\theta = 0$.

(b) Show that the area of $\triangle OPQ$ is $r^2 \sin\theta \cos\theta$.

(c) Hence show that, for this value of θ,

$$\frac{\text{area of sector } OPQ}{\text{area of triangle } OPQ} = \frac{3}{2\cos\theta}$$ [E]

107 (a) Solve the equation

$$2x^2 + x - 2 = 0$$

giving your answers correct to 2 decimal places.

(b) The quadratic equation $x^2 - x + r = 0$ is satisfied when $x = 2$. Find the value of r. [E]

108 (a) $\text{f}(x) \equiv x^4 + px^3 - 8x^2 + qx + 16$, where p and q are constants. $\text{f}(x)$ is exactly divisible by $(x - 4)$ and $(x + 2)$.

(i) Find and simplify two simultaneous equations satisfied by p and q.

(ii) Solve your equations to find p and q.

(b) Solve the simultaneous equations

$$3x - y = 6$$

$$x^2 + xy - y^2 = 4$$ [E]

109

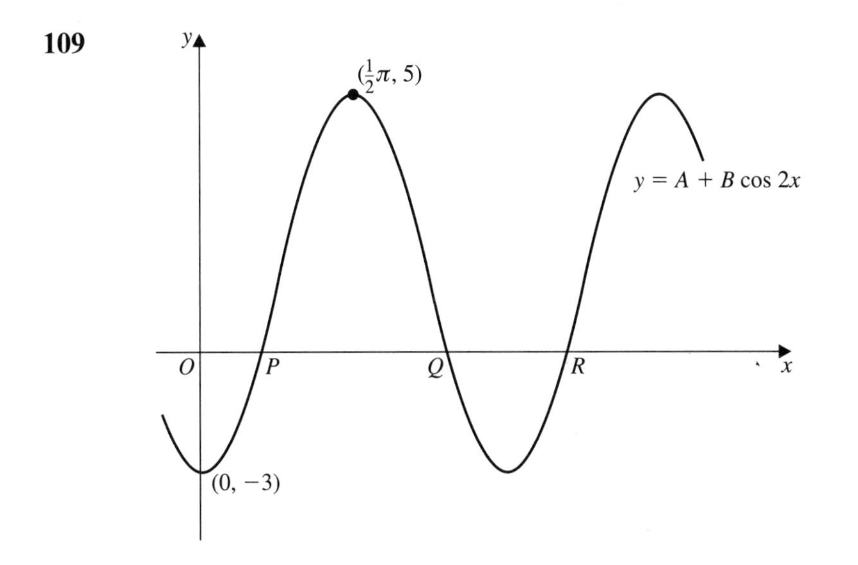

The graph shows a sketch of part of the curve with equation $y = A + B\cos 2x$, where A and B are constants and x is in radians. The curve passes through the points $(0, -3)$ and $(\frac{1}{2}\pi, 5)$ and meets the x-axis at the points P, Q and R, as shown.

(a) Find the values of A and B and hence show that $\dfrac{A}{B} = -\dfrac{1}{4}$.

(b) Hence, or otherwise, determine the x-coordinates of P, Q and R, giving your answers to two decimal places. [E]

110

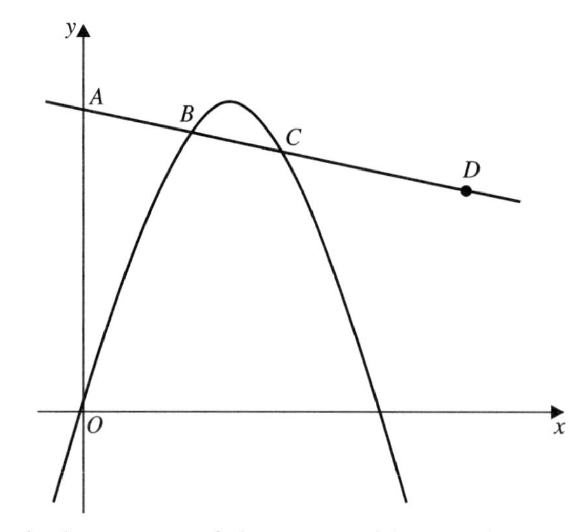

The graph shows part of the curve with equation $y = 16x - kx^2$, where k is a constant. The points A and D have coordinates (0, 18) and (6, 15) respectively.

(a) Calculate, giving your answer to 3 significant figures, the length of AD.

The line l passes through the points A and D and intersects the curve at the points B and C, as shown.

(b) Obtain an equation for l in the form $y = mx + c$, where m and c are constants.

Given also that C has coordinates (4, 16),

(c) show that $k = 3$

(d) calculate the coordinates of B. [E]

111 (a) Given that $(3x + 2)$ is a factor of $3x^3 + Ax^2 - 4x - 4$, show that $A = 5$.

(b) Factorise $3x^3 + 5x^2 - 4x - 4$ completely.

(c) Given that $0° \leqslant t \leqslant 360°$, find the values of t, to the nearest degree, for which

$$3\sin^3 t + 5\sin^2 t - 4\sin t - 4 = 0$$ [E]

Sequences and series

4

4.1 Sequences including those given by a general formula for the *n*th term

A sequence is a succession of numbers formed by following a rule. The numbers in a sequence are sometimes called **terms**. For example, the sequence of numbers produced by the rule $2n + 2$ is:

$$4, \quad 6, \quad 8, \quad 10, \quad 12 \ldots$$

first term second term

The *n*th term is also called the **general term** of the sequence because it can be used to find any term in the sequence. When the general term of a sequence is given in terms of n you can find the first, second, third and fourth terms of the sequence by substituting $n = 1$, $n = 2$, $n = 3$ and $n = 4$ in the general term.

Example 1

The *n*th term of a sequence is $2n - 3$. Find the first four terms and the 15th term.

Take $n = 1$, 2, 3 and 4 in turn to find the terms

$$-1, \quad 1, \quad 3, \quad 5,$$

which are the 1st, 2nd, 3rd and 4th terms of the sequence.

Taking $n = 15$ gives $30 - 3 = 27$, which is the 15th term of the sequence.

Example 2

The first four terms of a sequence are 3, 12, 27, 48. Find a possible *n*th term.

Rewrite the terms as:

$$3 \times 1, \quad 3 \times 4, \quad 3 \times 9, \quad 3 \times 16$$

That is: $3 \times 1^2, \quad 3 \times 2^2, \quad 3 \times 3^2, \quad 3 \times 4^2$

From this pattern, you can see that $3n^2$ is a possible *n*th term for the sequence with first four terms 3, 12, 27, 48.

Notice that the nth terms of the sequences discussed in examples 1 and 2 get progressively larger as n gets larger. Such sequences are called **divergent**.

Example 3

Discuss the sequences whose nth terms are:

(a) $(-1)^n$ (b) $(\frac{1}{2})^n$ (c) $\cos(60n°)$.

(a) The terms of the sequence are $-1, 1, -1, 1, \ldots$ The terms **oscillate** between -1 and 1 successively.

(b) The terms of the sequence are $\frac{1}{2}, \frac{1}{4}, \frac{1}{8}, \frac{1}{16}, \frac{1}{32}, \ldots$ Each term is half of the immediately preceding term. As n increases, succeeding terms get progressively nearer to zero.

Sequences whose nth term approaches a finite number as n approaches infinity are called **convergent** – they converge on or get closer to a number. The number they converge on is sometimes called the **limit** or the limiting value.

(c) If you use your calculator you will find the first six terms of the sequence are $\frac{1}{2}, -\frac{1}{2}, -1, -\frac{1}{2}, \frac{1}{2}, 1, \frac{1}{2}, -\frac{1}{2}, -1, \ldots$ and these terms then repeat from the 7th term to the 12th term, from the 13th term to the 18th term and so on. A sequence that repeats in a set number of terms is called a **periodic sequence**. Here period means the number of terms before the sequence repeats. This sequence is periodic over six successive terms.

Representing a sequence graphically

It is often useful to represent the first few terms of a sequence such as $u_1, u_2, u_3, u_4, \ldots$ graphically. You can do this by plotting points whose coordinates are $(1, u_1), (2, u_2), (3, u_3), (4, u_4), \ldots$ These graphs show the first few terms of the three sequences discussed in example 3. The graphs can help you understand the behaviour of a sequence.

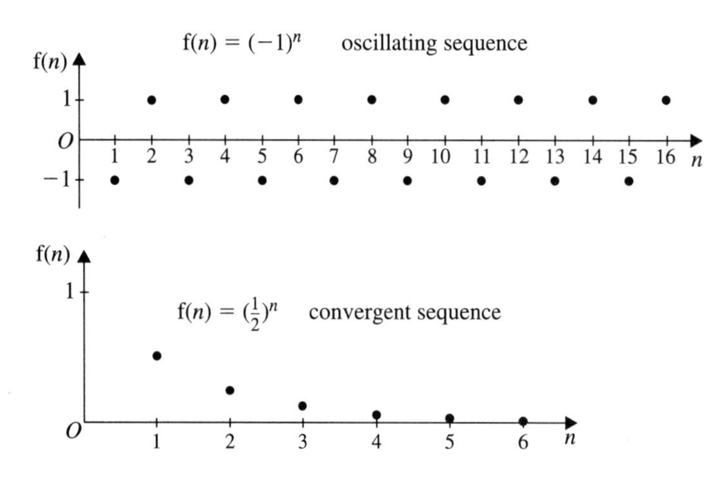

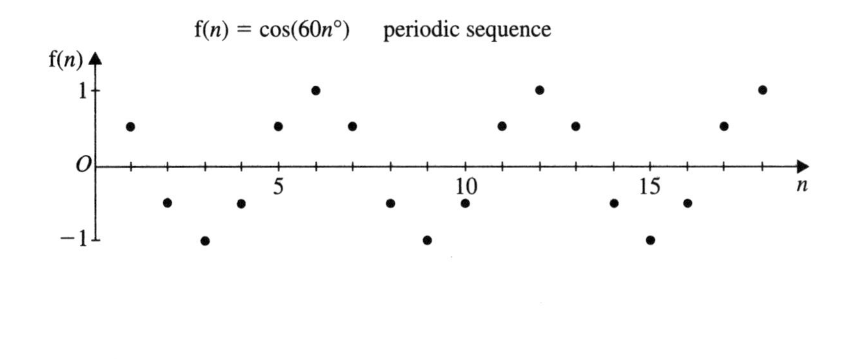

Exercise 4A

1 Write down the first four terms of the sequences whose nth terms are

(a) $3n$ (b) $5n - 4$ (c) $2n^2 - 3$ (d) 2^{n-1}

2 Write down possible 5th and 6th terms for the sequence whose first four terms are given. Find a possible nth term.

(a) 3, 5, 7, 9, . . . (b) 3, 9, 27, 81, . .
(c) $\frac{1}{2}, \frac{2}{3}, \frac{3}{4}, \frac{4}{5}, \ldots$ (d) 5, 9, 17, 33, . . .
(e) 2, 6, 12, 20, . . .

3 The 1st, 2nd and 3rd terms of a sequence are 1, 2 and 5. When asked to give the 4th term, Anne says it is 10, Brian says it is 14 and Cleo says it is 26. Find possible 5th and 6th terms for each of the three students' sequences.

4 The nth term of a sequence is n^{-2}. Write down the first four terms. Is this sequence convergent? Give a reason for your answer.

5 Using your calculator, discuss the sequences whose nth terms are
(a) $\cos(180n° - 60°)$ (b) $\sin(45n°)$ (c) $(\sin 45°)^n$.

6 Find the first four terms of the sequence whose nth term is

$$\frac{n-1}{n+1}$$

Investigate what happens to the nth term of this sequence as n increases.

7 Investigate the following sequences to determine whether or not they converge. State the limiting value of those sequences which converge.

(a) $4, 3\frac{1}{2}, 3\frac{1}{3}, \ldots, 3 + \frac{1}{n}, \ldots$ (b) $3, 2, 1, \ldots, 4 - n, \ldots$
(c) $3, 2, \frac{4}{3}, \ldots, 3(\frac{2}{3})^{n-1}, \ldots$ (d) $3, -2, \frac{4}{3}, \ldots, 3(-\frac{2}{3})^{n-1}$
(e) $3, -1, 3, \ldots, 1 + 2(-1)^{n+1}, \ldots$

8 In the following sequences the terms successively increase, or decrease, by a fixed value. Determine the missing terms and find a general term.

(a) 8, ___, ___, 29, . . .

(b) -1, -7, ___, ___, . . .

(c) 5, ___, -11, ___, . . .

9 At the start of its motion, a pendulum swings through an angle of 30°. Each successive angle of swing is 1% less than the size of the angle in the immediate last swing. Starting with 30°, find the first four terms in the angle-of-swing sequence. Find also the 20th and the 100th angles of swing.

10 The sequence $u_1, u_2, u_3 \ldots u_n, \ldots$ is such that $u_1 = 1$ and

$$u_n = u_{n-1} + \tfrac{3}{2}(2)^n$$

Find the values of u_2, u_3 and u_4. By further investigation, find the smallest value of n for which $u_n > 1000$.

4.2 Arithmetic series

A sequence consists of a set of terms that follow each other in definite order, such as:

$$u_1, u_2, u_3, \ldots, u_n \ldots$$

If you write down the *sum* of the first n terms of the sequence you obtain:

$$u_1 + u_2 + u_3 + \ldots + u_n$$

This is called a **finite series** of n terms. This series is often written as

$$\sum_{r=1}^{n} u_r$$

which means 'the sum of the terms obtained by substituting 1, 2, 3, . . . , n in turn for r in u_r'. The sigma sign Σ is a Greek capital letter S to stand for sum. The numbers below and above it show the lower and upper limits between which the variable is being summed.

If the terms continue infinitely without stopping then

$$u_1 + u_2 + u_3 + \ldots + u_n + \ldots$$

is called an **infinite series**. It is written

$$\sum_{r=1}^{\infty} u_r$$

Example 4

Write down the first four terms and the nth term of the series

$$\sum_{r=1}^{n}(4r-1).$$

Taking $r = 1, 2, 3$ and 4 in $4r - 1$ gives: $(4 - 1)$, $(8 - 1)$, $(12 - 1)$ and $(16 - 1)$ as the first four terms. That is, 3, 7, 11 and 15 are the first four terms and the nth term of the series is $4n - 1$.

Notice that, in this series, each new term is obtained by adding 4 to the previous term.

For example,

$$\text{3rd term} = 4 + \text{2nd term} = 4 + 7 = 11$$
$$\text{4th term} = 4 + \text{3rd term} = 4 + 11 = 15$$

and so on. This property of the series is important and series having this property are called arithmetic series.

An **arithmetic series** is a series in which each term is obtained from the previous term by adding to it or taking from it a constant quantity. The constant quantity is called the **common difference**, because the difference between any two consecutive terms (terms that follow one another) is the same.

Example 5

An arithmetic series has first term 54 and common difference -3. Find the second, third and fourth terms of the series and the nth term.

$$\text{Second term} = 54 - 3 = 51$$
$$\text{Third term} = (54 - 3) - 3 = 54 - 6 = 48$$
$$\text{Fourth term} = [(54 - 3) - 3] - 3 = 54 - 9 = 45$$

Notice this pattern from the first four terms:

$$\begin{array}{ccccccc} \text{1st term} & + & \text{2nd term} & + & \text{3rd term} & + & \text{4th term} \\ 54 & + & 54 - (3 \times 1) & + & 54 - (3 \times 2) & + & 54 - (3 \times 3) \end{array}$$

From this you can deduce that the nth term is

$$54 - 3(n - 1) = 54 - 3n + 3$$

So the nth term is $57 - 3n$.

You can check this because $n = 1$ gives first term as 54, $n = 2$ makes second term 51, etc.

The arithmetic series with first term a and common difference d is:

$$a + (a + d) + (a + 2d) + (a + 3d) + \dots$$

By studying the pattern of successive terms you can see that

- **the nth term is $a + (n - 1)d$**

Example 6

Find the common difference and the 10th term of the arithmetic series given by

$$\sum_{r=1}^{n} (4r - 17)$$

By taking $r = 1$, 2 and 3 in turn you find that the first three terms of the series are:

$$(4 - 17), (8 - 17) \text{ and } (12 - 17)$$

that is:

$$-13, -9 \text{ and } -5$$

The common difference is

$$\begin{aligned} \text{2nd term} - \text{1st term} &= -9 - (-13) \\ &= -9 + 13 \\ &= 4 \end{aligned}$$

The 10th term is found by taking $r = 10$ and is

$$(4 \times 10 - 17) = 40 - 17 = 23$$

The sum of an arithmetic series

Example 7

Find the sum of the first 10 terms of the arithmetic series $6 + 10 + \ldots$

First notice that the common difference is 4 and that the last term (the 10th term in this case) is $6 + 9 \times 4 = 42$. The series is:

$$6 + 10 + 14 + \ldots + 38 + 42$$

Written backwards the series is:

$$42 + 38 + 34 + \ldots + 10 + 6$$

Adding the corresponding terms in these two series gives:

$$48 + 48 + 48 + \ldots + 48 + 48$$

That is, *twice* the sum of the series is $10 \times 48 = 480$. The sum of the series is 240.

The sum of *any* arithmetic series

You can use the method just shown to find the sum S_n of the first n terms of the general arithmetic series

$$S_n = a + (a + d) + (a + 2d) + \ldots + (L - d) + L$$

where L is the nth term (L for 'last').

Rewriting the series in the reverse order gives:

$$S_n = L + (L - d) + (L - 2d) + \dots + (a + d) + a$$

Add these two together to get:

$$2S_n = (a + L) + (a + L) + \dots + (a + L)$$

That is,

$$2S_n = n(a + L) \quad \text{and} \quad S_n = \frac{n}{2}(a + L).$$

Now $L = a + (n - 1)d$ because L is the nth term of the series.

Substituting for L in the expression for S_n gives:

■ $$\boldsymbol{S_n = \frac{n}{2}[2a + (n - 1)d]}$$

You can use this formula to find the sum of any arithmetic series to n terms if you know the first term and the common difference of the series.

Example 8

Find the sum of the first 22 terms of the arithmetic series $6 + 4 + 2 + \dots$

Use the formula $S_n = \frac{n}{2}[2a + (n - 1)d]$, with $n = 22$, $a = 6$ and $d = -2$. This gives:

$$\begin{aligned} S_{22} &= 11[12 + 21(-2)] \\ &= 11 \times [-30] \\ &= -330 \end{aligned}$$

The sum of the first 22 terms is -330.

Example 9

Evaluate $\sum_{r=1}^{16}(5r - 1)$.

The first three terms are found by taking $r = 1$, 2 and 3, and $r = 16$ gives the last term. This gives the arithmetic series $4 + 9 + 14 + \dots + 79$, which has 16 terms. Use the formula $S_n = \frac{n}{2}(a + L)$ with $n = 16$, $a = 4$ and $L = 79$.

$$S_{16} = 8(4 + 79) = 664$$

Example 10

The 3rd and 7th terms of an arithmetic series are 71 and 55. Find:

(a) the first term and the common difference

(b) the sum of the first 45 terms.

(a) For the arithmetic series with first term a and common difference d,

$$3\text{rd term} = a + 2d = 71$$
$$7\text{th term} = a + 6d = 55$$

Solving these equations simultaneously gives:

$$4d = -16,\ d = -4$$

$$a = 55 - (-24) = 79$$

The first term is 79 and the common difference is -4.

(b) Use the formula $S_n = \frac{n}{2}[2a + (n-1)d]$ with $n = 45$, $a = 79$ and $d = -4$.

$$S_{45} = \frac{45}{2}[158 + 44 \times (-4)] = \frac{45}{2}[-18] = -405$$

The sum of the first 45 terms is -405.

Example 11

The sum of the first n terms of a series is $n^2 + 3n$.

(a) Find the nth term of the series.

(b) Show that the series is arithmetic and find the common difference.

(a) The nth term of the series $= S_n - S_{n-1}$

$$\begin{aligned} &= n^2 + 3n - (n-1)^2 - 3(n-1) \\ &= n^2 + 3n - (n^2 - 2n + 1) - 3n + 3 \\ &= n^2 + 3n - n^2 + 2n - 1 - 3n + 3 \\ &= 2n + 2 \end{aligned}$$

(b) Substituting $n = 1, 2, 3, \ldots$ in turn in the nth term, $2n + 2$, gives the series:

$$4 + 6 + 8 + 10 + \ldots$$

This is clearly an arithmetic series with first term 4 and common difference 2.

4.3 The sum of the first *n* natural numbers

The set of natural numbers $\mathbb{N} = \{1, 2, 3, 4, \ldots\}$ can be formed into the series

$$1 + 2 + 3 + \ldots + n$$

for the first n natural numbers.

This series can also be written as $\sum_{r=1}^{n} r$, and it clearly is an arithmetic series with first term 1 and common difference 1.

Using the formula $S_n = \frac{n}{2}[2a + (n-1)d]$ you have

$$\sum_{r=1}^{n} r = \frac{n}{2}[2 + (n-1)1] \text{ because } a = d = 1$$

$$= \frac{n(n+1)}{2}$$

■ **The sum of the first *n* natural numbers is $\frac{1}{2}n(n+1)$.**

Example 12

Find the sum of the first 1000 natural numbers.

Using the formula $S_n = \frac{n(n+1)}{2}$, when $n = 1000$:

$$S_{1000} = \frac{1000 \times 1001}{2} = 500\,500$$

Example 13

Find the sum of all those integers between 1 and 500 which are multiples of 3.

Here, you require the sum S of the series

$$S = 3 + 6 + 9 + \ldots + 498$$

that is:

$$S = 3[1 + 2 + 3 + \ldots + 166]$$

$$= 3 \times \frac{166 \times 167}{2}$$

$$= 41\,583$$

Notice also that you could also take

$S = 3 + 6 + 9 + \ldots + 498$ and use the formula

$$S_n = \frac{n}{2}[2a + (n-1)d]$$

with $a = 3$, $d = 3$ and $n = 166$ to give

$$S_{166} = \tfrac{166}{2}[6 + (165 \times 3)] = 41\,583$$

as before.

Exercise 4B

In questions 1–5 find the common difference and the sum of the arithmetic series to the number of terms stated.

1	$1 + 4 + 7 + \dots$	17 terms
2	$-3 + 1 + 5 + \dots$	21 terms
3	$19 + 12 + 5 + \dots$	16 terms
4	$0.25 + 0.5 + 0.75 + \dots$	30 terms
5	$£50 + £75 + £100 + \dots$	66 terms

Evaluate each of the following:

6 $\sum_{r=1}^{23} r$ **7** $\sum_{r=1}^{32} (r - 3)$ **8** $\sum_{r=7}^{13} (3r - 2)$

9 Find the sum of all positive even numbers less than 200.

10 Find the sum of all positive odd numbers between 100 and 400.

11 An arithmetic series has 1st term 10 and 17th term 122. Find the common difference and the sum of the first 20 terms.

12 In an arithmetic series the 20th term is 200. The common difference of this series is -6. Find the first term and the sum of the first 19 terms.

13 In an arithmetic series the 8th term is 3 and the 3rd term is 8. Find the first term, the common difference and the sum of the first 10 terms.

14 An arithmetic series has first term 5 and common difference 3. Starting with the first term, find the least number of terms that have a sum greater than 1500. Starting with the first term, find also the number of terms that have a sum of 3925.

15 The boring of a well costs £5 for the first metre depth, £11 for the second metre and £17 for the third metre. The costs for successive 1 metre depths continue in the arithmetic series which has £5 as the first term and £6 as its common difference. Find the cost of boring a well of depth (a) 50 m (b) 100 m.

16 The sum of the first n terms of a series is $3n^2 - 2n$. Find the 1st, 10th and nth terms of the series.

17 The sum of the first n terms of an arithmetic series is S_n. $S_{20} = 45$ and $S_{40} = 290$. Find the first term and the common difference of the series. Find also the sum of the first 60 terms of the series.

18 The 3rd, 4th and 5th terms of an arithmetic series are $(4 + x)$, $2x$ and $(8 - x)$. Find the value of x, and the sum of the first 24 terms of the series.

19 The 5th and 7th terms of an arithmetic series are $\frac{1}{6}$ and $\frac{1}{2}$. Find the rth term and the sum of the first 18 terms of the series.

20 The first and last terms of an arithmetic series are -12 and 22. The sum of all the terms is 260. Find the number of terms in the series.

21 Given that $\sum_{r=1}^{2n}(4r - 1) = \sum_{r=1}^{n}(3r + 59)$, find the value of n.

22 An arithmetic series has common difference d where $d > 0$. Three consecutive terms of the series, $x - d$, x, and $x + d$, have a sum of 24 and a product of 120. Calculate the value of d.

23 The sums of the first four terms and the last four terms of an arithmetic series are 2 and -18. The 5th term is -2. Find the common difference and the number of terms in the series.

24 Find the sum of all the integers between 0 and 200 that are not divisible by 4.

25 The sum of the 1st and 2nd terms of an arithmetic series is x and the sum of the $(n - 1)$th and nth terms is y. Show that the sum of the first n terms is $\frac{n}{4}(x + y)$. Find an expression for the common difference in terms of x, y and n.

4.4 Geometric series

A series in which each term is obtained by multiplying the previous term by a fixed number r, where r can take any value except 0, 1, or -1, is called a **geometric series.**

The number r is called the **common ratio** of the series because the ratio of any term to the term before it in the series is constant.

Example 14

Find the common ratios and the nth terms of these geometric series:

(a) $1 + 3 + 9 + \ldots$
(b) $125 + 25 + 5 + \ldots$
(c) $16 - 8 + 4 - \ldots$

(a) The common ratio is 3 because each term is 3 times the previous term. The series can be written as:

$$1 + 3 + 3^2 + \dots$$

The nth term is 3^{n-1}.

(b) The common ratio is $\frac{1}{5}$ because each term is one-fifth the previous term. The series can be written as:

$$5^3 + 5^2 + 5 + \dots$$

The nth term is 5^{4-n}.

(c) The common ratio is $-\frac{1}{2}$ because each term is $-\frac{1}{2}$ times the previous term. The series can be written as:

$$2^4 - 2^3 + 2^2 - \dots$$

The nth term $= 2^4 \times (-\frac{1}{2})^{n-1}$

$= (-1)^{n-1} \times 2^{5-n}$

For each nth term that you find, check by putting $n = 1, 2, 3 \dots$ that the correct series is generated.

Here is the standard way of representing a general geometric series:

$$a + ar + ar^2 + \dots$$

The first term is a and the common ratio is r.

■ **The nth term of the series $a + ar + ar^2 + \dots$ is ar^{n-1}**

The sum of a geometric series

The sum of a geometric series can be written:

$$S = a + ar + ar^2 + \dots + ar^{n-1}$$

Muliplying by r gives: $rS = \quad ar + ar^2 + \dots + ar^{n-1} + ar^n$

Subtracting the expression for rS from the expression for S gives:

$$S - rS = a - ar^n$$

That is: $S(1 - r) = a(1 - r^n)$

Dividing by $1 - r$ gives:

■ $$\mathbf{S = \frac{a(1 - r^n)}{1 - r}}$$

This is the standard formula for the sum of the first n terms of a geometric series with first term a and common ratio r. It may also be written as

■ $$\mathbf{S = \frac{a(r^n - 1)}{r - 1}}$$

4.5 The sum to infinity of a convergent geometric series

The sum to n terms of the geometric series with first term a and common ratio r is:

$$\frac{a(1-r^n)}{1-r} = \frac{a}{1-r} - \frac{ar^n}{1-r}$$

If r lies between -1 and 0 or between 0 and 1, then r^n gets smaller and smaller as n increases. [For example, $(\frac{1}{2})^2 = \frac{1}{4}$, $(\frac{1}{2})^3 = \frac{1}{8}$, $(\frac{1}{2})^4 = \frac{1}{16} \ldots$] This means that you can make the term $\frac{ar^n}{1-r}$ as small as you like provided you take a large enough value of n. As n approaches infinity so $\frac{ar^n}{1-r}$ approaches zero. In other words the values of $\frac{ar^n}{1-r}$ form a decreasing sequence which has the **limiting value** zero as n approaches infinity.

The two statements $-1 < r < 0$ and $0 < r < 1$, can be combined by writing $|r| < 1$. This reads 'mod r is less than 1'. Mod r, written $|r|$, means the absolute value of r, or the actual numerical value of r. For example, if $r = 3$, $|r| = 3$ but if $r = -3$ then $|r| = 3$ too.

- **If $|r| < 1$, the geometrical series $a + ar + ar^2 + \ldots + ar^{n-1} + \ldots$ has the sum to n terms**

$$S_n = \frac{a(1-r^n)}{1-r}$$

This converges to the value $\frac{a}{1-r}$ as n approaches infinity.

- **The series has a sum to infinity of $\frac{a}{1-r}$, provided that $|r| < 1$.**

Example 15

Find to the nearest integer the sum of the first 11 terms of the geometric series with 1st term 5 and 2nd term 6.

$$a = 5,\ ar = 6 \quad \text{and so} \quad r = \frac{ar}{a} = \frac{6}{5}$$

The common ratio of the series is $\frac{6}{5} = 1.2$. With $a = 5$, $r = 1.2$ and $n = 11$, using the formula

$$S_n = \frac{a(1-r^n)}{1-r}$$

gives: $$S_{11} = \frac{5(1-1.2^{11})}{1-1.2} = 161 \text{ (nearest integer)}$$

Example 16

For a geometric series with first term 12 and common ratio $\frac{2}{3}$ find the 5th term, the sum of the first 8 terms and the sum to infinity. Give answers to 3 significant figures where necessary.

$a = 12,\ r = \frac{2}{3}$

5th term is: $ar^4 = 12(\frac{2}{3})^4 = 2.37$ (3 s.f.)

The sum to 8 terms is: $\dfrac{12[1-(\frac{2}{3})^8]}{1-\frac{2}{3}} = 34.6$ (3 s.f.)

The sum to infinity is: $\dfrac{12}{1-\frac{2}{3}} = 36$

Example 17

Evaluate $\sum_{n=1}^{7} 16(-\frac{1}{2})^n$ and find the sum to infinity of the series.

The series is $-8 + 4 - 2 + \dots$ to 7 terms.

This is a geometric series with $a = -8$, $r = -\frac{1}{2}$ and $n = 7$.

The sum to 7 terms is: $\dfrac{(-8)[1-(-\frac{1}{2})^7]}{1-(-\frac{1}{2})} = \dfrac{-129}{24} = -5.375$

The sum to infinity is: $\dfrac{a}{1-r} = \dfrac{-8}{1-(-\frac{1}{2})} = -\dfrac{16}{3} = -5\frac{1}{3}$

Example 18

Find the least number of terms of the geometric series with 1st term 50 and 2nd term 47 for which the sum exceeds 800.

$$a = 50,\ ar = 47 \quad \text{and so} \quad r = \frac{47}{50} = 0.94$$

The first term is 50 and the common ratio is 0.94.

Using $S_n = \dfrac{a(1-r^n)}{1-r}$ the sum to n terms is

$$S_n = \frac{50[1-0.94^n]}{1-0.94}$$

and you require the smallest integer value of n for which $S_n > 800$. That is,

$$1 - 0.94^n > \frac{800 \times 0.06}{50} = 0.96$$

which gives $0.94^n < 0.04$.

Using a method of trial and improvement and a calculator you will find that

$$0.94^{52} = 0.040\,054\,2$$

and $$0.94^{53} = 0.037\,650\,9$$

This gives the value of n required as 53.

Note: Book P2 shows you another method of solving equations of the type $a^x = b$ using logarithms in place of the trial and improvement approach used here.

Exercise 4C

Where necessary, give answers to 3 s.f.

1 Find the 10th term and the sum of the first 10 terms of the geometric series with first term u_1 and common ratio r when:

(a) $u_1 = 10$ and $r = 2$

(b) $u_1 = 10$ and $r = -2$

(c) $u_1 = 10$ and $r = \frac{1}{2}$

2 Find the sum of the first 12 terms of the geometric series $u_1 + u_2 + u_3 + \ldots + u_{12}$, when:

(a) $u_1 = 3, u_2 = 6$

(b) $u_1 = 3, u_2 = -6$

(c) $u_1 = 3, u_2 = 2$

(d) $u_1 = 3, u_2 = -2$

3 A geometric series has first term a and common ratio r. Find the nth term and the sum to n terms when:

(a) $a = 4, r = \frac{1}{2}$ and $n = 8$

(b) $a = -4, r = 2$ and $n = 10$

(c) $a = 48, r = -\frac{1}{3}$ and $n = 8$

(d) $a = -100, r = 0.9$ and $n = 13$

4 State which series in question 3 has no sum to infinity and explain why.

Find the sum to infinity of the three convergent series in question 3.

5 Find the first three terms of a geometric series which has a 4th term of -3 and a 7th term of 81. Find also the sum of the first 12 terms of this series.

6 Write down the sum of the first n terms of the following geometric series and find the sum to infinity in those cases where the series is convergent.

(a) $\frac{3}{10} + \frac{3}{100} + \frac{3}{1000} + \dots$

(b) $16 - 8 + 4 - \dots$

(c) $8 - 12 + 18 - \dots$

7 The sum to infinity of a geometric series is 5 and the first term is 7. Find the common ratio and the sum of the first 15 terms.

8 The sum to infinity of a geometric series is 9. The common ratio is positive and the sum of the first two terms is 5. Find the first term, the common ratio and the sum of the first 12 terms.

9 A geometric series has the following properties. The 1st and 2nd terms have a sum of -4 and the 4th and 5th terms have a sum of 108. Find the 1st term and the common ratio of the series. Explain why the series has no sum to infinity.

10 A state's population at the end of each year is 2% greater than at the beginning of the year. Find the number of years required for the population to double.

11 The first term of a geometric series is 500 and the common ratio is 0.93. The nth term of the series is u_n. Find the least value of n for which (a) $u_n < 40$ (b) $u_n < 30$.

12 The 1st and the 4th terms of a geometric series are 27 and 8 respectively. Find the 7th term and the common ratio. Find also the sum to infinity.

13 By the end of a year, the value of a shop has increased by 3% of its value at the start of the year. At the start of 1994 a shop was valued at £45 000. Estimate, to the nearest £100, the value of the shop at the end of the year (a) 2000 (b) 2200.

14 Find, to 3 significant figures, the values of:

(a) $\sum_{n=4}^{9} (1.5)^n$ (b) $\sum_{n=2}^{10} 4(\frac{3}{4})^{n-1}$

15 Find the value of:

$$\sum_{n=1}^{10} [3(\tfrac{2}{3})^n - 2]$$

giving your answer to 3 significant figures.

4.6 Further examples on arithmetic and geometric series

Example 19

Evaluate $\sum_{n=1}^{11}(1.2^n + 1.2n)$.

You can make this easier by separating it into two summations:

$$\sum_{n=1}^{11} 1.2^n + \sum_{n=1}^{11} 1.2n$$

The first is a geometric series with first term 1.2 and common ratio 1.2 which has a sum of

$$\frac{1.2(1 - 1.2^{11})}{1 - 1.2} = 38.58$$

The second is an arithmetic series with first term 1.2 and common difference 1.2 which has a sum of

$$\tfrac{11}{2}\,[2.4 + 10 \times 1.2] = 79.2$$

Hence the sum of the series is $79.2 + 38.58 = 117.78$.

Example 20

The 1st, 2nd and 3rd terms of an arithmetic series are a, b and a^2, where a is negative. The 1st, 2nd and 3rd terms of a geometric series are a, a^2 and b. Find: (a) the values of a and b (b) the sum to infinity of the geometric series (c) the sum of the first 40 terms of the arithmetic series.

(a) As a, b and a^2 are consecutive terms in an arithmetic series,

$$b - a = a^2 - b$$

which gives:

$$2b = a + a^2 \qquad \text{(i)}$$

As a, a^2 and b are consecutive terms in a geometric series,

$$\frac{a^2}{a} = \frac{b}{a^2}$$

which gives:

$$a^3 = b \qquad \text{(ii)}$$

Eliminating b from equations (i) and (ii) gives:

$$2a^3 = a + a^2$$

Since a is not zero, $\quad 2a^2 - a - 1 = 0$

Factorising: $\quad (2a + 1)(a - 1) = 0$

It follows, since a is negative, that $a = -\frac{1}{2}$ and $b = -\frac{1}{8}$.

(b) The first term of the geometric series is $-\frac{1}{2}$ and the common ratio is $-\frac{1}{2}$.

The sum to infinity $= \dfrac{-\frac{1}{2}}{1-(-\frac{1}{2})} = -\frac{1}{3}$

(c) The arithmetic series has first term $-\frac{1}{2}$ and common difference $(-\frac{1}{8}+\frac{1}{2}) = \frac{3}{8}$.

The sum of the first 40 terms $= 20[-1 + 39(\frac{3}{8})]$

$= 272.5$

Exercise 4D

1 The numbers p, 10 and q are three consecutive terms of an arithmetic series. The numbers p, 6 and q are three consecutive terms of a geometric series. Show that $p^2 - 20p + 36 = 0$ and hence find the values of p and of q for which the geometric series converges.
Find the sum to infinity of the geometric series in this case.

2 A geometric series has all terms positive. The sum of the first four terms is 15 and the sum to infinity of the series is 16. Find the sum of the first eight terms.

3 A ball is dropped from a height of 10 m and bounces on a horizontal floor to a height of 8 m. On each successive bounce the height reached is 0.8 times the height reached in the previous bounce. Find the total distance travelled by the ball before it comes to rest.

4 The first term of an arithmetic series is -10.5 and the common difference is 1.5. The sum of n terms is -42. Find the two values of n which satisfy these conditions.

5 A geometric series has 1st term $\sqrt{2}$ and 2nd term $\sqrt{6}$. Find the 12th term and the sum of the first 12 terms.

6 The sum of n terms of an arithmetic series is 36. The first term is 1 and the nth term is 11. Find the value of n and the common difference of the series.

7 The first three terms of a geometric series are $4x$, $x+1$ and x. Given that x is negative, find the sum to infinity of the series.

8 For the arithmetic series $9.5 + 9.1 + \ldots$, find:

(a) the 50th term

(b) the least value of n for which the sum of the first n terms is negative.

9 A geometric series has 4th term 10 000 and 8th term 1. Find two series which satisfy these data and find the sum to infinity of each of these series.

10 Evaluate:

$$\sum_{n=2}^{9} [(0.9)^n + (0.9n + 4)]$$

giving your answer to 2 decimal places.

11 The sum of the first n terms of a series is $(4n+5)^2$. Find the nth term of the series.

12 The first and last terms of an arithmetic series are -12 and 22 respectively. The sum of all n terms in the series is 260.

(a) Find the value of n.

(b) Find the sum of the first 30 terms of the series with the same first term and common difference. [E]

13 The sum of the first two terms of a geometric series is -4 and the sum of the fourth and fifth terms is 108. Calculate the common ratio and the fourth term. [E]

14 Calculate the number of positive terms of the arithmetic series $100 + 97 + 94 + \ldots$ and find their sum. [E]

15 Find the sum of all the integers between 0 and 500 inclusive which are *not* divisible by 4.

SUMMARY OF KEY POINTS

1 The nth term of the arithmetic series $a + (a+d) + (a+2d) + \ldots$ is $[a + (n-1)d]$.

2 The sum of the arithmetic series $a + (a+d) + (a+2d) + \ldots + [a + (n-1)d]$ is $\frac{n}{2}[2a + (n-1)d]$.

3 The sum of the first n natural numbers is $\frac{1}{2}n(n+1)$.

4 The nth term of the geometric series $a + ar + ar^2 + \ldots$ is ar^{n-1}.

5 The sum of the finite geometric series $a + ar + ar^2 + \ldots + ar^{n-1}$ is

$$\frac{a(1-r^n)}{1-r} = \frac{a(r^n-1)}{r-1}$$

6 If $|r| < 1$, then the infinite geometric series $a + ar + ar^2 + \ldots$ has sum $\frac{a}{1-r}$

Differentiation

5

Differential calculus is the area of mathematics concerned with the rate at which things change. For example, the speed of a car is the rate at which the distance it travels changes with time.

This chapter starts by looking at the gradient of a straight line graph, which represents a rate of change.

5.1 The gradient of a graph

Gradient of a straight line graph

A straight line graph has a constant gradient or slope – one which is the same at any point on the line. The gradient can be found either from the equation of the line or by calculation from the coordinates of two points on the line. The gradient of the line between two points (x_1, y_1) and (x_2, y_2) is:

$$\frac{\text{change in } y}{\text{change in } x} = \frac{y_2 - y_1}{x_2 - x_1} = m$$

where m is a fixed number called a constant.

For the example of speed given above:

$$\text{speed} = \frac{\text{distance travelled}}{\text{time}}$$

The speed is the rate of change of distance travelled with respect to time.

In the same way a gradient, written:

$$\text{gradient} = \frac{\text{change in } y}{\text{change in } x} = m$$

can be thought of as the rate of change of y with respect to x.

The constant m is the rate of change of y with respect to x.

Gradient of a curve

A curve does not have a constant gradient – its direction is continuously changing, so its gradient will continuously change too. The gradient of a continuous curve with equation $y = \mathrm{f}(x)$ at any point on the curve is defined as the gradient of the tangent to the curve at this point.

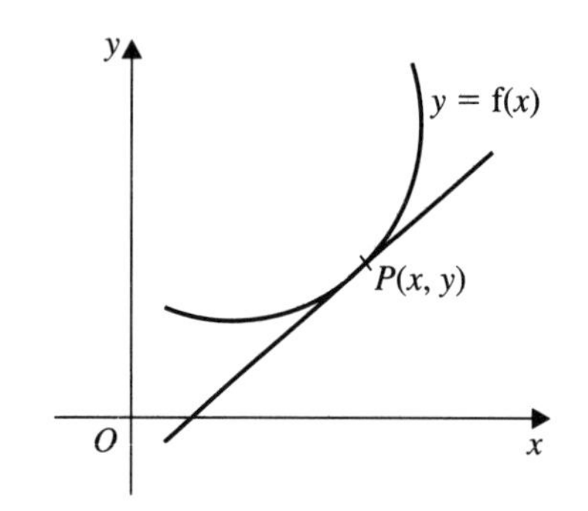

The **tangent** to a continuous curve at any point is a straight line which just touches the curve at this point. The gradient of the tangent to the curve $y = \mathrm{f}(x)$ at the point $P(x, y)$ is the rate at which y is changing with respect to x.

5.2 Using geometry to approximate to a gradient

Look at this curve:

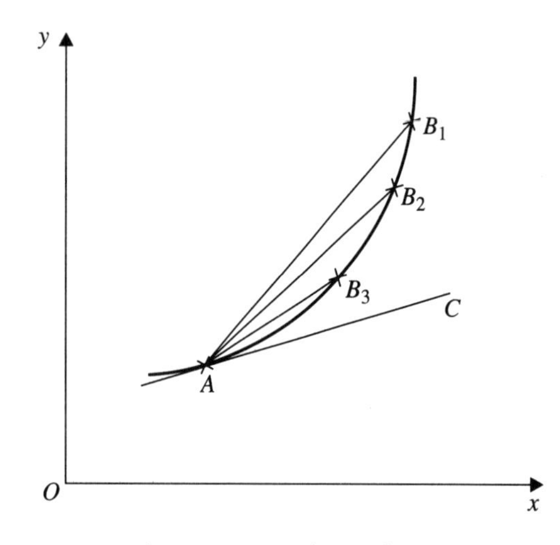

AC is the tangent to the curve at the point A.

A is a fixed point on the curve. Look at the chords AB_1, AB_2, AB_3, . . . (A chord is a line joining two points on a curve.) For points B_1, B_2, B_3 . . . that are closer and closer to A the sequence of chords AB_1, AB_2, AB_3 . . . move closer to becoming the tangent AC.

The gradients of the chords AB_1, AB_2, AB_3, . . . move closer to becoming the gradient of the tangent AC. So these gradients move closer to becoming the gradient of the curve at point A.

A numerical approach to rates of change

Here is how this idea can be applied to a real example. Look at the section of the graph of $y = x^2$ for $3 \leqslant x \leqslant 4$.

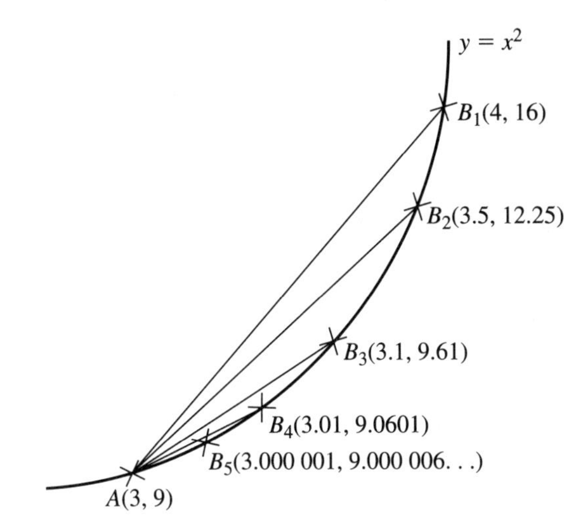

The gradient of the chord AB_1 is:

$$\frac{16-9}{4-3} = 7$$

Here are the gradients of the chords AB_1, AB_2, AB_3, AB_4 and AB_5 :

Chord	x changes from	y changes from	$\frac{\text{change in } y}{\text{change in } x}$ = gradient
AB_1	3 to 4	9 to 16	$\frac{16-9}{4-3} = 7$
AB_2	3 to 3.5	9 to 12.25	$\frac{12.25-9}{3.5-3} = 6.5$
AB_3	3 to 3.1	9 to 9.61	$\frac{9.61-9}{3.1-3} = 6.1$
AB_4	3 to 3.01	9 to 9.0601	$\frac{0.0601}{0.01} = 6.01$
AB_5	3 to 3.000 001	9 to 9.000 006 . . .	$\frac{0.000\,006\ldots}{0.000\,001} = 6$

The ratio given for the chord AB_5 is approximate.

At the point (3.000 001, 9.000 006 . . .) the line AB_5 is virtually the same as the tangent to the curve at A. Its gradient is 6.

To convince yourself further using a numerical approach you could take a further series of points on the curve at $x = 2$, 2.5, 2.9, 2.99 and 2.999 999 (so that you are approaching $x = 3$ from below) and show that they also give gradients which progressively approach the value of 6.

This numerical approach shows that the gradient of the curve $y = x^2$ at the point $A(3, 9)$ is 6.

5.3 A general approach to rates of change

Using a step-by-step approach similar to that in section 5.2 it is possible to find a numerical value for the gradient or rate of change at *any* point on a curve. The method can be generalised by taking a point $P(x, y)$ on the curve $y = f(x)$.

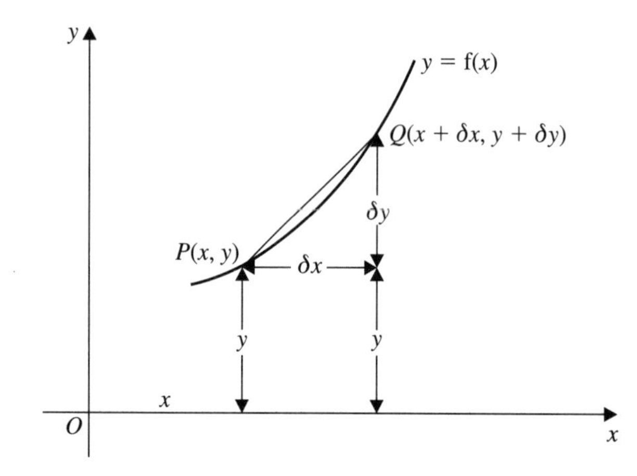

The point $Q(x + \delta x, y + \delta y)$ is very close to P on the curve. The small change from P in the value of x is δx and the corresponding small change in the value of y is δy. It is important to understand that δx is read as 'delta x' and is a *single symbol.* It is often called the **increment** (or change) in x and δy is called the increment in y.

The gradient of the chord PQ is:

$$\frac{(y + \delta y) - y}{(x + \delta x) - x} = \frac{\delta y}{\delta x}$$

As the equation of the curve is $y = \mathrm{f}(x)$, the coordinates of P can also be written as $[x, \mathrm{f}(x)]$ and the coordinates of Q as $[(x + \delta x), \mathrm{f}(x + \delta x)]$.

The value of δx can be made as small as you like. The smaller the value of δx, the smaller the value of δy and the closer point Q will be to point P. As shown in section 5.2 the ratio $\dfrac{\delta y}{\delta x}$ approaches a definite limit as δx gets smaller and approaches zero. This limit is the gradient of the tangent at P which is the gradient of the curve at P. It is called the **rate of change of y with respect to x** at the point P. This is denoted by $\dfrac{\mathrm{d}y}{\mathrm{d}x}$.

$$\frac{\mathrm{d}y}{\mathrm{d}x} = \lim_{\delta x \to 0}\left(\frac{\delta y}{\delta x}\right)$$

$$= \lim_{\delta x \to 0}\left[\frac{\mathrm{f}(x + \delta x) - \mathrm{f}(x)}{(x + \delta x) - x}\right]$$

$$= \lim_{\delta x \to 0}\left[\frac{\mathrm{f}(x + \delta x) - \mathrm{f}(x)}{\delta x}\right]$$

The symbol $\dfrac{\mathrm{d}y}{\mathrm{d}x}$ is called the **derivative** or the **differential coefficient** of y with respect to x. Read aloud it sounds like 'dee y by dee x'. In words $\lim\limits_{\delta x \to 0}\left(\dfrac{\delta y}{\delta x}\right)$ is: 'the limit of $\dfrac{\delta y}{\delta x}$ as δx tends to zero'. 'Tends to' is another way of saying 'approaches'.

If $y = \mathrm{f}(x)$, you can also use the notation:

$$\frac{\mathrm{d}y}{\mathrm{d}x} = \mathrm{f}'(x)$$

In this case f′ is often called the **derived function** of f.

The procedure used to find $\dfrac{\mathrm{d}y}{\mathrm{d}x}$ from y is called **differentiating y with respect to x**.

Example 1

Find $\frac{dy}{dx}$ for the function $y = x^2$.

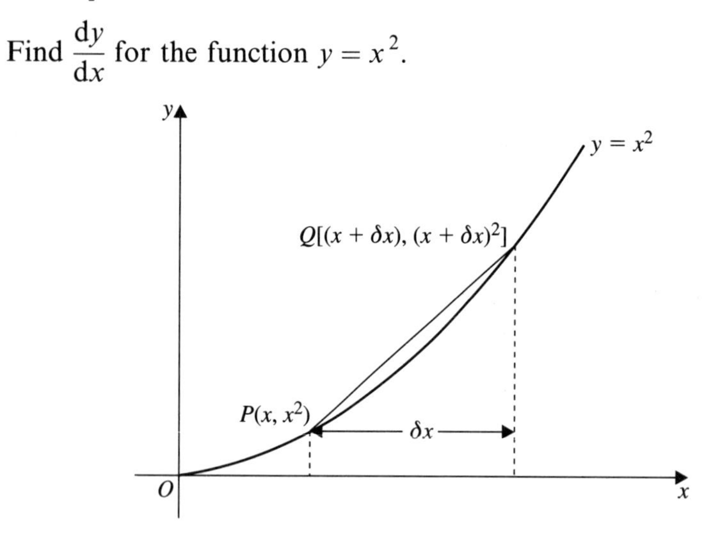

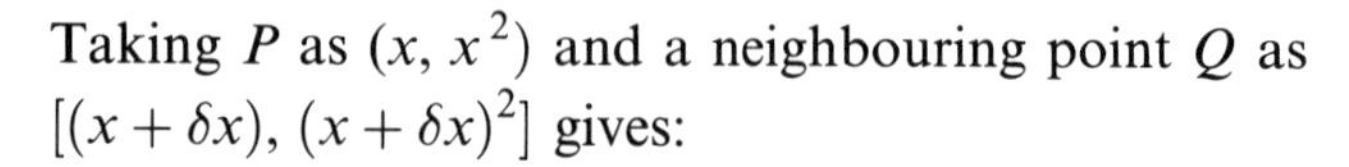
Taking P as (x, x^2) and a neighbouring point Q as $[(x + \delta x), (x + \delta x)^2]$ gives:

$$\frac{dy}{dx} = \lim_{\delta x \to 0} \left[\frac{(x + \delta x)^2 - x^2}{(x + \delta x) - x} \right]$$

$$= \lim_{\delta x \to 0} \left[\frac{x^2 + 2x\,\delta x + (\delta x)^2 - x^2}{\delta x} \right]$$

$$= \lim_{\delta x \to 0} \left[\frac{2x\,\delta x + (\delta x)^2}{\delta x} \right]$$

$$= \lim_{\delta x \to 0} (2x + \delta x)$$

As δx approaches zero ($\delta x \to 0$) the limiting value is $2x$. That is, the derivative of x^2 with respect to x is $2x$. (You could also say that the derived function of x^2 with respect to x is $2x$.) More often this is written:

$$\frac{dy}{dx} = 2x$$

For the curve $y = x^2$, we have generally $\frac{dy}{dx} = 2x$. So when $x = 3$, $\frac{dy}{dx} = 6$.

Look back at section 5.2 to see how the numerical approach used there agrees with the more general result used in example 1.

Example 2

Find $\frac{dy}{dx}$ for the function $y = \frac{1}{x}$.

For neighbouring points $P\left(x, \frac{1}{x}\right)$ and $Q\left(x + \delta x, \frac{1}{x + \delta x}\right)$:

$$\frac{dy}{dx} = \lim_{\delta x \to 0}\left[\frac{\left(\frac{1}{x + \delta x} - \frac{1}{x}\right)}{(x + \delta x) - x}\right] = \lim_{\delta x \to 0}\left[\frac{\left(\frac{x - x - \delta x}{x(x + \delta x)}\right)}{\delta x}\right]$$

So:
$$\frac{dy}{dx} = \lim_{\delta x \to 0}\left[\frac{-\delta x}{x\,\delta x(x + \delta x)}\right] = \lim_{\delta x \to 0}\left[\frac{-1}{x(x + \delta x)}\right]$$

The limit as δx approaches zero is:

$$\frac{dy}{dx} = \frac{-1}{x^2} = -x^{-2}$$

So for $y = \frac{1}{x}$ (or x^{-1}),

$$\frac{dy}{dx} = -\frac{1}{x^2} \quad (\text{or } -x^{-2})$$

5.4 A general formula for $\frac{dy}{dx}$ when $y = x^n$

If you extend the results obtained in examples 1 and 2 for further functions of x such as x^3 and $x^{\frac{1}{2}}$, you will find that $\frac{dy}{dx} = 3x^2$ and $\frac{dy}{dx} = \frac{1}{2}x^{-\frac{1}{2}}$ respectively.

- **In general, when $y = x^n$, where n is any real number:**

$$\frac{dy}{dx} = nx^{n-1}$$

Here are some other useful results which you can use. Their proofs are not shown.

- **When $y = kx^n$ where k is a constant:**

$$\frac{dy}{dx} = nkx^{n-1}$$

- **When $y = u \pm v$ where u and v are functions of x**

$$\frac{dy}{dx} = \frac{du}{dx} \pm \frac{dv}{dx}$$

Example 3

Find $\frac{dy}{dx}$ for each of the following:

(a) $y = x^3 - x^7$

(b) $y = 2x^4 - 3$

(c) $y = \frac{1}{x^3} - \frac{2}{x}$

Apply the formula:

(a) $\frac{dy}{dx} = 3x^{3-1} - 7x^{7-1} = 3x^2 - 7x^6$

(b) $\frac{dy}{dx} = 2(4x^{4-1}) - 0 = 8x^3.$

(Notice that $3 = 3 \times 1 = 3x^0$. So $\frac{dy}{dx} = 3(0x^{0-1}) = 0$. In other words a constant such as 3 has no rate of change and its derivative is zero.)

(c) Write this first as $y = x^{-3} - 2x^{-1}$.

Then:
$$\begin{aligned}\frac{dy}{dx} &= -3x^{-3-1} - 2(-1x^{-1-1}) \\ &= -3x^{-4} + 2x^{-2} \\ &= -\frac{3}{x^4} + \frac{2}{x^2}\end{aligned}$$

Example 4

You may need to find the value of $\frac{dy}{dx}$ at a particular point.

Find the value of $\frac{dy}{dx}$ at the point where $x = 3$ on the curve whose equation is $y = (2x - 1)(3x + 2)$.

Multiply out the brackets giving:

$$\begin{aligned}y &= 2x(3x + 2) - 1(3x + 2) \\ &= 6x^2 + 4x - 3x - 2 \\ &= 6x^2 + x - 2\end{aligned}$$

Differentiate with respect to x:

$$\begin{aligned}\frac{dy}{dx} &= 6(2x^{2-1}) + 1x^{1-1} - 0 \\ &= 12x + 1\end{aligned}$$

At the point where $x = 3$, $\frac{dy}{dx} = 36 + 1 = 37$

Exercise 5A

1 Find the y-coordinate of the point Q on the curve $y = x^3$ when the x-coordinate is

(a) 2 (b) 1.5 (c) 1.1 (d) 1.01 (e) 1.0001

The point P has coordinates (1, 1).

Find the gradients of the chords joining P to Q in each case.

Use them to find an estimate for the gradient of the tangent to the curve at P.

2 Repeat question 1 for the curve with equation

(a) $y = x^{-2}$ (b) $y = \sqrt{x}$ (c) $y = x^{-\frac{1}{3}}$.

3 In each of the following y is given as a function of x. Find the derived function $\frac{dy}{dx}$.

(a) x^4 (b) x^{-3} (c) -2

(d) $3x$ (e) $4x^3$ (f) $2x^{-5}$

(g) $\frac{1}{2x}$ (h) $\frac{3}{2x^2}$ (i) $x^2 - x^{-2}$

(j) $3x^{-1} - 2x^{-2}$ (k) $(x-1)(x+2)$ (l) $x(x^2-3)$

(m) $x^2(x^{-1} - x^{-2})$ (n) $(2x^2-3)^2$ (o) $\frac{x^3-1}{2x}$

4 Find the gradient of the curve with equation $y = f(x)$ at the point where $x = a$ when:

(a) $f(x) = x^6$, $a = -1$ (b) $f(x) = -3x^2$, $a = 2$

(c) $f(x) = 1 - 3x^2$, $a = \frac{1}{2}$ (d) $f(x) = \frac{1}{\sqrt{x}}$, $a = 4$

(e) $f(x) = \sqrt{(12x)}$, $a = \sqrt{3}$ (f) $f(x) = x^2(2x-1)$, $a = -1$

(g) $f(x) = \frac{x-4}{x^2}$, $a = -2$

5 Here is a sketch of the curve $y = x^2 - 2x$, which cuts the x-axis at the origin O and the point A, and passes through the point $B(1, -1)$. Find the gradient of the curve at the points A, B and O.

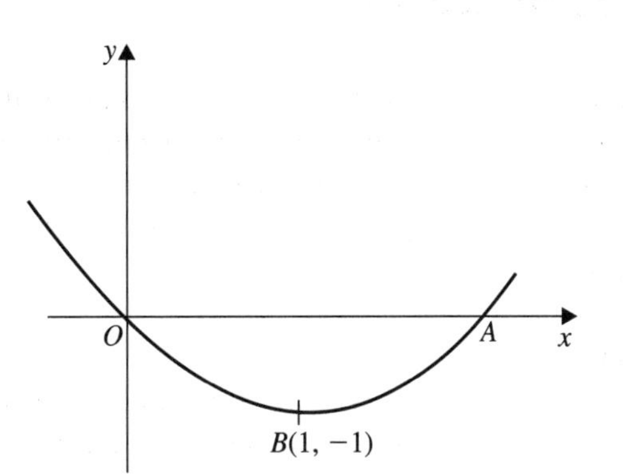

6 The points P, Q and R on the curve with equation $y = 1 - x^3$ have x-coordinates 2, -1 and -2. The curve crosses the x-axis at S. Find the gradient of the curve at P, Q, R and S.

7 Find the y-coordinate and the value of $\frac{dy}{dx}$ at the point P whose x-coordinate is -1 on the curve with equation $y = 2x^2 + \frac{1}{x}$.

8 The curve with equation $y = 100x - x^2$ represents the path of an arrow fired from the origin O and landing on the horizontal ground again at the point A. Find the coordinates of A and the gradient of the path of the arrow at O and A. Find also the coordinates of the point H at which $\frac{dy}{dx} = 0$. Explain what is happening to the arrow at this point.

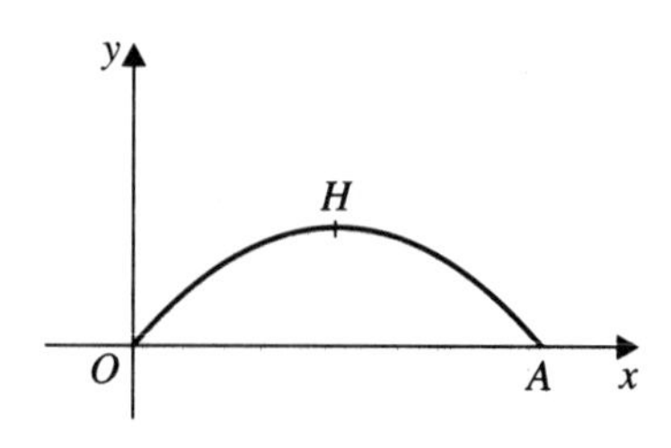

9 A curve is given by the equation

$$y = 3x^2 + 3 + \frac{1}{x^2}, \text{ where } x > 0$$

At the points A, B and C on the curve $x = 1$, 2 and 3 respectively. Find the gradient at A, B and C. [E]

10 Calculate the x-coordinates of the points on the curve with equation $y = 7x^2 - x^3$ at which the gradient is equal to 16. [E]

11 Find the x-coordinates of the two points on the curve with equation $y = x^3 - 11x + 1$ where the gradient is 1. Find the corresponding y-coordinates. [E]

12 The curve with equation $y = ax^2 + bx + c$ passes through the point (1, 2). The gradient of the curve is zero at the point (2, 1). Find the values of a, b and c. [E]

5.5 Increasing and decreasing functions

A function f which increases as x increases in the interval from $x = a$ to $x = b$ is called an **increasing function** in the interval (a, b). Similarly a function f which decreases as x increases in the interval from $x = c$ to $x = e$ is called a **decreasing function** in the interval (c, e).

Example 5

The function $f(x) = x^5$ is increasing for all real values of x except at $x = 0$. This means that the tangent to the curve at any point (except the origin O) makes an acute angle with Ox. Notice also that $\frac{dy}{dx} = 5x^4$ and that $5x^4 > 0$ for all real values of x except $x = 0$.

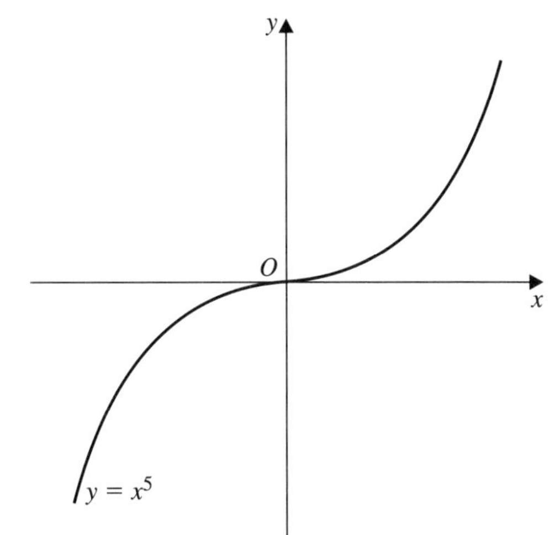

Example 6

Here is the graph of the function $f(x) = -(x-1)^2$:

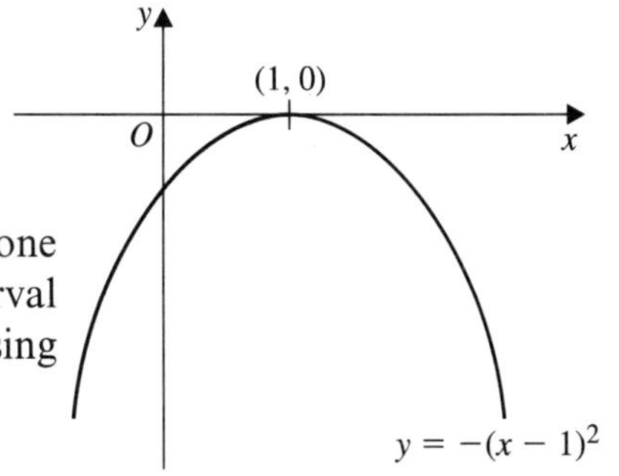

Like many other functions, this function is increasing in one interval (in the interval $x < 1$) and decreasing in another interval (the interval $x > 1$). At $x = 1$ the function is neither increasing nor decreasing. At this point it is said to be **stationary**.

5.6 Second derivatives

For the function f, given by $y = f(x)$, the first derivative is $f'(x)$ or $\frac{dy}{dx}$ and we write

$$\frac{dy}{dx} = f'(x)$$

The **second derivative** of y with respect to x is obtained by differentiating $f'(x)$ and is denoted by $f''(x)$; we write:

$$\frac{d}{dx}\left(\frac{dy}{dx}\right) = f''(x)$$

$\frac{d}{dx}\left(\frac{dy}{dx}\right)$ is written in shorthand form as $\frac{d^2y}{dx^2}$ and read as 'dee two y by dee x squared'. $\frac{d^2y}{dx^2}$ is called the **second derivative of y with respect to x**.

This notation can be extended; for example:

$$\frac{d}{dx}\left(\frac{d^2y}{dx^2}\right) = \frac{d^3y}{dx^3}$$

and this is called the **third derivative** of y with respect to x.

Example 7
Given that $y = x^{\frac{3}{2}} + 3x^{\frac{5}{2}}$, find $\frac{dy}{dx}$ and $\frac{d^2y}{dx^2}$.

Differentiating:

$$\frac{dy}{dx} = \tfrac{3}{2}x^{\frac{1}{2}} + 3\left(\tfrac{5}{2}x^{\frac{3}{2}}\right) = \tfrac{3}{2}x^{\frac{1}{2}} + \tfrac{15}{2}x^{\frac{3}{2}}$$

Differentiating again:

$$\frac{d^2y}{dx^2} = \tfrac{3}{2}\left(\tfrac{1}{2}x^{-\frac{1}{2}}\right) + \tfrac{15}{2}\left(\tfrac{3}{2}x^{\frac{1}{2}}\right)$$

$$= \tfrac{3}{4}x^{-\frac{1}{2}} + \tfrac{45}{4}x^{\frac{1}{2}}$$

5.7 Turning points

The graph following shows the curve with equation $y = f(x)$ and you can see that the function f is increasing in some intervals and decreasing in others. For example, f decreases from A to B, increases from B to C and increases also from C to E.

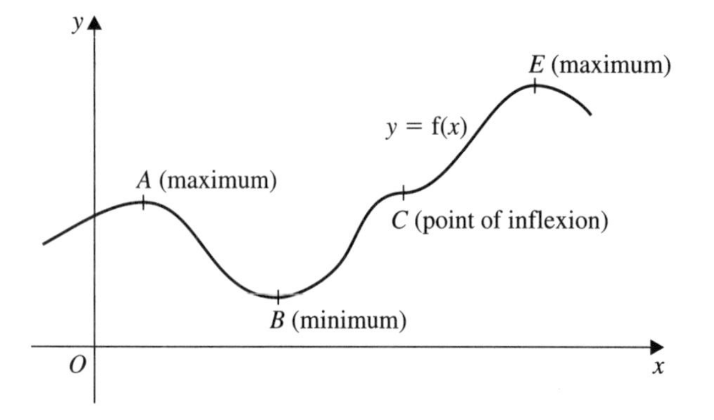

Suppose that the x-coordinates of A, B, C and E are a, b, c and e. Near point A, f(x) is increasing for $x < a$ and decreasing for $x > a$. At the point A, the curve reaches 'the top of a hill'. The value f(a) is called a **maximum** value of f. Also at A, $f'(a) = 0$; that is, the gradient of the curve is zero at A because the tangent to the curve at A is parallel to the x-axis.

Similarly at B, $f'(b) = 0$. The function f(x) is decreasing for $x < b$ and increasing for $x > b$. At the point B the curve reaches 'the lowest point'. The value f(b) is called a **minimum** value of f. Again, the tangent at B to the curve is parallel to the x-axis.

Between the points B and E, the curve is increasing, except at C where it levels out and has a tangent parallel to the x-axis. The point C is called an **inflexion** or a **point of inflexion**. At C, $f'(c) = 0$.

At any turning point on the curve $y = f(x)$, $f'(x) = 0$. The turning point may be a maximum, a minimum or a point of inflexion. You can find out what type a turning point is by considering the sign of $\frac{dy}{dx}$ on either side of the turning point **or** by evaluating $\frac{d^2y}{dx^2}$ at the turning point.

You should note that:

- **For the maximum points at A and E the value of $f''(x)$ is *negative*.**
- **At the minimum point B, $f''(x)$ is *positive*.**
- **At the point of inflexion C, $f''(x) = 0$**

However, although it is *always* true that if $f'(x) = 0$ and $f''(x) < 0$, then the curve with equation $y = f(x)$, has a maximum at x, and if $f'(x) = 0$ and $f''(x) > 0$, then the curve has a minimum at x, it is not sufficient for a point of inflexion to know that $f'(x) = 0$ and $f''(x) = 0$. For example, the curve with equation $y = x^4$ looks like this:

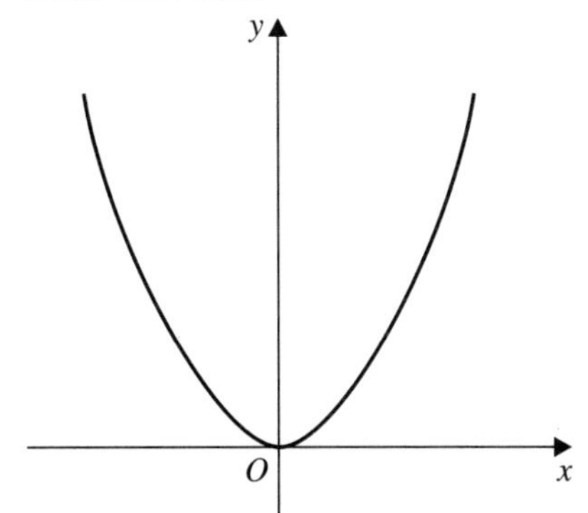

Now if $f(x) = x^4$, then $f'(x) = 4x^3$ and $f''(x) = 12x^2$.

Putting $f'(x) = 0$ gives $4x^3 = 0 \Rightarrow x = 0$ so there is a turning point at (0, 0). However, at $x = 0$, $f''(x) = 12 \times 0^2 = 0$. But, as you can see, the turning point at (0, 0) is not a point of inflexion. In fact:

- **the condition for a point of inflexion is that $f'(x) = 0$, $f''(x) = 0$ but $f'''(x) \neq 0$.**

So the second derivative is very useful when you want to find out what kind of stationary point you have identified on a curve: maximum, minimum or point of inflexion. Suppose we have $y = f(x)$ and $f'(x_1)$, $f'(x_2)$and $f'(x_3)$ are all zero. If the corresponding values of y are y_1, y_2 and y_3 respectively, then (x_1, y_1), (x_2, y_2) and (x_3, y_3) are stationary points of $y = f(x)$.

The turning points of a curve are sometimes called **stationary points**. Points of inflexion, maxima, and minima are all **stationary points**.

If $f''(x_1) < 0$, then y_1 is a maximum value of $f(x)$.

If $f''(x_2) > 0$, then y_2 is a minimum value of $f(x)$.

If $f''(x_3) = 0$ and $f'''(x_3) \neq 0$, then y_3 is an inflexion point.

Example 8

Find the coordinates of the stationary points on the curve with equation $y = (x+2)(x-1)^2$. Sketch the curve, showing the stationary points and the coordinates of the points at which the curve meets the axes.

$$(x+2)(x-1)^2 = (x+2)(x^2 - 2x + 1)$$

$$= x^3 - 2x^2 + x + 2x^2 - 4x + 2$$

$$y = x^3 - 3x + 2$$

Differentiating with respect to x gives:

$$\frac{dy}{dx} = 3x^2 - 3$$

Stationary points occur where $\frac{dy}{dx} = 0$, that is:

$$3(x^2 - 1) = 0$$

$$3(x-1)(x+1) = 0 \Rightarrow x = 1 \text{ or } x = -1$$

Substituting for x in $(x+2)(x-1)^2$ gives $y = 0$ or $y = 4$. The coordinates of the stationary points are (1, 0) and (−1, 4). The curve meets the x-axis when $(x+2)(x-1)^2 = 0$, that is, at $x = -2$ and at $x = 1$. The curve meets the y-axis when $x = 0$, that is at (0, 2).

Here is a sketch of $y = (x+2)(x-1)^2$:

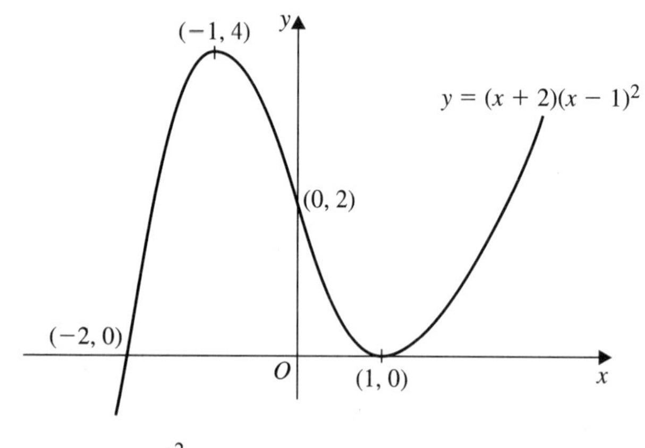

Notice further that $\frac{d^2y}{dx^2} = 6x$.

At $x = -1$, $\frac{d^2y}{dx^2} = -6 < 0$ and (−1, 4) is a *maximum* turning point.

At $x = 1$, $\frac{d^2y}{dx^2} = 6 > 0$ and (1, 0) is a *minimum* turning point.

Example 9
Sketch the curve with equation $y = x^3$ and show that the curve has a point of inflexion at the origin O (0, 0).

For positive x, x^3 increases as x increases and also for negative x, x^3 increases as x increases.

At O, $\frac{dy}{dx} = 3x^2 = 0$ *and* $\frac{d^2y}{dx^2} = 6x = 0$.

But $\frac{d^3y}{dx^3} = 6 \neq 0$, thus the curve has a point of inflexion at O.

A sketch of the curve looks like this:

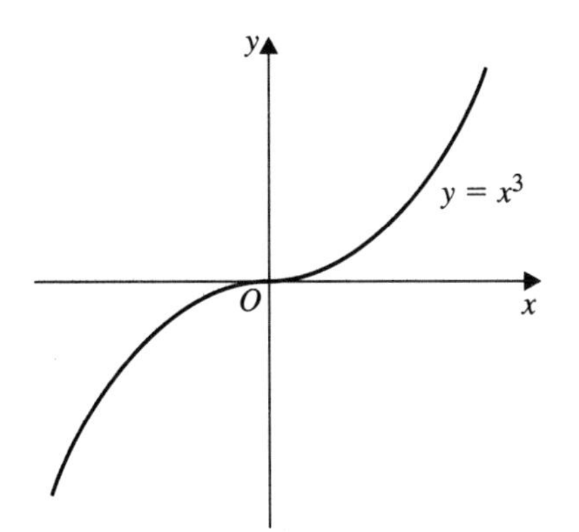

5.8 Tangents and normals to curves

The gradient at any point on the curve f(x) is

$$\frac{dy}{dx} = f'(x)$$

At $P(a,b)$
$$\frac{dy}{dx} = f'(a)$$

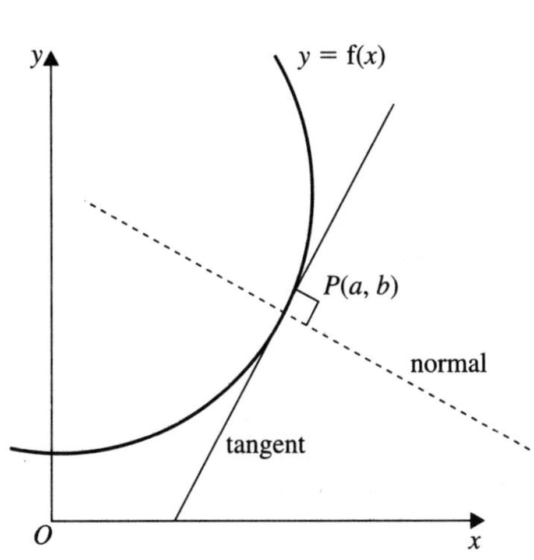

From section 5.1 you know that the gradient of the tangent to the curve at P is the same as the gradient of the curve at P.

■ **The equation of the tangent at P to the curve is**
$$\boldsymbol{y - b = \mathrm{f}'(a)(x - a)}$$

By definition the normal to the curve at P is perpendicular to the tangent. The gradient of the normal is therefore $-\dfrac{1}{\mathrm{f}'(a)}$, as the product of the gradients of lines at right angles is -1.

■ **The equation of the normal is**

$$\boldsymbol{y - b = -\frac{1}{\mathrm{f}'(a)}(x - a)}$$

Example 10
Find the equation of the tangent and of the normal at the point $P(2, 8)$ on the curve with equation $y = x^3$.

Differentiating: $$\frac{\mathrm{d}y}{\mathrm{d}x} = 3x^2$$

At P, $x = 2$ so $\dfrac{\mathrm{d}y}{\mathrm{d}x} = 3(2^2) = 12$.

Gradient of tangent at P is 12, and the tangent has equation

$$y - 8 = 12(x - 2)$$

so: $$y - 8 = 12x - 24$$

that is: $$12x - y - 16 = 0$$

Gradient of normal at P is $-\frac{1}{12}$, and the normal at P has equation

$$y - 8 = -\tfrac{1}{12}(x - 2)$$

or: $$12y - 96 = -x + 2$$

that is: $$x + 12y - 98 = 0$$

5.9 Using differentiation to solve practical problems

There are many practical problems that can be solved by mathematics where you want to find the maximum or minimum value of some quantity. This kind of problem can often be solved by differentiation.

When you first look at the problem, it may appear that the variable you want to maximise or minimise, y, depends on more than one other variable. You must use the data given in the problem to find a way of expressing y in terms of only one other variable x. Once $y = \mathrm{f}(x)$ is set up, you can differentiate and test for stationary points as usual, to find the maximum (or minimum) value of y. Example 11 is typical of the kind of practical problems you will meet.

Example 11

A rectangular sheet of cardboard is 12 cm long and 7.5 cm wide. Equal squares, each of side x cm, are cut from each corner. The flaps are then folded to make an open box in the form of a cuboid. The volume of the box is $V\,\text{cm}^3$.

Show that $V = 4x^3 - 39x^2 + 90x$.

Given that x can vary, find the maximum value of V and the value of x when this occurs.

If a square of side x cm is cut from each corner then the length of the box is $(12 - 2x)$ cm and its width is $(7.5 - 2x)$ cm.
Volume of box is given by

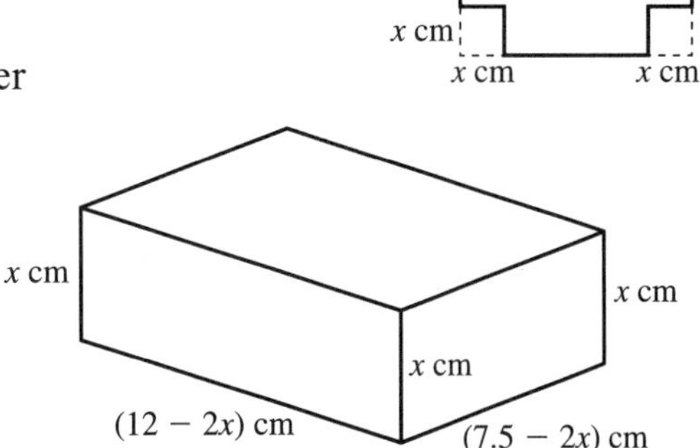

$$V = x(12 - 2x)(7.5 - 2x)$$

$$= x[90 - 24x - 15x + 4x^2]$$

$$= 4x^3 - 39x^2 + 90x, \text{ as required.}$$

The maximum value of V occurs when $\dfrac{\mathrm{d}V}{\mathrm{d}x} = 0$, so

$$\frac{\mathrm{d}V}{\mathrm{d}x} = 12x^2 - 78x + 90 = 0$$

Dividing by 6: $\quad 2x^2 - 13x + 15 = 0$

Factorising: $\quad (2x - 3)(x - 5) = 0$

So: $\quad x = 1\frac{1}{2} \quad \text{or} \quad x = 5$

You can see that $x = 5$ cannot apply in this practical problem because $(7.5 - 2x)$ would be negative.

At $\quad x = 1\frac{1}{2},\ V = 1\frac{1}{2} \times 9 \times 4\frac{1}{2} = 60\frac{3}{4}$

Notice that $\dfrac{\mathrm{d}^2V}{\mathrm{d}x^2} = 24x - 78 = -42 < 0$ when $x = 1\frac{1}{2}$ so $60\frac{3}{4}$ is the maximum value of V and occurs when $x = 1\frac{1}{2}$.

Example 12

A cylindrical tin, closed at both ends, is made of thin sheet metal. Find the dimensions of a tin like this that holds $1000\,\text{cm}^3$ and has a minimum total surface area.

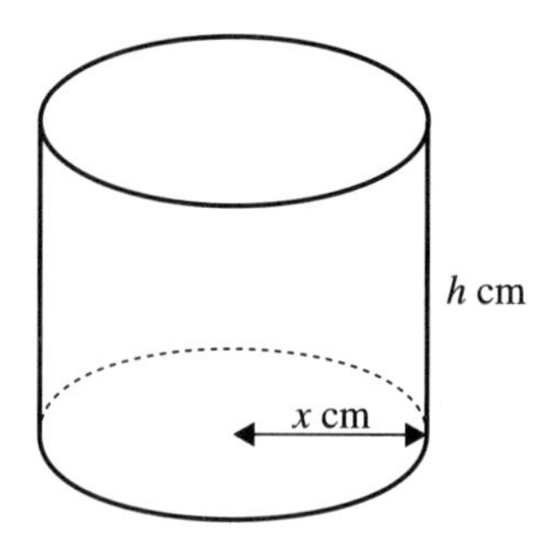

Let the tin have radius x centimetres and height h centimetres.

Then: $$\text{volume} = \pi x^2 h = 1000 \Rightarrow h = \frac{1000}{\pi x^2}$$

The curved surface area is $2\pi xh$ square centimetres. The area of the two circular ends is $2\pi x^2$ square centimetres. Let the total surface area of the tin be y square centimetres.

Then: $$y = 2\pi xh + 2\pi x^2$$

Both x and h vary. To solve the equation change it to an equation in two variables by substituting $h = \dfrac{1000}{\pi x^2}$ in the equation for y. This gives an equation in only two variables, x and y:

$$y = 2\pi x\left(\frac{1000}{\pi x^2}\right) + 2\pi x^2$$

So: $$y = 2000x^{-1} + 2\pi x^2$$

The surface area y is now expressed as a function of x. Differentiating with respect to x gives:

$$\frac{\mathrm{d}y}{\mathrm{d}x} = -2000x^{-2} + 4\pi x$$

For a stationary value of y, $\dfrac{\mathrm{d}y}{\mathrm{d}x} = 0$.

So: $$\frac{2000}{x^2} = 4\pi x$$

Rearranging gives: $$x^3 = \frac{2000}{4\pi} = \frac{500}{\pi}$$

Taking the cube root, the value of x is 5.42. Substituting this value for x in the equation $h = \dfrac{1000}{\pi x^2}$ gives the value of h as 10.84. When $\dfrac{\mathrm{d}y}{\mathrm{d}x} = 0$, $x = 5.42$ and $h = 10.84$.

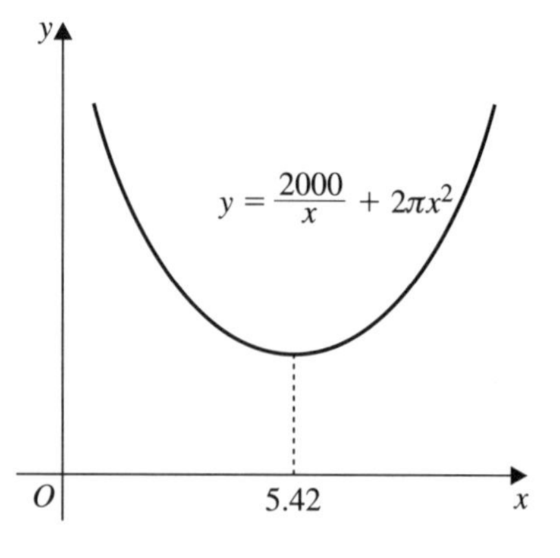

If you look at the sketch of the curve $y = \dfrac{2000}{x} + 2\pi x^2$, you can see that the only stationary point for $x > 0$ is a minimum. So the cylindrical tin of minimum surface area which has a volume of $1000\,\text{cm}^3$ has a radius of 5.42 cm and a height of 10.84 cm.

Exercise 5B

1 Draw a sketch of the function $f(x) = -x^3$, $x \in \mathbb{R}^+$. Find $f'(x)$ and explain why f is a decreasing function.

2 For each of the following functions, find the set of values of x for which (i) the function is increasing, (ii) the function is decreasing. The domain of each is $x \in \mathbb{R}$. Sketch the function in each case.

(a) $f(x) = (x - 2)^2$ (b) $f(x) = x^2 - 4$
(c) $f(x) = -4x^2$ (d) $f(x) = x(2 - x)$

3 Find the least value of the function $f(x) = x^2 - 3x + 3$. Find the greatest value of the function $f(x) = -2x^2 + 6x - 6$.

4 Find the greatest value of $4x - x^2$ and state the value of x for which this occurs. If x takes all real values, find the range of the function $f(x) = 4x - x^2$.

5 Find the coordinates of the points on the following curves at which the gradient is zero. Describe the nature of each stationary point.

(a) $y = x^2 + x^3$ (b) $y = x^3 - 3x + 2$
(c) $y = x + x^{-1}$ (d) $y = x^2 + 27x^{-1}$
(e) $y = (1 - x^2)(1 - 4x)$ (f) $y = x^3 - 3x^2 + 6$

6 Find the coordinates of the points on the curve $y = 2x^3 + 3x^2 - 12x + 6$ at which y has a stationary value. For each stationary value you find, say whether the value of y is a minimum or a maximum.

7 Show that the curve $y = 3x - x^3$ cuts the x-axis at three points. Find the turning points for this curve and so sketch the curve. Write all the information you have found on your sketch.

8 For the curves $y = 2x^3$ and $y = 4x - 5x^2$, find the values of x when the gradients of the two curves are equal.

9 A curve has equation $y = x^3 - 4x^2 + 4x + 3$.

(a) Find the coordinates of the points on this curve at which the gradient is -1.

(b) Find also the set of values of x for which y is decreasing.

10 Find an equation of the tangent and of the normal at (p, q) to the curve with equation $y = f(x)$ when

(a) (p, q) is $(1, 2)$ and the curve has equation $y = 2x^2$

(b) (p, q) is $(-1, -4)$ and the curve has equation $y = \frac{4}{x}$

(c) (p, q) is $(4, 2)$ and the curve has equation $y = x^{\frac{1}{2}}$

(d) (p, q) is $(8, \frac{1}{2})$ and the curve has equation $y = x^{-\frac{1}{3}}$

11 $f(x) \equiv 12 - 4x + 2x^2$

(a) Find an equation of the tangent and the normal at the point where $x = -1$ on the curve with equation $y = f(x)$.

(d) Show that the minimum value of $f(x)$ is 10. [E]

12 Given that $y = x^{-1}(x - 1)^2$ find the maximum and minimum values of y.

13 A stone is thrown vertically upwards and after t seconds its height y metres above its starting point is given by $y = 100t - 5t^2$. Calculate the greatest height reached by the stone and the value of t when this happens.

14 Given that $y = x^2 + 16x^{-1}$, find the maximum or minimum value of y and the value of x at which this occurs.

15 The sum of two variable positive numbers is 100. Let the numbers be x and $100 - x$ and let their product be y. Using differentiation find the maximum value of y.

16 The product of two variable positive numbers is 100. What is their least possible sum?

17 The organisers of a rave concert intend to use a large, flat expanse of land with a long, straight wall as one boundary. The rectangular enclosure is to be made from a total of 1000 metres of fencing on three sides with the wall for the fourth side. Given that the length of fencing on the side opposite the wall is $2x$ metres, show that the area y square metres of the enclosure is given by $y = 1000x - 2x^2$. Using differentiation find the maximum value of y.

18 A rectangular box with no lid is made from thin cardboard. The base is $2x$ centimetres long and x centimetres wide and the volume is 48 cubic centimetres. Show that the area, y square centimetres, of cardboard used is given by $y = 2x^2 + 144x^{-1}$. As x varies, find the value of x for which y is stationary and so find the minimum value of y.

19 A closed rectangular box is made of thin hardboard $3x$ centimetres long and x centimetres wide. The volume of the box is 288 cubic centimetres. Express the surface area, y square centimetres, of the box in terms of x and so find the value of x for which y is least.

20 A closed cylindrical olive oil tin is made of thin sheet metal of area $24\pi\,\text{cm}^2$. Find the great possible capacity of the tin and the radius of the tin in this case.

21 A rectangular sheet of thin cardboard is 80 cm by 50 cm. A square of side x centimetres is cut away from each corner of the sheet which is then folded to form an open rectangular box of volume y cubic centimetres. Show that

$$y = 4000x - 260x^2 + 4x^3$$

Given that x varies, find the greatest volume of the box.

22 A rectangle of length x centimetres, where x varies, has a constant area of 12 square centimetres. Express the perimeter of the rectangle, y, in terms of x. Find the least possible value of y.

23 A circular ink stain gets larger as time goes on. Its radius x centimetres at time t seconds is given by $x = 0.1t^2$. Find the rate of change of x with respect to t when $t = 0.5$ and when $t = 1.5$. Express A square centimetres, the area of the stain at time t seconds, in terms of t and so find the rate of change of A with respect to t when $t = 0.5$ and when $t = 1.5$.

24 At time t seconds the radius x centimetres of an expanding sphere is given by $x = \frac{1}{2}t$. Express the volume, V cubic centimetres, and the surface area, A square centimetres, in terms of t. From this find the rate of change of V and the rate of change of A with respect to t at the instant when $t = 3$.

25 The diameter of a closed cylindrical tin is equal to the height of the tin. Show that, for a fixed volume, a tin made to this specification requires the least amount of sheet metal.

26 A curve has equation $y = (x - 6)(x + 6)(3 - 2x)$. Find the set of values of x for which $\frac{\mathrm{d}y}{\mathrm{d}x} > 0$. On a sketch of the curve write in the coordinates of the turning points and also the coordinates of the points where the curve crosses the coordinate axes.

27 A water tank is being filled so that the volume of water in the tank at the end of t seconds is $V\,\text{m}^3$ where

$$100V = 15t^2 + 16t$$

Calculate the rate at which the tank is filling when $t = 6$.

28 Find the value of $\frac{dy}{dx}$ at $x = 2.5$ given that $y = \frac{x^2}{4} - \frac{4}{x^2}$.

Find also the value of $\frac{d^2y}{dx^2}$ at $x = 2.5$.

29 Find the coordinates of those points on the curve with equation $y = 4x - \frac{16}{3}x^3$ where the tangent to the curve is parallel to the x-axis. [E]

30 Given that $y = 3 + 3x^2 - x^3$, find the value of x when y takes a maximum value and state the value of this maximum. Find also an equation of the tangent and the normal at the point where $x = 4$.

SUMMARY OF KEY POINTS

1 The gradient of a curve $y = f(x)$ at the point $x = a$ is the gradient of the tangent to the curve at $x = a$.

2 $\frac{dy}{dx}$ is the derivative of y with respect to x. It represents the rate of change of y with respect to x.

3 If $y = f(x)$ then $\frac{dy}{dx} = f'(x)$.

4 If $y = x^n$ then

$$\frac{dy}{dx} = nx^{n-1}$$

5 If $y = kx^n$, where k is a constant, then

$$\frac{dy}{dx} = nkx^{n-1}$$

6 If $y = u \pm v$, where u and v are functions of x, then

$$\frac{dy}{dx} = \frac{du}{dx} \pm \frac{dv}{dx}$$

7 A function which increases in the interval from $x = a$ to $x = b$ is called an increasing function in the interval (a, b).

8 A function which decreases as x increases in the interval from $x = c$ to $x = e$ is called a decreasing function in the interval (c, e).

9 If $y = \mathrm{f}(x)$, then $\dfrac{\mathrm{d}^2y}{\mathrm{d}x^2} = \mathrm{f}''(x)$ is called the second derivative of y with respect to x.

10 A point on the curve $y = \mathrm{f}(x)$ for which $\mathrm{f}'(x) = 0$ is called a turning (or stationary) point of the curve.

11 There are three types of turning point: a maximum, a minimum and a point of inflexion.

12 If $\mathrm{f}'(x) = 0$ and $\mathrm{f}''(x) < 0$, the turning point is a maximum.

13 If $\mathrm{f}'(x) = 0$ and $\mathrm{f}''(x) > 0$, the turning point is a minimum.

14 If $\mathrm{f}'(x) = 0$, $\mathrm{f}''(x) = 0$ and $\mathrm{f}'''(x) \neq 0$, the turning point is a point of inflexion.

15 The equation of the tangent to the curve $y = \mathrm{f}(x)$ at the point on the curve with coordinates (a, b) is

$$y - b = \mathrm{f}'(a)(x - a)$$

16 The equation of the normal to the curve $y = \mathrm{f}(x)$ at the point on the curve with coordinates (a, b) is

$$y - b = -\frac{1}{\mathrm{f}'(a)}(x - a)$$

Integration

6

Chapter 5 on differentiation is about the rate at which things change. Differentiation is the procedure used to find $\frac{dy}{dx}$, or the rate of change of a function y with respect to x.

You need a good understanding of the ideas in chapter 5 before working on chapter 6. This chapter looks at the reverse or inverse process to differentiation, called **integration**.

6.1 Integration as the inverse of differentiation

Chapter 5 on differentiation shows that if $y = x^3$, then $\frac{dy}{dx} = 3x^2$. That is, for a given function y the process of differentiation gives a unique derivative $\frac{dy}{dx}$. However, differentiating $y = x^3 - 5$ or $y = x^3 + 2.5$ gives the same result for $\frac{dy}{dx}$ in each case: $3x^2$. So if you know what $\frac{dy}{dx}$ is and you want to find y, there is more than one possible answer.

In this way differentiation and its inverse process, integration, are quite different. Differentiation produces a unique answer, but the result of integration is a whole set of solutions. If you know that $\frac{dy}{dx} = 3x^2$, then you can only say that

$$y = x^3 + C$$

where C is a constant. The solutions $y = x^3, y = x^3 - 5$ and $y = x^3 + 2.5$ are *all* members of this set of solutions. Here are graphs of these solutions:

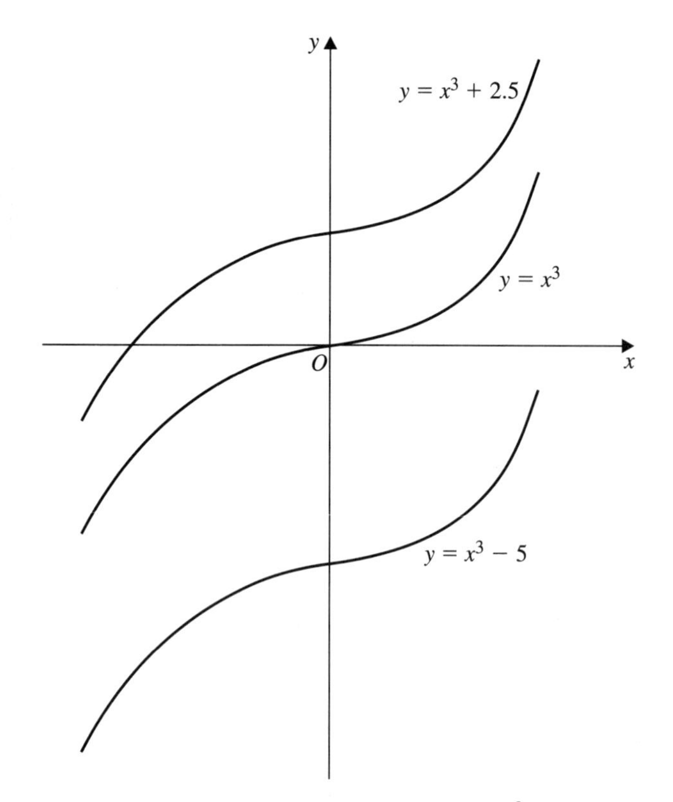

For different values of C, the equation $y = x^3 + C$ represents a set or family of curves. All of them are 'parallel' to one another, and no two curves ever pass through the same point.

This inverse process of moving from the differential relation $\frac{dy}{dx} = f'(x)$ to the general solution $y = f(x) + C$ is called **indefinite integration** – the constant C could take any value so the process is indefinite. A special symbol $\int$, looking like a stretched S, is used for integration. In this notation:

$$\int f'(x)\, dx = f(x) + C$$

In words this is: 'the integral of the function f' with respect to x equals the function f plus a constant'. As in the notation for differentiation, the dx stands for 'with respect to x'.

6.2 Integration of x^n where $n \in \mathbb{Q}$, $n \neq -1$

Working backwards from the standard formulae for differentiation in chapter 5 (p. 135) gives the following indefinite integrals, where k and C are constants.

- $\displaystyle\int x\,\mathrm{d}x = \tfrac{1}{2}x^2 + C$
- $\displaystyle\int k\,\mathrm{d}x = kx + C$
- $\displaystyle\int x^n\,\mathrm{d}x = \frac{1}{n+1}x^{n+1} + C$, where $n \neq -1$
- $\displaystyle\int (u \pm v)\,\mathrm{d}x = \int u\,\mathrm{d}x \pm \int v\,\mathrm{d}x$

You can check the first three results by differentiating the right-hand sides of the equations.

Example 1

Find the following indefinite integrals:

(a) $\displaystyle\int x^7\,\mathrm{d}x$ (b) $\displaystyle\int x^{-3}\,\mathrm{d}x$ (c) $\displaystyle\int x^{\frac{3}{2}}\,\mathrm{d}x$

(a) Using the formula for $\int x^n\,\mathrm{d}x$ with $n = 7$ gives:

$$\int x^7\,\mathrm{d}x = \tfrac{1}{8}x^8 + C$$

(b) Using the formula for $\int x^n\,\mathrm{d}x$ with $n = -3$ gives:

$$\int x^{-3}\,\mathrm{d}x = -\tfrac{1}{2}x^{-2} + C$$

(c) Using the formula for $\int x^n\,\mathrm{d}x$ with $n = \frac{3}{2}$ gives:

$$\int x^{\frac{3}{2}}\,\mathrm{d}x = \frac{1}{\frac{5}{2}}x^{\frac{5}{2}} + C$$
$$= \tfrac{2}{5}x^{\frac{5}{2}} + C$$

Example 2

Find $\displaystyle\int \left(2x^2 - \frac{1}{x}\right)^2 \mathrm{d}x$

Multiply out the bracket to obtain:

$$\int \left(2x^2 - \frac{1}{x}\right)^2 \mathrm{d}x = \int \left(4x^4 - 4x + \frac{1}{x^2}\right) \mathrm{d}x$$
$$= \int 4x^4\,\mathrm{d}x - \int 4x\,\mathrm{d}x + \int x^{-2}\,\mathrm{d}x$$
$$= \tfrac{4}{5}x^5 - \tfrac{4}{2}x^2 - x^{-1} + C$$
$$= \tfrac{4}{5}x^5 - 2x^2 - x^{-1} + C$$

Remember to check your final result by differentiation to make sure that you have used the formula for $\int x^n\,\mathrm{d}x$ correctly.

Example 3

Find (a) $\int x^{\frac{2}{3}}\,dx$ (b) $\int x^{-\frac{5}{3}}\,dx$.

(a) Here you have
$$\int x^{\frac{2}{3}}\,dx = \frac{1}{\frac{2}{3}+1}x^{\frac{2}{3}+1} + C$$
$$= \tfrac{3}{5}x^{\frac{5}{3}} + C$$

(b) Using the formula for $\int x^n\,dx$, with $n = -\frac{5}{3}$

$$\int x^{-\frac{5}{3}}\,dx = \frac{x^{-\frac{5}{3}+1}}{-\frac{5}{3}+1} + C$$
$$= \frac{x^{-\frac{2}{3}}}{-\frac{2}{3}} + C$$
$$= -\tfrac{3}{2}x^{-\frac{2}{3}} + C$$

Exercise 6A

1 Find the following integrals:

(a) $\int x\,dx$ (b) $\int x^4\,dx$ (c) $\int 6x^3\,dx$

(d) $\int x^{-2}\,dx$ (e) $\int 2x^{-4}\,dx$ (f) $\int 9x^2\,dx$

(g) $\int (4x+5)\,dx$ (h) $\int x(x-1)\,dx$ (i) $\int x^{-1}(x-x^2)\,dx$

(j) $\int (x+1)^2\,dx$ (k) $\int (2-x)^2\,dx$ (l) $\int \left(x-\frac{1}{x}\right)^2\,dx$

(m) $\int (x^{-2}-x^2)\,dx$ (n) $\int (1+x^{-2})^2\,dx$

(o) $\int (3x^5-2x^{-3})\,dx$ (p) $\int 15x^{-4}\,dx$

(q) $\int (3x-2x^{-1})^2\,dx$ (r) $\int (3x-2)(2x-3)\,dx$

(s) $\int (5x^2-3)(5x^2+3)\,dx$ (t) $\int x(\sqrt{x}-2)^2\,dx$

2 Integrate the following functions with respect to x:

(a) $x^{\frac{1}{2}}$ (b) $x^{\frac{3}{5}}$ (c) $x^{-\frac{1}{2}}$ (d) $x^{-\frac{3}{4}}$

(e) $5x^{\frac{3}{2}}$ (f) $\sqrt{(x^3)}$ (g) $(\sqrt{x})^{-3}$ (h) $4x^{\frac{1}{3}}$

(i) $3x^{\frac{5}{4}}$ (j) $-\frac{1}{2}x^{-\frac{5}{4}}$ (k) $(3x)^{\frac{1}{2}}$ (l) $(4x)^{-\frac{1}{2}}$

3 Find the following integrals:

(a) $\int (x^{-4} + 2)\, dx$ (b) $\int (2x - x^{-2})\, dx$

(c) $\int (2x^{-\frac{1}{2}} + x^{-2})\, dx$ (d) $\int \frac{3}{x^3}\, dx$

(e) $\int \frac{1}{3x^2}\, dx$ (f) $\int \frac{5}{8x^4}\, dx$

(g) $\int (3x^{-1} - x)^2\, dx$ (h) $\int x(4 + x^{-3})\, dx$

(i) $\int \frac{3x^2 - 4}{2x^2}\, dx$ (j) $\int \frac{(3x-2)(2x+3)}{\sqrt{x}}\, dx$

6.3 The solution of the differential equation $\frac{dy}{dx} = f(x)$

The process of getting from $\frac{dy}{dx} = f'(x)$ to $y = f(x) + C$ is often called **indefinite integration** because of the inclusion of the constant C which could take any value. This is, so far, the same question as asking you to find $\int f'(x)\, dx$. The statement $\frac{dy}{dx} = f'(x)$ is usually called a **differential equation** and the statement found by integration, $y = f(x) + C$, is called the **general solution of the differential equation**.

Here are two other differential equations:

$\frac{d^2y}{dx^2} + y = \frac{dy}{dx}$, called a **second order** differential equation, because the highest differential coefficient is $\frac{d^2y}{dx^2}$.

$\frac{d^3y}{dx^3} + \frac{d^2y}{dx^2} = x$, called a **third order** differential equation, because the highest differential coefficient is $\frac{d^3y}{dx^3}$.

The differential equation $\frac{dy}{dx} = f'(x)$ is a **first order** equation and this is the only type that you will need to solve in the P1 course.

Example 4

Find the general solution of the differential equation

(a) $\dfrac{dy}{dx} = x^4$ (b) $\dfrac{dy}{dx} = (\sqrt{x} + 1)^2$

(a) $\dfrac{dy}{dx} = x^4$

Integrating with respect to x:

$$y = \tfrac{1}{5}x^5 + C$$

and this is the required general solution.

(b) You must first expand the brackets:

$$(\sqrt{x} + 1)^2 = (\sqrt{x})^2 + 2\sqrt{x} + 1$$

$$= x + 2x^{\frac{1}{2}} + 1$$

$$\frac{dy}{dx} = x + 2x^{\frac{1}{2}} + 1$$

$$y = \int x\,dx + 2\int x^{\frac{1}{2}}\,dx + \int 1\,dx$$

$$y = \tfrac{1}{2}x^2 + \tfrac{4}{3}x^{\frac{3}{2}} + x + C$$

and this is the required general solution.

6.4 Boundary conditions

In section 6.2, you have seen how to obtain the general solution of $\dfrac{dy}{dx} = f'(x)$ as $y = f(x) + C$, where the constant C is often called an **arbitrary constant**.

If you are given a condition such as 'the value of y is 6 at $x = -1$' then you can find the value of this arbitrary constant C. A condition of this type is called a **boundary condition**. A boundary condition gives you extra information which you can substitute in the general solution to find the value of C. The solution which includes the value of C is called a **particular solution** of the differential equation.

Example 5

A curve passes through the point (2,–1). The gradient of the curve at any point is equal to $2 - 6x^2$. Find the equation of the curve.

You are told that $\frac{dy}{dx} = 2 - 6x^2$, because the gradient is $\frac{dy}{dx}$.

Integrating gives:

$$y = 2x - 2x^3 + C, \text{ the general solution}$$

(Check by differentiating that the integration is correct.)

You know that $y = -1$ at $x = 2$. Substituting these values in the equation gives:

$$-1 = 4 - 16 + C$$

and:

$$C = 11$$

So the equation of the curve is $y = -2x^3 + 2x + 11$ and this is also the particular solution of the differential equation $\frac{dy}{dx} = 2 - 6x^2$ for $y = -1$ at $x = 2$.

Example 6

For the curve with equation $y = f(x)$, it is known that $f'(x)$ is proportional to $x^2 + 1$ and that the curve passes through the points (3, 0) and (0, 36). Find the equation of the curve.

Since $f'(x)$ is proportional to $x^2 + 1$, you can say that

$$\frac{dy}{dx} = k(x^2 + 1), \text{ where } k \text{ is a constant.}$$

So:

$$y = \int kx^2\,dx + \int k\,dx$$

that is:

$$y = \tfrac{1}{3}kx^3 + kx + C, \text{ where } C \text{ is a constant}$$

Now the curve passes through the point (3, 0).

That is:

$$0 = 9k + 3k + C$$

$\therefore$

$$12k + C = 0$$

The curve also passes through the point (0, 36).

So:

$$36 = 0 + 0 + C$$

That is

$$C = 36$$

Hence $12k = -36$ and $k = -\frac{36}{12} = -3$.

The equation of the curve is

$$y = \tfrac{1}{3}kx^3 + kx + C$$

and you know that $k = -3$ and $C = 36$.

So the equation of the curve is

$$y = -x^3 - 3x + 36.$$

6.5 Definite integrals

With indefinite integration we have:

$$\int f'(x)\,dx = f(x) + C$$

where C is an arbitrary constant.

The **definite integral** $\int_a^b$ is defined by:

$$\int_a^b f'(x)\,dx = \Big[f(x)\Big]_a^b = f(b) - f(a)$$

provided that f' is the derived function of f throughout the interval (a, b).

Notice the use of the **square brackets** around $f(x)$. This notation for definite integrals is standard. The numbers a and b are called the **limits** of the definite integral. In words $\int_a^b g(x)\,dx$ is: 'the integral of $g(x)$ with respect to x between a and b'.

Example 7

Evaluate $\int_4^9 \sqrt{x}\,dx$.

First write $\sqrt{x}$ in index form: $x^{\frac{1}{2}}$.

Using the formula for integrating x^n gives:

$$\int_4^9 \sqrt{x}\,dx = \left[\frac{1}{\frac{3}{2}}x^{\frac{3}{2}}\right]_4^9 = \tfrac{2}{3}\left[9^{\frac{3}{2}} - 4^{\frac{3}{2}}\right]$$

$$= \tfrac{2}{3}(27 - 8) = \tfrac{38}{3} = 12\tfrac{2}{3}$$

Example 8

Evaluate $\int_{-2}^{3} (2x - 1)(3x + 1)\,dx$.

First remove the brackets:

$$(2x - 1)(3x + 1) = 2x(3x + 1) - 1(3x + 1)$$

$$= 6x^2 + 2x - 3x - 1$$

$$= 6x^2 - x - 1$$

$$\int_{-2}^{3}(6x^2 - x - 1)\,dx = \int_{-2}^{3} 6x^2\,dx - \int_{-2}^{3} x\,dx - \int_{-2}^{3} 1\,dx$$

$$= \left[2x^3 - \frac{x^2}{2} - x\right]_{-2}^{3}$$

$$= \left[2(3)^3 - \frac{3^2}{2} - 3\right] - \left[2(-2)^3 - \frac{(-2)^2}{2} - (-2),\right]$$

$$= \left[54 - \tfrac{9}{2} - 3\right] - [-16 - 2 + 2]$$

$$= 46\tfrac{1}{2} + 16$$

$$= 62\tfrac{1}{2}$$

Exercise 6B

1 In each of the following $\frac{dy}{dx}$ is given in terms of x, and $y = 1$ at $x = 0$. Find y in terms of x.

(a) $\frac{dy}{dx} = 2x - 1$

(b) $\frac{dy}{dx} = x^3 + x$

(c) $\frac{dy}{dx} = x^4 + 2x$

(d) $\frac{dy}{dx} = -2x^{\frac{3}{4}}$

(e) $\frac{dy}{dx} = (3x - 4)^2$

(f) $\frac{dy}{dx} = (x - 1)(3x - 2)$

2 Evaluate each of the following definite integrals:

(a) $\int_0^2 3x^2\,dx$

(b) $\int_2^4 x^{-2}\,dx$

(c) $\int_1^4 \sqrt{x}\,dx$

(d) $\int_{-1}^{1} (2x - 1)^2\,dx$

(e) $\int_{-1}^{0} x^9\,dx$

(f) $\int_2^5 \frac{3}{x^2}\,dx$

3 Show that $\int_0^2 (4 - x^2)^2\,dx = \frac{256}{15}$.

Evaluate the definite integrals in questions 4–7.

4 $\int_1^2 (x - 1)(x - 2)\,dx$

5 $\int_4^9 (2x^{\frac{1}{2}} + 3x^{-\frac{1}{2}})\,dx$

6 $\int_{-1}^{2} x^3(4 - 5x)\,dx$

7 $\int_0^2 (x^{\frac{1}{2}} - 2)^2\,dx$

8 Find the value of $\int_{-1}^{1} x^n\,dx$ for $n = 6, 7, 8$ and 9.

9 Given that $\frac{dy}{dx} = x^3 + x^{-3}$ and that $y = 4$ at $x = 1$, find (a) y in terms of x (b) the value of y at $x = 2$.

10 For the curve $y = f(x)$, $f'(x) = 2x - \frac{x^2}{2}$. The curve passes through the point (0, 1). Find the value of f(3).

11 Evaluate: (a) $\int_4^9 x^{\frac{1}{2}}(2x - 3)\,dx$ (b) $\int_{-3}^{-1} \frac{(x-1)}{x^4}\,dx$

12 Given that $\frac{dy}{dx} = 6x - 3x^2$, and $x \geqslant 1$, and also that $y = 6$ at $x = 1$, find the value of x for which y is greatest and so find the greatest positive value of y.

6.6 Finding the area of a region bounded by lines and a curve

Consider first two simple examples.

Example 9

Here is a sketch of the shaded region bounded by the line $y = h$, the ordinates $x = a$ and $x = b$ and the x-axis:

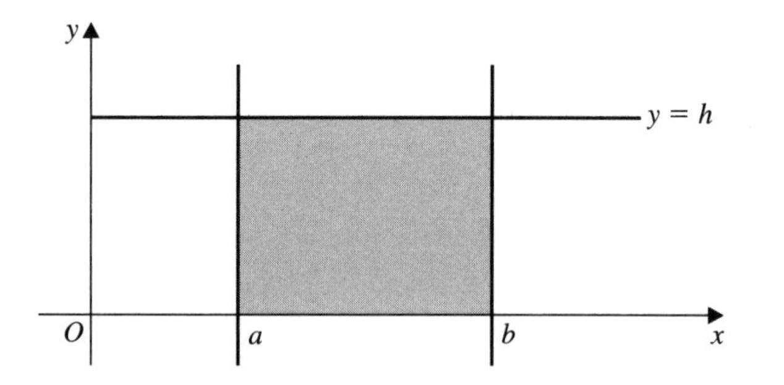

This region is a rectangle of length $(b - a)$ and width h. So the area of the region is $h(b - a)$. Notice also that:

$$\int_a^b h\,dx = \Big[hx\Big]_a^b = h\Big[x\Big]_a^b$$

$$= h(b - a)$$

The area of the region is $h(b - a)$.

Example 10

Here is a sketch of the shaded region bounded by the line $y = x$, the ordinates $x = a$ and $x = b$ and the x-axis.

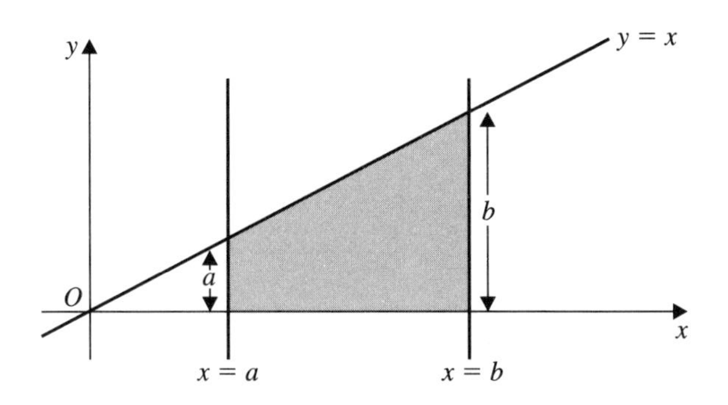

This region is a trapezium whose parallel sides are of length a and b. The perpendicular distance between these parallel lines is $(b - a)$ and the area of the trapezium is:

$$\frac{(a+b)}{2}(b-a) = \frac{b^2 - a^2}{2}$$

Notice also that:

$$\int_a^b x \, dx = \tfrac{1}{2}\left[x^2\right]_a^b$$

$$= \tfrac{1}{2}(b^2 - a^2)$$

The area of the region is $\frac{1}{2}(b^2 - a^2)$.

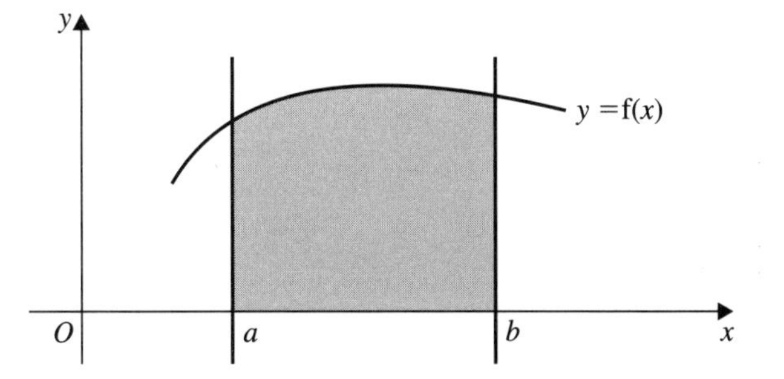

- **In general, the area of the region bounded by the curve $y = f(x)$, the ordinates $x = a$ and $x = b$ and the x-axis can be found by evaluating the definite integral $\int_a^b f(x)\, dx$, when it exists.**

Examples 9 and 10 show that this formula is valid for two simple cases when $f(x)$ is constant and when $f(x)$ is a linear function. A proof of the general case is not required in an advanced course. You are expected to assume the formula and to use it for the functions that you are able to integrate.

Notice also that the area of the region bounded by the curve with equation $x = g(y)$, the lines $y = \alpha$ and $y = \beta$ and the y-axis can be found by evaluating the definite integral $\int_{\alpha}^{\beta} g(y)\,dy$ when it exists. The region would look like this:

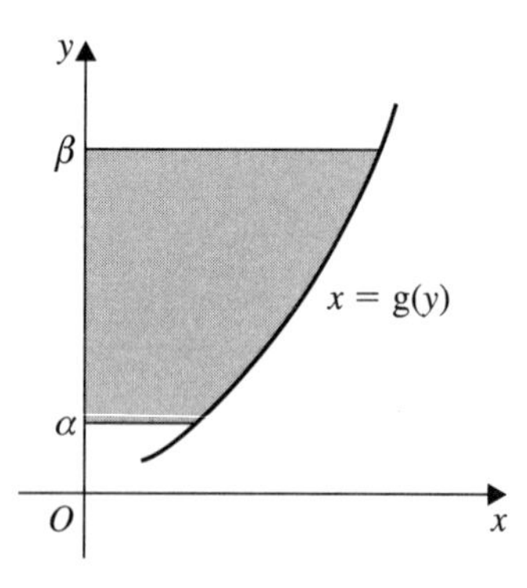

Example 11

Find the area of the finite region bounded by the curve with equation $y = 2x^3$, the lines $x = 2$, $x = 4$ and the x-axis.

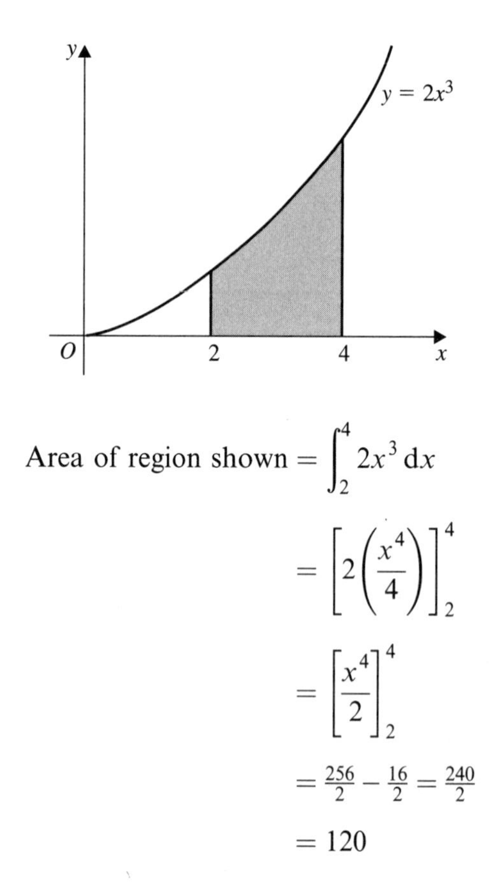

$$\text{Area of region shown} = \int_2^4 2x^3\,dx$$

$$= \left[2\left(\frac{x^4}{4}\right)\right]_2^4$$

$$= \left[\frac{x^4}{2}\right]_2^4$$

$$= \tfrac{256}{2} - \tfrac{16}{2} = \tfrac{240}{2}$$

$$= 120$$

The required area is 120 sq. units.

Example 12

Find the area of the finite region bounded by the curve $y = x^2 - 4$ and the x-axis.

The curve cuts the x-axis where $x^2 - 4 = 0$, that is at (2, 0) and (−2, 0). The curve cuts the y-axis at (0, −4). Here is a sketch of the curve, which is similar to $y = x^2$ but with its lowest point at (0, −4).

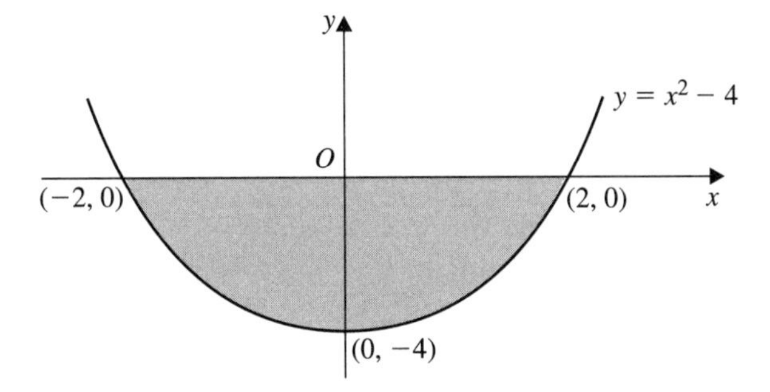

The shaded region is bounded by the x-axis and the curve.

The area of the shaded region is:

$$\int_{-2}^{2} (x^2 - 4)\,dx = \left[\frac{x^3}{3} - 4x\right]_{-2}^{2}$$
$$= \tfrac{8}{3} - 8 - (-\tfrac{8}{3} + 8)$$
$$= -\tfrac{32}{3}$$
$$= -10\tfrac{2}{3}$$

Notice that the negative sign means the region is below the x-axis. It does not mean that the area is negative. There is no such thing as a negative area!

Example 13

Here is a sketch of the curve $y = 2x - x^2$ which meets the line $y = -2x$ at the origin O and the point P. Determine the coordinates of P and so find the area of the shaded region.

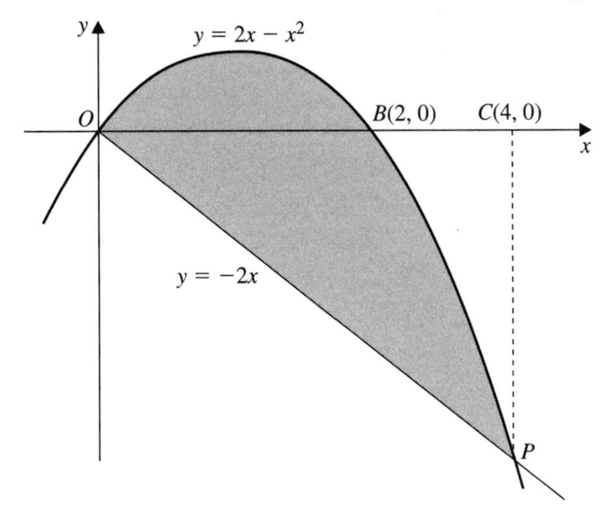

Find the coordinates of P by solving the simultaneous equations $y = -2x$ and $y = 2x - x^2$.

Replacing y by $-2x$ in the equation of the curve gives:

$$-2x = 2x - x^2$$

That is: $x^2 - 4x = 0$

Factorising gives: $x(x - 4) = 0$

So $x = 0$ at O. Also $x = 4$ and $y = -8$ at P.

The coordinates of P are $(4, -8)$.

Notice that the curve crosses the x-axis at the point $B(2, 0)$. The perpendicular from P to the x-axis meets the x-axis at the point $C(4, 0)$. You can now find the areas of three separate regions in the sketch. Here they are in separate sketches, labelled A_1, A_2 and A_3:

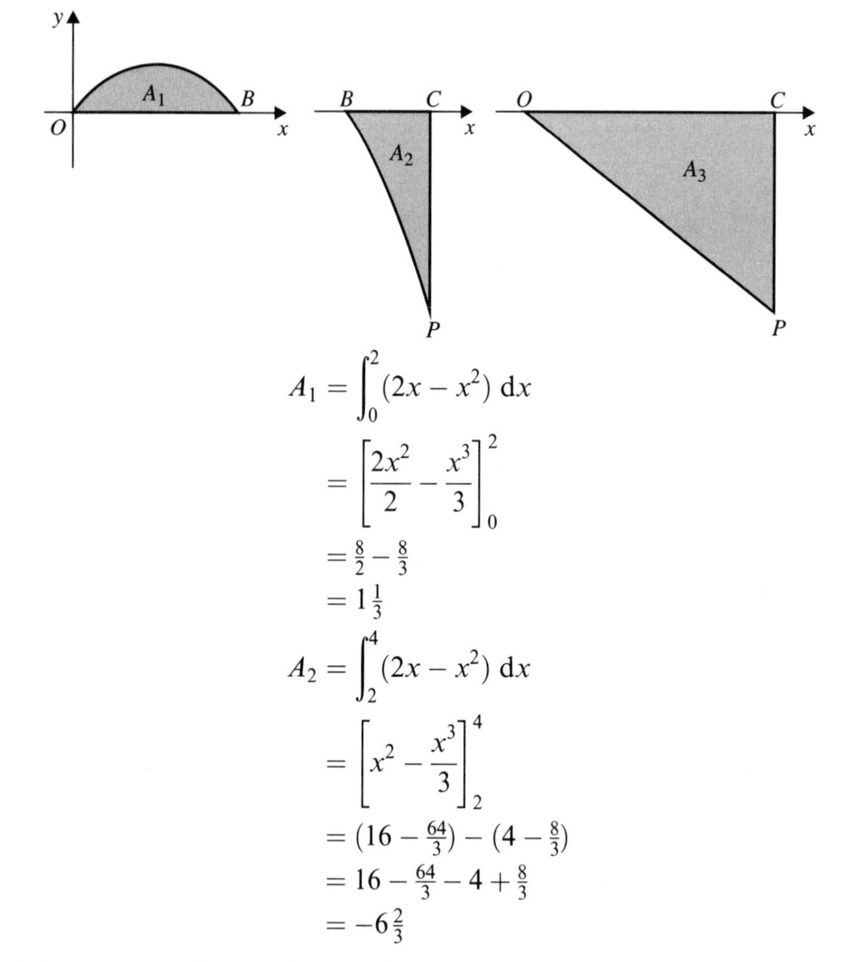

$$A_1 = \int_0^2 (2x - x^2)\,dx$$

$$= \left[\frac{2x^2}{2} - \frac{x^3}{3}\right]_0^2$$

$$= \frac{8}{2} - \frac{8}{3}$$

$$= 1\tfrac{1}{3}$$

$$A_2 = \int_2^4 (2x - x^2)\,dx$$

$$= \left[x^2 - \frac{x^3}{3}\right]_2^4$$

$$= (16 - \tfrac{64}{3}) - (4 - \tfrac{8}{3})$$

$$= 16 - \tfrac{64}{3} - 4 + \tfrac{8}{3}$$

$$= -6\tfrac{2}{3}$$

The area of triangle OCP is A_3.

$$A_3 = \tfrac{1}{2} \times 4 \times 8 = 16$$

The area of the shaded region is:

$$A_3 + A_1 - |A_2| = 16 + 1\tfrac{1}{3} - 6\tfrac{2}{3} = 10\tfrac{2}{3}$$

Notice that you have to find the areas of the regions above the x-axis and regions below the x-axis separately.

Example 14

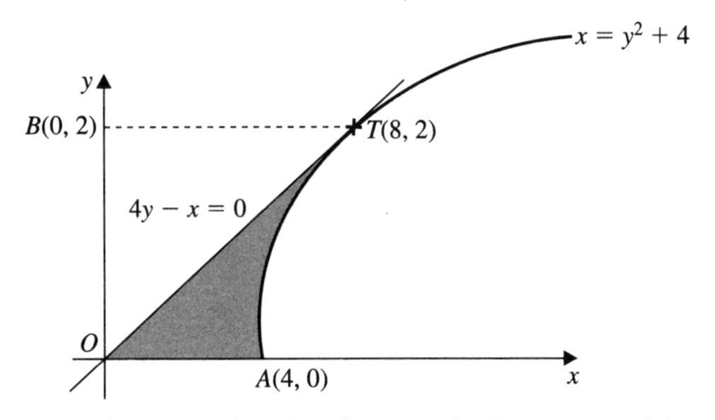

The diagram shows a sketch of part of the curve with equation $x = y^2 + 4$. The tangent at $T(8, 2)$ to the curve passes through the origin O. The coordinates of A and B are (4, 0) and (0, 2). Find the area of the shaded region.

$$\text{The area of } \triangle BTO = \tfrac{1}{2} \times 8 \times 2 \text{ units}^2$$
$$= 8 \text{ units}^2$$

$$\text{Area of whole region } OATB = \int_{y=0}^{y=2} x \, \mathrm{d}y$$
$$= \int_0^2 (y^2 + 4) \, \mathrm{d}y$$
$$= \left[\tfrac{1}{3}y^3 + 4y\right]_0^2$$
$$= \tfrac{8}{3} + 8 = 10\tfrac{2}{3} \text{ units}^2$$

$$\text{Area of shaded region} = \text{area of } OATB - \text{area of } \triangle BTO$$
$$= \left(10\tfrac{2}{3} - 8\right) \text{units}^2$$
$$= 2\tfrac{2}{3} \text{ units}^2$$

Exercise 6C

1 In each of the following cases find the area of the finite region bounded by the curve $y = \mathrm{f}(x)$, the x-axis and the lines $x = a$ and $x = b$.

	$\mathrm{f}(x)$	a	b
(a)	$6x^2 - 5$	1	3
(b)	$x^{\frac{5}{2}}$	1	4
(c)	$1 - x^2$	-1	1
(d)	$\frac{1}{3}x^3$	-3	0

2 In each of the following cases find the area of the finite region bounded by the curve $x = g(y)$, the y-axis and the lines $y = p$ and $y = q$.

	$g(y)$	p	q
(a)	$y^2 + 2$	2	4
(b)	$y^{\frac{1}{2}}$	9	16
(c)	$(2y - 3)^2$	3	4
(d)	$3y^{-2}$	-3	-1

3 Find the area of the region bounded by the x-axis and the part of the curve $y = 4x - 3 - x^2$ from $x = 1$ to $x = 3$.

4 Find the area of the region bounded by the x-axis, the curve $y = x - 8x^{-2}$ and the ordinates $x = 2$ and $x = 4$.

5 Find the area of the finite region bounded by the curve $y = 2 + x - x^2$ and the x-axis.

6 Find the area of the region bounded by the x-axis, the curve $y = (x - 6)^2$ and the ordinates $x = 2$ and $x = 5$. Explain the sign of the area obtained. Without further integration, state the value of $\int_2^5 (12x - x^2 - 36)\,dx$.

7 Sketch the curve $y = x^2$ and the line $y = 5x - 4$. Show that the curve and line intersect at the points P and Q whose x-coordinates are 1 and 4. Find the area of the finite region bounded by the line PQ and the curve.

8 Evaluate the integrals $\int_{-1}^{0} x^3\,dx$, $\int_0^1 x^3\,dx$ and $\int_{-1}^{1} x^3\,dx$.
Interpret these results geometrically by sketching graphs.

9 Sketch the curve $y = (1 - x)(x - 4)$ and find the coordinates of the points where the curve intersects the x-axis. Shade in the regions whose areas are represented by $\int_1^4 (1 - x)(x - 4)\,dx$ and $\int_4^5 (1 - x)(x - 4)\,dx$. Without actually integrating, explain why:

$$\int_4^5 (1 - x)(x - 4)\,dx = \int_0^1 (1 - x)(x - 4)\,dx$$

10 Calculate the area of the finite region bounded by the curves $y = x^2$ and $y = \sqrt{x}$.
(Hint: the curves are symmetrical about $y = x$ in the first quadrant.)

11 Find the area of the region bounded by the curve $y^2 = x$ and the line $y = x$.

12 Sketch the curve $x = 4y - y^2$. Evaluate the integral $\int_0^4 (4y - y^2)\,dy$ and show in your diagram the region whose area is given by your evaluation of the integral.

13 The region R is bounded by the curve $y = x^2 + 2$, the x- and y-axes and the line joining the point (2, 6) to the point (26, 0), as shown in the diagram. Find the area of R. Show also that the line joining (2, 6) to (26, 0) is the normal to the curve at (2, 6).

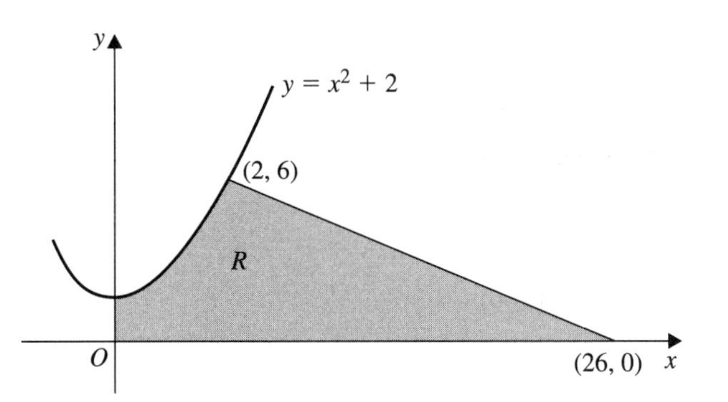

14 Find the area of the region bounded by the line $y = 2x$ and the curve $y = \sqrt{(20x)}$.

15 Calculate the area of the region bounded by the curve $y = 12 - 4x + 2x^2$ and the lines $x = -2$, $x = 3$ and $y = 5$.

16 Given that $\int_2^4 \left(3t^2 - 2t - \frac{k}{t^2}\right) dt = 40$, find the value of the constant k.

17 The line $y = 2$ meets the curve $y = 6 - x^2$ in two points A and B, as shown. Find the area of the shaded region.

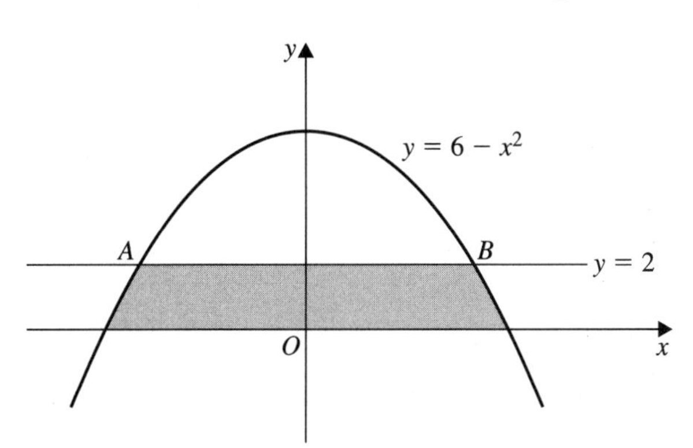

18 For the curve $y = (x - 1)(x + 2)^2$ find:

(a) the coordinates of the turning points, distinguishing between a maximum and a minimum point

(b) the area of the region bounded by the curve, the y-axis and the positive x-axis.

19 Sketch the curve with equation $y = x(4 - x^2)$. Evaluate

(a) $\int_0^2 x(4 - x^2)\,dx$ (b) $\int_{-2}^0 x(4 - x^2)\,dx$ and show by shading on your sketch the regions whose areas are equal to your answers for (a) and (b).

20 Find the area of the finite region bounded by the curve $x = y^2 - 1$, the y-axis and the line $y = 2$.

21 Calculate the area of the finite region between the curve $x = y^2 + 3y$ and the line $x = 4y$.

22 Given that $\frac{dy}{dx} = 2 + x^{-2}$ and $y = 2$ at $x = 1$, find:

(a) y in terms of x (b) the value of y at $x = -1$

23 The tangent to the curve $y = 4 - x^2$ at $P(1, 3)$ meets the x-axis at Q.

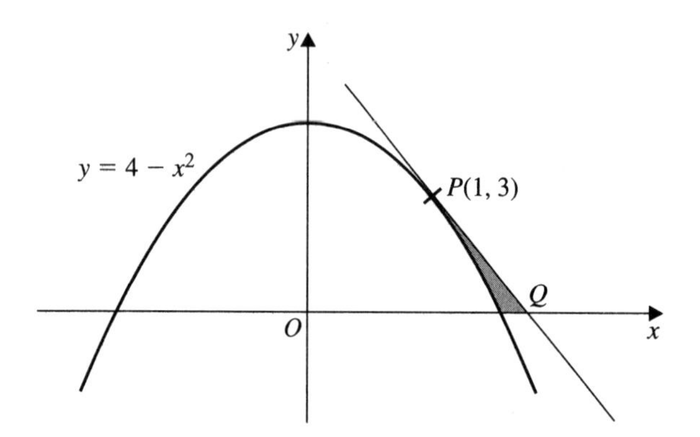

(a) Find the x-coordinate of Q.

(b) Find the area of the shaded region.

24 Find an equation of the normal to the curve $y = 2 + 5x - x^2$ at the point where $x = 3$.

Show that the normal meets the curve again at (1, 6).

Find the area of the region bounded by the curve and the normal.

SUMMARY OF KEY POINTS

1 $y = \mathrm{f}(x) + C$, where C is an arbitrary constant, is called the indefinite integral of $\frac{\mathrm{d}y}{\mathrm{d}x} = \mathrm{f}'(x)$

2 $\int x^n \, \mathrm{d}x = \frac{1}{n+1} x^{n+1} + C$, where $n \neq -1$.

3 $\int (u + v) \, \mathrm{d}x$, where u and v are functions of x, is $\int u \, \mathrm{d}x + \int v \, \mathrm{d}x$.

4 A statement such as $\frac{\mathrm{d}y}{\mathrm{d}x} = \mathrm{f}'(x)$, $\frac{\mathrm{d}^2 y}{\mathrm{d}x^2} + y = \frac{\mathrm{d}y}{\mathrm{d}x}$, $\frac{\mathrm{d}^3 y}{\mathrm{d}x^3} + \frac{\mathrm{d}^2 y}{\mathrm{d}x^2} = x$, is called a differential equation.

5 The differential equation $\frac{\mathrm{d}y}{\mathrm{d}x} = \mathrm{f}'(x)$ is a first order equation.

6 The solution of a differential equation which contains one or more arbitrary constants is called the general solution of the equation.

7 The conditions which allow you to evaluate the arbitrary constant(s) in the general solution of a differential equation are the boundary conditions.

8 A solution to a differential equation where the value of the arbitrary constant is known is called a particular solution.

9 The definite integral

$$\int_a^b \mathrm{f}'(x) \, \mathrm{d}x = [\mathrm{f}(x)]_a^b = \mathrm{f}(b) - \mathrm{f}(a)$$

provided that f′ is the derived function of f throughout the interval (a, b).

10 The area of the region bounded by the curve $y = \mathrm{f}(x)$, the ordinates $x = a$ and $x = b$ and the x-axis can be found by evaluating the definite integral $\int_a^b \mathrm{f}(x) \, \mathrm{d}x$, when it exists.

Proof

7

The word **proof** has many meanings, including the standard of the strength of an alcoholic drink, the trial impression taken by a printer before final printing and the demonstration of the truth, or otherwise, of some proposition, or theorem, by logical argument. It is the last of these that concerns mathematicians.

In mathematics you learn techniques and processes which are then presented in a logical sequence in order to reach an end result, called the answer. Often the answer may be given to you or you know it from memory or past experience.

A **direct proof** has a well defined starting point, followed by a series of valid, logical steps which lead to the required conclusion. You will be expected to exhibit sound understanding and correct use of mathematical language when writing a proof. In section 7.1 you will be given direct proofs of some well known formulae, which you have met already during the course.

7.1 Some standard proofs

Example 1

Given that a and b are unequal real numbers, prove that $a^2 + b^2 > 2ab$.

As a and b are unequal and real, $(a - b)$ is either a positive number (if $a > b$) or a negative number (if $a < b$).

So $(a - b)^2$ is a positive number.

That is:

$$(a - b)^2 > 0$$

$$a^2 - 2ab + b^2 > 0$$

so:

$$a^2 + b^2 > 2ab$$

Example 2

Prove that the sum of the first n positive integers is $\frac{1}{2}n(n + 1)$

What you are required to prove is that

$$\sum_{r=1}^{n} r = \frac{1}{2}n(n + 1)$$

Consider the series S, where S is given by

$$S = 1 + 2 + 3 + \ldots + (n-1) + n = \sum_{r=1}^{n} r$$

Also $\quad S = n + (n-1) + (n-2) + \ldots + 2 + 1$

when written backwards

Adding:

$$2S = (n+1) + (n+1) + (n+1) + \ldots + (n+1) + (n+1)$$

$$= n(n+1), \text{ as there are } n \text{ terms, each of which is } (n+1)$$

So:

$$S = \tfrac{1}{2}n(n+1), \text{ as required.}$$

Example 3

A geometric series has first term a and common ratio r. Prove that the sum of the first n terms is $\dfrac{a(r^n - 1)}{r-1}$.

Consider the geometric series S, where S is given by

$$S = a + ar + ar^2 + \ldots + ar^{n-2} + ar^{n-1}$$

$$rS = \quad ar + ar^2 + \ldots + ar^{n-2} + ar^{n-1} + ar^n$$

Taking the first line from the second:

$$rS - S = ar^n - a$$

Factorising gives: $\quad S(r-1) = a(r^n - 1)$

So: $\quad S = \dfrac{a(r^n - 1)}{r-1}$, as required.

7.2 Proof by a direct method

In section 7.1, three proofs have been given in examples 1, 2 and 3. In each case, the starting point of the proof had to be decided. After this, a few logical steps were required in order to move forward correctly to the result. Each example is an illustration of a proof by a direct method. You will meet many problems in mathematics which start with 'Given that statement A is true, prove (or show) that statement B is true'. In other cases, you will know that a starting statement such as 'Given A' implies a second statement B. For example, 'Given that $f(x) = x^2 - 4$, then $(x-2)$ is a factor of $f(x)$' is such a situation. You can write

$$\{f(x) = x^2 - 4\} \Rightarrow \{x - 2 \text{ is a factor of } f(x)\}.$$

The symbol $\Rightarrow$ stands for 'implies'.

'If A, then B' is written as $A \Rightarrow B$. This is the same as saying $B \Leftarrow A$ which reads 'B is implied by A'. The symbol $\Leftarrow$ stands for 'is implied by'.

For $A \Rightarrow B$ you can also say:

A is a **sufficient** condition for B.
B is a **necessary** condition for A.

These are equivalent to 'if A, then B'.

If A and B are logically equivalent statements, then you can write

$$A \Leftrightarrow B$$

and you say that 'if and only if A, then B' or 'A is a **necessary and sufficient** reason for B'.

In general, $A \Leftarrow B$ is the **converse** of $A \Rightarrow B$. You should understand that a statement and its converse are not necessarily both true, nor both false. It is unsound logically to assume that if a statement is true, so is its converse. Consider the following examples.

Example 4
The statement $\{x = 4\} \Rightarrow \{3x = 12\}$ is true.

The statement $\{3x = 12\} \Rightarrow \{x = 4\}$ is also true.

So: $\{x = 4\} \Leftrightarrow \{3x = 12\}$ is true

Example 5
The statement $\{x = 4\} \Rightarrow \{x^2 = 16\}$ is true, but $\{x^2 = 16\} \Rightarrow \{x = 4\}$ is NOT true, the correct statement being:

$$\{x^2 = 16\} \Rightarrow \{x = \pm 4\}$$

So here the statement $\{x = 4\} \Rightarrow \{x^2 = 16\}$ is true but its converse is not.

Example 6
Prove that the equation $x^2 + 2px + q = 0$, where p and q are constants, has distinct real roots if, and only if, $p^2 > q$.

That is, you are required to prove that

$$\{x^2 + 2px + q = 0 \text{ has real, distinct roots}\} \Leftrightarrow \{p^2 > q\}$$

$$x^2 + 2px + q = 0$$

So: $$x^2 + 2px = -q$$

Add p^2 to both sides:

$$x^2 + 2px + p^2 = p^2 - q$$

So: $$(x + p)^2 = p^2 - q \qquad (*)$$

Now if the equation has real, distinct roots, $p^2 - q$ must have real, distinct square roots. This is only true if $p^2 - q > 0$.

So: $$p^2 - q > 0$$

and $$p^2 > q$$

Conversely, if $p^2 > q$ then $p^2 - q > 0$. So $p^2 - q = k$, say, where k is a positive number.

So, from (*): $(x + p)^2 = k$ where $k > 0$

so: $$x + p = \pm\sqrt{k}$$

The equation $x^2 + 2px + q = 0$ therefore has roots given by

$$x = p + \sqrt{k}$$

and $$x = p - \sqrt{k}$$

Since $k > 0$ these are real and distinct.

That is, $\{p^2 > q\} \Rightarrow \{x^2 + 2px + q = 0$ has real distinct roots$\}$.

Summing up then, we have now proved that:

$\{$the equation $x^2 + 2px + q = 0$ has real, distinct roots$\} \Leftrightarrow \{p^2 > q\}$

You should note in this proof that the argument must be valid in both directions, that is:

> **if A, then B and**
> **if B, then A**
> **after which you can say A is a necessary and sufficient condition for B.**

Exercise 7A

1 Prove that the sum of the first n terms of the arithmetic series with first term a and common difference d is S, where

$$S = \frac{n}{2}[2a + d(n - 1)]$$

2 In $\triangle ABC$, $AB = 7$ cm, $AC = 5$ cm and $BC = x$ cm. Let $\angle BAC$ be denoted by θ. Given that the area of $\triangle ABC$ is 10 cm^2, prove that $\sin\theta = \frac{4}{7}$.
Hence find the two possible sizes of θ in degrees to one decimal place.

3 Prove that for all real values of x

$$x^2 + 1 \geqslant 2x$$

4 Given that $\frac{\mathrm{d}y}{\mathrm{d}x} = x^{-\frac{1}{2}}$ and that $y = 3$ when $x = 4$, prove that $y = 2x^{\frac{1}{2}} - 1$.

5 Explain why each of the following assertions is not necessarily correct:

(a) $\{\sin x = \frac{3}{5}\} \Rightarrow \{\cos x = \frac{4}{5}\}$

(b) $\{x^4 = 16\} \Rightarrow \{x = -2\}$

(c) $\{x^2 - 5x = 14\} \Rightarrow \{x = -2\}$

6 Investigate whether, or not, the converse statements of (a), (b) and (c) in question **5** are correct.

7 The following assertions are given: if they are true, prove them to be correct, if they are false provide a counter-example to illustrate the falsehood:

(a) $\{x^2 + 3x - 28 \leqslant 0\} \Leftrightarrow \{-7 \leqslant x \leqslant 4\}$

(b) $\{y = x^3 - 3x^2\} \Leftrightarrow \left\{\frac{\mathrm{d}y}{\mathrm{d}x} = 3x^2 - 6x\right\}$

(c) $\{x < y\} \Leftrightarrow \left\{\frac{1}{x} > \frac{1}{y}\right\}$

8 The $\triangle ABC$ is such that A is (1, 1), B is (−2, 5) and C is (4, 5). Prove that $\triangle ABC$ is isosceles and determine its area.

9 Prove that real roots of the equation $x^2 + 8x + k = 0$ do not exist if $k > 16$.

10 State a condition for the roots of the equation $x^2 + ax + b = 0$ to be real. Prove that your condition is true.

11 Prove that the equation $ax^2 + bx + 1 = 0$ has identical roots if and only if $b^2 - 4a = 0$.

12 Prove that the tangent to the curve with equation $y = x^2$ at the point where $x = 1$ has equation $y = 2x - 1$.

13 Find the condition, in terms of k, for the equation $x^2 - 2kx - 5k = 0$ to have real, distinct roots. Is the condition you have found both necessary and sufficient? Justify your answer.

14 Prove that there is no real value of λ for which the equation

$$x^2 + (\lambda - 3)x + \lambda^2 + 4 = 0$$

has real roots.

15 Prove that $(x - a)$ is a factor of $x^3 - 3a^2x + 2a^3$. Hence factorise $x^3 - 3a^2x + 2a^3$ completely.

16 Prove that -2 is a root of the equation $x^3 - 6x - 4 = 0$.
Find the other two roots of the equation in surd form.

17 Prove that the sum to infinity of the geometric series with first term $-\frac{4}{5}$ and common ratio $-\frac{4}{5}$ is $-\frac{4}{9}$.

18 Prove that the area, in units^2, of the shaded region shown is $\frac{8}{3}$.

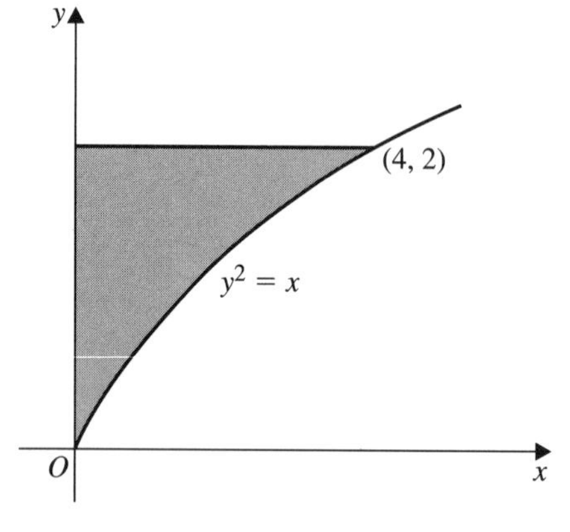

19 Prove that

$$\frac{\sin\theta}{1+\sin\theta} \equiv \frac{\tan\theta}{\cos\theta} - \tan^2\theta$$

20 Prove that $\dfrac{3-\sqrt{5}}{3+\sqrt{5}}$ can be expressed in the form $A - B\sqrt{5}$ where the rational numbers A and B are to be found.

21 Prove that $\dfrac{\sqrt{3}+\sqrt{2}}{\sqrt{3}-\sqrt{2}}$ can be expressed in the form $m + n\sqrt{6}$ where the integers m and n are to be found.

22 Prove that the equation $2^{2x} - 6(2^x) + 8 = 0$ is satisfied by two integer values of x only, and find these integers.

23 The points $P(-1, -2)$, $Q(5, 0)$, $R(4, 3)$ and $S(1, 2)$ are the vertices of a quadrilateral $PQRS$. Prove that angles PQR and QRS are 90°. Find the area of the quadrilateral, given that the units are cm.

24 Prove that the lines with equations $3x - 2y = 19$, $2x + 3y = 4$ and $2x + 5y = 0$ pass through a common point P.
Find the equation of the line through P which is perpendicular to the line $2x + 5y = 0$.

25 Given that $(x - 1)$ and $(x + 2)$ are factors of the expression $3x^3 - x^2 + Ax + B$,

(a) show that $A = -10$

(b) find the value of B

(c) find the third factor of the expression.

26 Given that $f(x) = \left(\frac{x^2 + 1}{x}\right)^2$, show that $f'(x) = 2x - 2x^{-3}$.

Hence find the equation of the normal to the curve $y = f(x)$ at the point where $x = -1$.

27 Show that the equation $4 - 5\cos\theta = 2\sin^2\theta$ has just two roots in the interval $[0, 360°]$ and find the value of these roots *exactly*.

28 Prove that the sum of the first 200 positive integers is 20 100. Hence find the sum of those integers N, such that $1 < N < 200$ and N is *not* a multiple of 3.

29

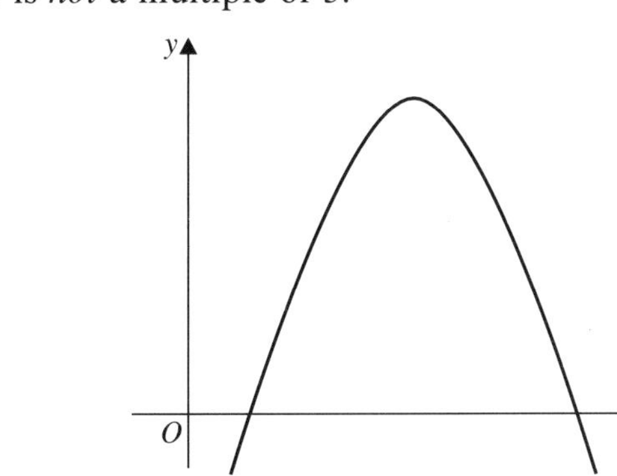

The diagram shows part of the curve with equation $y = px^2 + qx + r$, where p, q and r are constants. Prove that each of the following are true for this curve:

(a) $p < 0$ (b) $r < 0$ (c) $q^2 - 4pr > 0$

30 Given that $-5 \leqslant x \leqslant 6$, prove that

$$-8\tfrac{1}{12} \leqslant 3x^2 - 7x - 4 \leqslant 106$$

SUMMARY OF KEY POINTS

1. A direct proof has a well defined starting point, followed by a series of valid, logical steps which lead to the required conclusion.
2. The symbol $\Rightarrow$ stands for 'implies'.
3. The symbol $\Leftarrow$ stands for 'is implied by'.
4. If $A \Leftrightarrow B$ then A and B are logically equivalent statements.

Review exercise 2

1 Find the positive constants a and b such that 0.25, a, 9 are in geometric progression and 0.25, a, $9 - b$ are in arithmetic progression. [E]

2 Write down the first five terms of the sequence defined by

$$u_{r+1} = -6u_r,\ u_1 = -\tfrac{1}{3}$$

3 Find the greatest value of $5x - 4x^2 - 1$.

4 Find the area of the region bounded by the curve $y = 1 - x + x^2$, the lines $x = -1$ and $x = 2$ and the x-axis.

5 The first three terms of a geometric series are $3(q + 5)$, $3(q + 3)$ and $(q + 7)$ respectively.

(a) Calculate the value of q.

(b) Find the common ratio, r, of the geometric series.

(c) Find the sum to infinity of the geometric series. [E]

6 Evaluate $\int_{-1}^{2} (x - 1)(2x + 5)\,dx$.

7 (a) The first term of a geometric series is 22 and the second term is 19. Find

(i) the sum, to 3 significant figures, of the first 17 terms of the series

(ii) the exact sum to infinity of the series.

(b) The first term of an arithmetic series is -50 and the common difference is 4. The sum of the first N terms of this series is -336. Form an equation in N and solve it to find the possible values of N. [E]

8 A cylindrical vessel, closed at both ends, is made of thin material and contains a volume of $16\pi\,\text{m}^3$. Given that the total exterior surface area of the vessel is a minimum, find its height and base radius. [E]

9 Find the area of the finite region between the curve $y = 8x - x^2$ and the x-axis.

10 (a) A geometric series has first term 3 and common ratio 0.8. Find the sum of the first 24 terms, giving your answer to 3 significant figures.

(b) Find $\sum_{r=1}^{52} \frac{r}{2}$. [E]

11 (a) The 3rd term of an arithmetic series is −20 and the 11th term is 20. Find the common difference and the 1st term of this series.

When the first k terms of this series are added together, their sum is zero. Find the non-zero value of k.

(b) The 1st and 7th terms of a geometric series of positive terms are 2 and 16 respectively.

Find, to 3 significant figures, the sum of the first 7 terms of this series. [E]

12 Given that $f(x) \equiv x^2(2x+1)(3-5x)$, find $f'(x)$.

13 Find the area of the finite region bounded by the curve $y = 12 - x - x^2$ and the x-axis.

14 (a) Find $\sum_{r=1}^{100} \frac{2r}{3}$.

(b) Find the sum to infinity of

$$\frac{7}{10} + \frac{7}{100} + \frac{7}{1000} + \frac{7}{10\,000} + \ldots.$$

giving your answer in the form $\frac{p}{q}$, where p and q are positive integers.

Hence, or otherwise, find the value of

$\sum_{r=1}^{\infty} \frac{k}{10^r}$ in terms of k. [E]

15 A geometric series has 4th term 40 and 9th term 1.25.

(a) Find the first term and the common ratio of the series.

(b) Show that the sum S_n of the first n terms is given by

$$S_n = 640\left[1 - \left(\tfrac{1}{2}\right)^n\right]$$ [E]

16 The radius of a circular oil slick is $(4 + 5t^2)$ cm at time t seconds.

Express the area $A\,\text{cm}^2$ at time t seconds in terms of t. Find the rate at which (a) the radius (b) the area of the slick is increasing at $t = 3$.

17 Evaluate $\int_1^8 (x^{\frac{1}{3}} - x^{-\frac{1}{3}})\,dx$.

18 A ball is dropped onto a horizontal plane and rebounds successively. The height above the plane reached by the ball after the first impact with the plane is 4 m. After each impact the ball rises to a height which is $\frac{3}{4}$ of the height reached after the previous impact. Calculate the total vertical distance travelled by the ball from the first impact until

(a) the fourth impact,

(b) the eighth impact,

(c) it comes to rest. [E]

19 A solid circular cylinder is to have a volume of 16π cm^3. Express the surface area A cm^2 of the cylinder in terms of the radius x cm of its base. Hence find the value of x for which A is a minimum. Find also the minimum value of A.

20 Find the area of the finite region bounded by the curve $y = -2x^2 - 8x - 6$ and the x-axis.

21 Given that

$$S_n = \sum_{r=1}^{n} 5^r$$

where n is a positive integer, show that

$$S_n = \tfrac{5}{4}(5^n - 1)$$ [E]

22 A rectangular box without a lid has a square base of side x cm. The total internal surface area is 40 cm^2 and its walls and base are thin. Show that the volume y cm^3 is given by

$$y = \tfrac{1}{4}x(40 - x^2)$$

Given that x can vary, find the maximum volume of the box.

23 Given that $y^{\frac{1}{2}} = x^{\frac{1}{3}} + 3$,

(a) show that $y = x^{\frac{2}{3}} + Ax^{\frac{1}{3}} + B$, where A and B are constants to be found.

(b) Hence find $\int y\,dx$.

(c) Using your answer from (b) determine the exact value of $\int_1^8 y\,dx$. [E]

24 (a) Find the sum of the integers which are divisible by 3 and lie between 1 and 400.

(b) Hence, or otherwise, find the sum of the integers, from 1 to 400 inclusive, which are *not* divisible by 3. [E]

25 Given that $\mathrm{f}(x) = \left(x - \frac{1}{x}\right)^3$, find the value of $\mathrm{f}'(3)$.

26 Find the area of the finite region bounded by the curve $y = 6x^2$ and the line $y = -5x - 1$.

27 A motorcycle has four gears. The maximum speed in bottom gear is $40\,\mathrm{km\,h^{-1}}$ and the maximum speed in top gear is $120\,\mathrm{km\,h^{-1}}$. Given that the maximum speeds in each successive gear form a geometric series, calculate, in $\mathrm{km\,h^{-1}}$ to one decimal place, the maximum speeds in the two intermediate gears. [E]

28 A savings scheme pays 5% per annum compound interest. A deposit of £100 is invested in this scheme at the start of each year.

(a) Show that at the start of the third year, after the annual deposit has been made, the amount in the scheme is £315.25.

(b) Find the amount in the scheme at the start of the fortieth year, after the annual deposit has been made. [E]

29 Given that $y = \dfrac{(x-3)(3x+2)}{\sqrt{x}}$, find $\dfrac{\mathrm{d}y}{\mathrm{d}x}$ and $\dfrac{\mathrm{d}^2y}{\mathrm{d}x^2}$.

30 (a) Find the coordinates of the turning points on the curve whose equation is:

$$y = x^3 - 9x^2 + 24x$$

(b) Evaluate $\displaystyle\int_2^4 (x^3 - 9x^2 + 24x)\,\mathrm{d}x$.

31 The second term of a geometric series is 80 and the fifth term of the series is 5.12.

(a) Show that the common ratio of the series is 0.4.

Calculate:

(b) the first term of the series

(c) the sum to infinity of the series, giving your answer as an exact fraction

(d) the difference between the sum to infinity of the series and the sum of the first 14 terms of the series, giving your answer in the form $a \times 10^n$, where $1 \leqslant a < 10$ and n is an integer.

32 A salesman is paid commission of £10 per week for each life insurance policy which he has sold. Each week he sells one new policy so that he is paid £10 commission in the first week, £20 commission in the second week, £30 commission in the third week and so on.

(a) Find his total commission in the first year of 52 weeks.

In the second year the commission increases to £11 per week on new policies sold, although it remains at £10 per week for policies sold in the first year. He continues to sell one policy per week.

(b) Show that he is paid £542 in the second week of his second year.

(c) Find the total commission paid to him in the second year.

[E]

33 Find the nature and the coordinates of the turning points on the curves with equations

(a) $y = (x + 2)(x - 1)^2$ (b) $y = (x + 4)(x - 1)^4$ [E]

34

$$f(x) \equiv x^3 - 5x^2 - 8x + 12$$

(a) Show that $(x + 2)$ is a factor of $f(x)$.

(b) Factorise $f(x)$ completely.

(c) Find the coordinates of the turning points of the curve with equation $y = f(x)$.

(d) Calculate the area of the finite region bounded by the curve, the x-axis and the lines with equations $x = -1$ and $x = +1$. [E]

35 The first three terms of a series, S, are $(m - 4)$, $(m + 2)$ and $(3m + 1)$.

(i) Given, also, that S is an arithmetic series,

(a) find m.

Using your value of m,

(b) write down the first four terms of the arithmetic series.

(ii) Given, instead, that S is a geometric series,

(a) find the two possible values of m

(b) write down the first four terms of each of the two geometric series obtained with your values of m

(c) state the value of the common ratio of each of the series.

One of these two geometric series has a sum to infinity.

(d) Find the sum to infinity of that series. [E]

36 Given that

$$\frac{\mathrm{d}x}{\mathrm{d}t} = 2t^2 - 10t + 12, \ t \geqslant 0$$

calculate:

(a) the values of t for which $\dfrac{\mathrm{d}x}{\mathrm{d}t} = 0$

(b) the value of x when $t = 4$, given further that $x = 2$ when $t = 1$.

37 A rectangular tank is made of thin sheet metal. The tank has a horizontal square base, of side x cm, and no top. When full the tank holds 500 litres.

(a) Show that the area, A cm^2, of sheet metal needed to make this tank is given by

$$A = x^2 + \frac{2\,000\,000}{x}$$

(b) Given that x can vary, find the minimum value of A and the value of x for which this occurs.

(c) Prove that this value of A is a minimum. [E]

38 John is given an interest-free loan to buy a second-hand car. He repays the loan in monthly instalments. He repays £20 the first month, £22 the second month and the repayments continue to rise by £2 per month until the loan is repaid. Given that the final monthly repayment is £114,

(a) show that the number of months it will take John to repay the loan is 48,

(b) find the amount, in pounds, of the loan. [E]

39 (i)

$$S_n = \sum_{r=1}^{n} (2r - 3)$$

(a) Write down the first three terms of this series.

(b) Find S_{50}.

(c) Find n such that $S_n = 575$.

(ii) The first three terms of a geometric series are $(x - 8)$, x and $(2x + 12)$ respectively.

(a) Find the two possible values of x and, hence, the two possible values of the common ratio of the series.

Given also that the common ratio is less than one,

(b) find the sum to infinity of the series. [E]

40 An open rectangular tank has a horizontal square base of side x metres. The height of the tank is y metres. The tank is made of thin sheet metal. The external surface area of the tank is S square metres and the volume is V cubic metres.

(a) Show that $S = 4\frac{V}{x} + x^2$.

Given that $V = 62.5$ and that x varies,

(b) find $\frac{\mathrm{d}S}{\mathrm{d}x}$.

(c) Determine the value of x for which $\frac{\mathrm{d}S}{\mathrm{d}x} = 0$ and find the corresponding value of S.

(d) Show that this value of S is the minimum possible value.

[E]

41

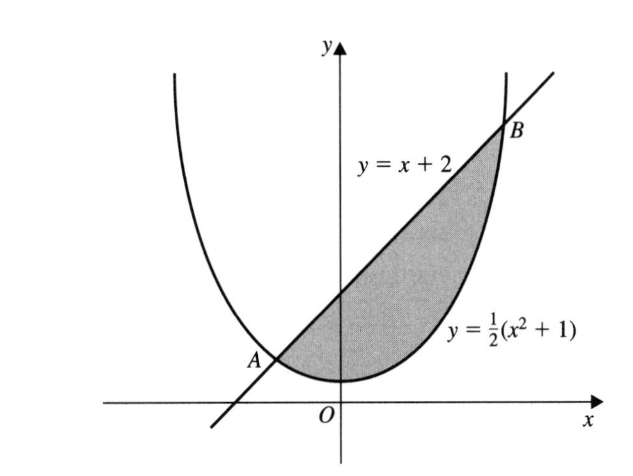

The diagram shows the curve $y = \frac{1}{2}(x^2 + 1)$ and the line $y = x + 2$, which meet at the points A and B.

(a) Find the coordinates of A and B.

(b) Determine the area of the shaded region.

42 A sequence of numbers $u_1, u_2, \ldots, u_n, \ldots$ is given by the formula

$$u_n = 3\left(\tfrac{2}{3}\right)^n - 1$$

where n is a positive integer.

(a) Find the values of u_1, u_2 and u_3.

(b) Find $\sum_{n=1}^{15} 3\left(\frac{2}{3}\right)^n$, and hence show that $\sum_{n=1}^{15} u_n = -9.014$ to 4 significant figures.

(c) Prove that $3u_{n+1} = 2u_n - 1$. [E]

43 (a) The sum of the first, sixth and seventh terms of an arithmetic progression is $-2\frac{1}{2}$. The sum of the first nine terms of the progression is zero. Calculate, for this progression, the value of

(i) the first term

(ii) the common difference

(iii) the fifth term.

(b) The first term of a geometric progression is a and the common ratio is r, where $r > 0$. The third term is 3. The first term of a second geometric progression is $\frac{a}{2}$ and the common ratio is $\frac{1}{r}$. The third term of the second progression is 24.

Calculate

(i) the value of r

(ii) the value of a

(iii) the sum of the first ten terms of the *second* progression. [E]

44 The diagram shows a square card $ABCD$ of side 8 cm. From each of the corners is removed a square of side x cm, and the remainder is folded along each of the dotted lines to form an open tray of depth x cm and volume V cm^3.

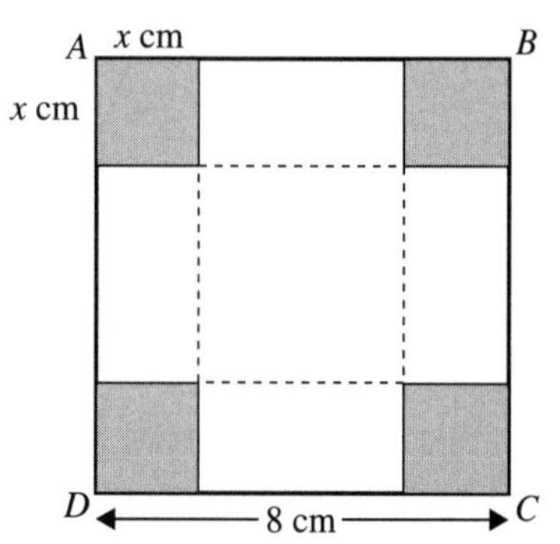

(a) Show that

$$V = 64x - 32x^2 + 4x^3$$

(b) Find $\frac{dV}{dx}$.

(c) Hence find the value of x for which $\frac{dV}{dx} = 0$.

(d) Show that this value of x gives a maximum value of V.

(e) Find the maximum value of V. [E]

45

$$y = 3x^{\frac{1}{2}} - 4x^{-\frac{1}{2}},\ x > 0$$

(a) Find $\frac{dy}{dx}$.

(b) Find $\int y\,dx$.

(c) Hence show that $\int_1^3 y\,dx = A + B\sqrt{3}$, where A and B are integers to be found. [E]

46 (i) The rth term of an arithmetic series is given by

$$2r - 5, \qquad r = 1, 2, 3, \ldots.$$

Write down the first four terms of this series and find

$$S_n = \sum_{r=1}^{n} (2r - 5)$$

in terms of n.

Given that $S_n = 165$ find the value of n.

(ii) In a geometric series, $(x + 1)$, $(x + 3)$ and $(x + 4)$ are the first, second and third terms respectively. Calculate the value of x and hence write down the numerical values of the common ratio and the first term of the series.

Calculate the numerical value of the sum to infinity of the series. [E]

47 The first three terms of a geometric series are $p(3q + 1)$, $p(2q + 2)$ and $p(2q - 1)$ respectively, where p and q are non-zero constants.

(a) Use algebra to show that one possible value of q is 5 and to find the other possible value of q.

(b) For each possible value of q, calculate the value of the common ratio of the series.

Given that $q = 5$ and that the sum to infinity of the geometric series is 896, calculate

(c) the value of p

(d) the sum, to 2 decimal places, of the first twelve terms of the series. [E]

48 A solid right cylinder has volume $40\pi\,\text{cm}^3$. The diameter of the cylinder is x cm and the height is h cm.

(a) Show that $x^2h = 160$.

The total surface area of the cylinder is $A\,\text{cm}^2$.

(b) Show that $A = \dfrac{\pi}{2}\left(x^2 + \dfrac{320}{x}\right)$.

(c) Find, to 3 significant figures, the value of x such that $\dfrac{\mathrm{d}A}{\mathrm{d}x} = 0$.

(d) Prove that this value of x gives a minimum value of A.

(e) Find, to 3 significant figures, the minimum value of A. [E]

49

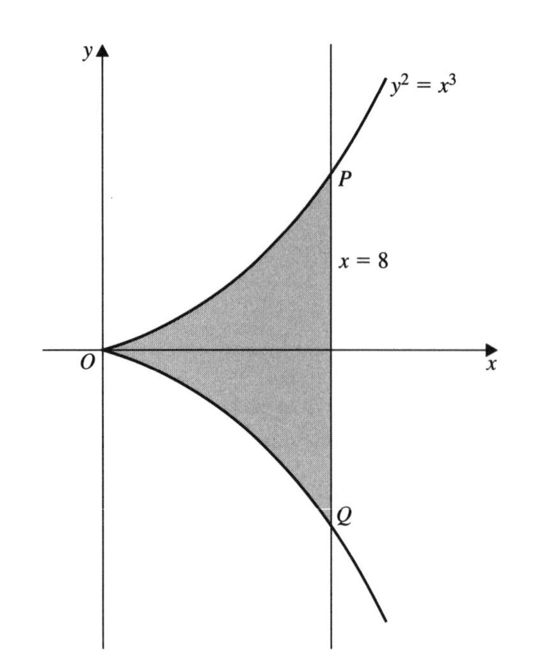

The curve $y^2 = x^3$ is shown and the line $x = 8$ which meets the curve at P and Q. Find the length of the line PQ and the area of the shaded region.

50 An investment of £2000 is made at the start of a year with a Finance Company. At the end of this year and at the end of each subsequent year the value of the investment is 11% greater than its value at the start of that year.

(a) Find, to the nearest £, the value of the investment at the end of

(i) the 5th year, (ii) the 10th year.

A client decides to invest £2000 at the start of each year. Write down a series whose sum is the total value of this annual investment at the end of 12 years.

(b) By finding the sum of your series, determine, to the nearest £, the value of the investment at the end of 12 years. [E]

51 (i) The eleventh term of an arithmetic series is 22 and the sum of the first 4 terms is 37.

(a) Find the first term and the common difference of this series.

(b) Find the sum of the first 21 terms of this series.

(ii) The first, second and third terms of a geometric series are x, $(x + 1)$ and $(x + 4)$, respectively.

(a) Find the value of x.

(b) Find the common ratio of the series.

(c) Find the seventh term of the series. [E]

52

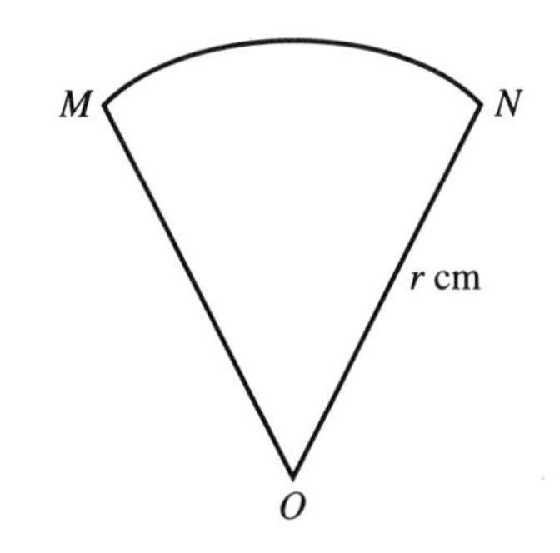

The diagram shows a minor sector OMN of a circle centre O and radius r cm. The perimeter of the sector is 100 cm and the area of the sector is A cm^2.

(a) Show that $A = 50r - r^2$.

Given that r varies, find

(b) the value of r for which A is a maximum and show that A is a maximum

(c) the value of $\angle MON$ for this maximum area

(d) the maximum area of the sector OMN. [E]

53

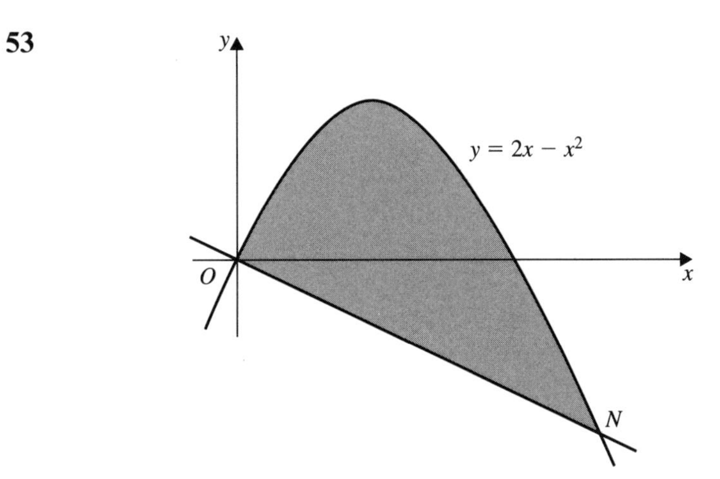

The diagram shows a sketch of the curve with equation $y = 2x - x^2$ and the line ON which is the normal to the curve at the origin O.

(a) Find an equation of ON.

(b) Show that the x-coordinate of the point N is $2\frac{1}{2}$ and determine its y-coordinate.

The shaded region shown is bounded by the curve and the line ON.

(c) Without using a calculator, determine the area of the shaded region. [E]

54 (i) The sum of the first five terms of an arithmetic series is 60 and the sum of the first ten terms of the series is $207\frac{1}{2}$.

(a) Calculate the first term of the series and the common difference.

The sum of the first n terms of the series is 500.

(b) Calculate the value of n.

(ii) A geometric series has first term $(2x - 6)$, second term $(2x + 2)$ and third term $(7x + 1)$.

(a) Find the two possible values of x.

Given that x is an integer,

(b) find the common ratio of the series

(c) calculate the sum of the first 10 terms of the series. [E]

55 The diagram shows a brick in the shape of a cuboid with base x cm by $2x$ cm and height h cm.

The total surface area of the brick is $300\,\text{cm}^2$.

(a) Show that $h = \dfrac{50}{x} - \dfrac{2x}{3}$.

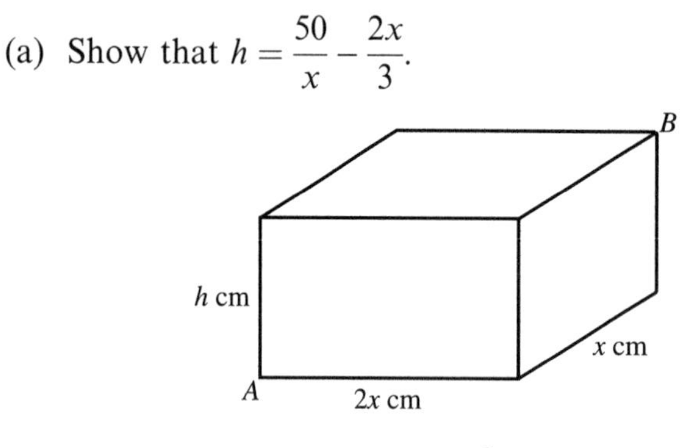

The volume of the brick is $V\,\text{cm}^3$.

(b) Express V in terms of x only.

Given that x can vary,

(c) find the maximum value of V. [E]

56 (i) In a geometric series, $(x - 1)$, $(x + 1)$ and $(x + 9)$ are consecutive terms. Find the value of the common ratio of the series.

(ii) The sum of the first n terms of an arithmetic series is 6014. Given that the sum of the first term and the nth term is 124,

(a) calculate the value of n,

(b) show that the value of the 49th term is 62.

Given also that the value of the 61st term is 77, find

(c) the value of the first term of the series,

(d) the common difference,

(e) the sum of the first 80 terms. [E]

57

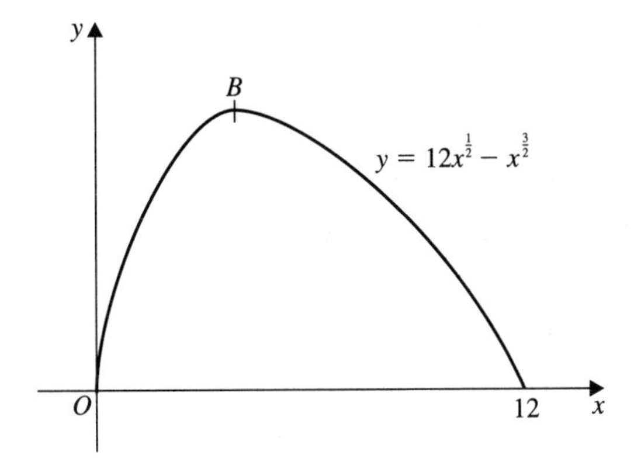

The diagram shows a sketch of the curve with equation

$$y = 12x^{\frac{1}{2}} - x^{\frac{3}{2}} \text{ for } 0 \leqslant x \leqslant 12$$

(a) Show that $\dfrac{dy}{dx} = \dfrac{3}{2}x^{-\frac{1}{2}}(4 - x)$.

At the point B on the curve the tangent to the curve is parallel to the x-axis.

(b) Find the coordinates of the point B.

(c) Find the area of the finite region bounded by the curve and the x-axis. [E]

58

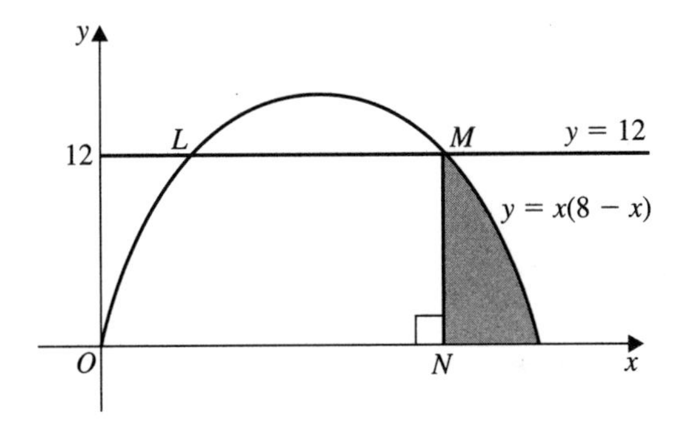

The diagram shows the curve C with equation $y = x(8 - x)$ and the line with equation $y = 12$ which meet at the points L and M.

(a) Determine the coordinates of the point M.

Given that N is the foot of the perpendicular from M onto the x-axis,

(b) calculate the area of the shaded region which is bounded by NM, the curve C and the x-axis. [E]

59 The third term of a geometric series is 15 and the common ratio of the series is 2.

(a) Calculate the value of the sixth term of the series.

(b) Calculate the sum of the first ten terms of the series.

The second and fifth terms of the series form the first two terms of an arithmetic series.

(c) Find the value of the ninth term of the arithmetic series.

(d) Find the sum of the first thirteen terms of the arithmetic series. [E]

60 (a) The seventh term of an arithmetic series is 6 and the eighteenth term is $22\frac{1}{2}$. Calculate

(i) the common difference of the series,

(ii) the first term of the series.

Given also that the sum of the first n terms of the series is 252,

(iii) find the value of n.

(b) The second term of a convergent geometric series is 24 and the sum of the first three terms is 126. Find the common ratio of the series. [E]

61

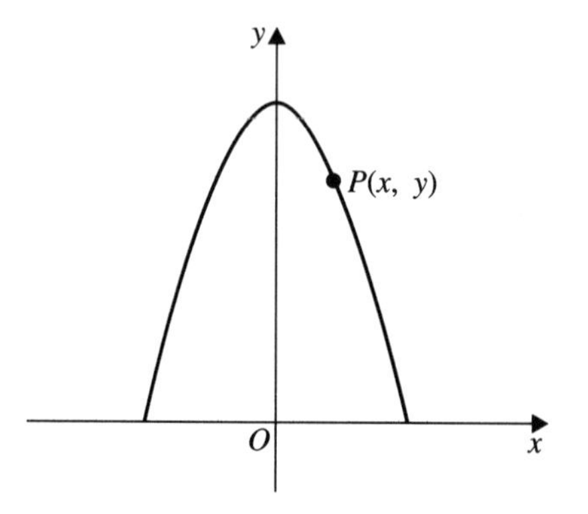

The diagram shows the part of the curve with equation $y = 5 - \frac{1}{2}x^2$ for which $y \geqslant 0$. The point $P(x, y)$ lies on the curve and O is the origin.

(a) Show that $OP^2 = \frac{1}{4}x^4 - 4x^2 + 25$.

Taking $f(x) \equiv \frac{1}{4}x^4 - 4x^2 + 25$,

(b) find the values of x for which $f'(x) = 0$.

(c) Hence, or otherwise, find the minimum distance from O to the curve, showing that your answer is a minimum. [E]

62

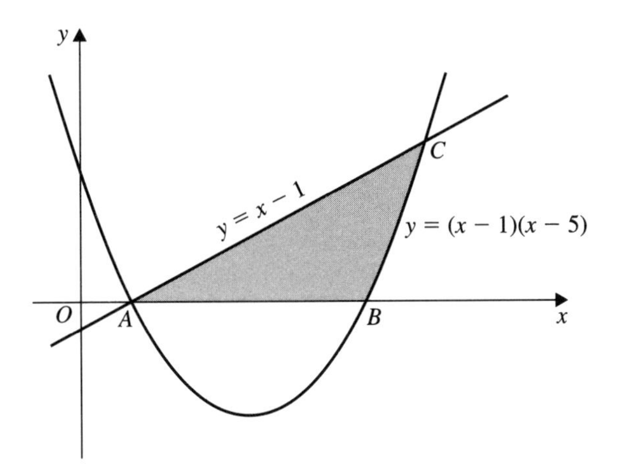

The diagram shows the line $y = x - 1$ meeting the curve with equation $y = (x - 1)(x - 5)$ at A and C. The curve meets the x-axis at A and B.

(a) Write down the coordinates of A and B and find the coordinates of C.

(b) Find the area of the shaded region bounded by the line, the curve and the x-axis. [E]

63

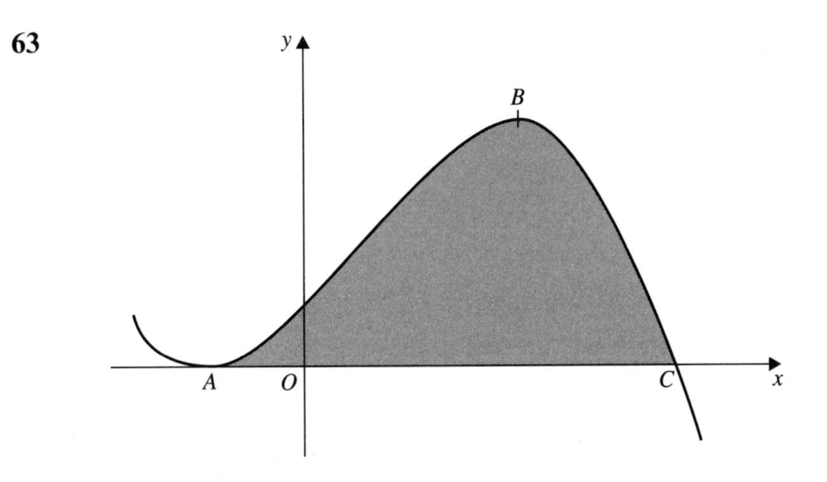

The diagram shows part of the curve with equation $y = 3 + 5x + x^2 - x^3$. The curve touches the x-axis at A and crosses the x-axis at C. The points A and B are stationary points on the curve.

(a) Show that C has coordinates (3, 0).

Using calculus and showing all your working, find

(b) the coordinates of A and B

(c) the area of the shaded region in the diagram bounded by the curve and the x-axis. [E]

64

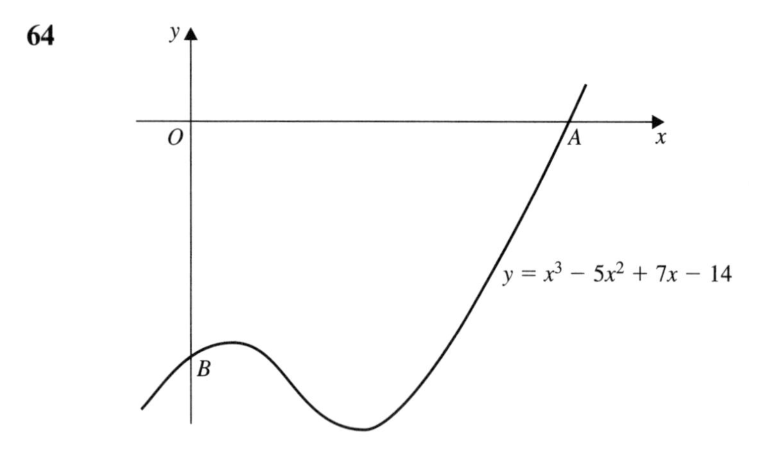

The curve C with equation $y = x^3 - 5x^2 + 7x - 14$ meets the x- and y-axes at the points A and B respectively, as shown in the diagram.

(a) State the coordinates of the point B.

(b) Determine, by calculation, the coordinates of the stationary points of the curve C.

(c) For the function f where $\mathrm{f}(x) = x^3 - 5x^2 + 7x - 14$, find the set of values of x for which f is decreasing. [E]

65 The sum of the first six terms of an arithmetic series is 90 and the eleventh term is five times the second term.

Find:

(a) the first term of the series

(b) the common difference of the series.

Given that the pth term is greater than 100,

(c) write down an inequality in p.

(d) Solve your inequality to determine the least value of p. [E]

66 The diagram shows a rectangular cake-box, with no top, which is made from thin card. The volume of the box is 500 cm^3. The base of the box is a square with sides of length x cm.

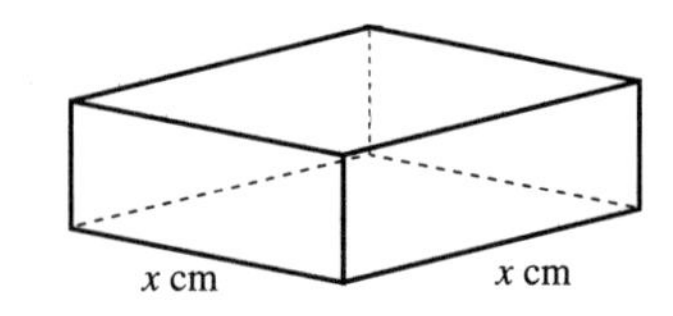

(a) Show that the area $A \text{ cm}^2$ of card used to make such an open box is given by

$$A = x^2 + \frac{2000}{x}$$

(b) Given that x varies, find the value of x for which $\frac{\mathrm{d}A}{\mathrm{d}x} = 0$.

(c) Find the height of the box when x has this value.

(d) Show that when x has this value, the area of card used is least. [E]

67

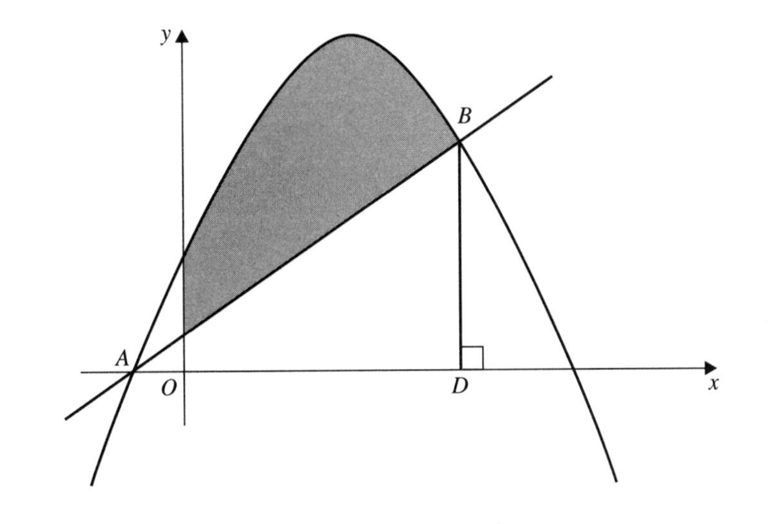

A and B are two points which lie on the curve C, with equation $y = -x^2 + 5x + 6$. The diagram shows C and the line l passing through A and B.

(a) Calculate the gradient of C at the point where $x = 2$.

The line l passes through the point with coordinates (2, 3) and is parallel to the tangent to C at the point where $x = 2$.

(b) Find an equation of l.

(c) Find the coordinates of A and B.

The point D is the foot of the perpendicular from B onto the x-axis.

(d) Find the area of the region bounded by C, the x-axis, the y-axis and BD.

(e) Hence find the area of the shaded region. [E]

68 (i) x^2, $(8x + 1)$ and $(7x + 2)$, where $x \neq 0$, are the second, fourth and sixth terms respectively of an arithmetic series.

(a) Find the value of x.

(b) Find the common difference and the first term of the series.

(c) Calculate the sum of the first 20 terms of the series.

(ii) The second term of a convergent geometric series is 5 and the sum to infinity is $-\frac{125}{6}$.

(a) Find the common ratio of the series.

(b) Calculate, to 3 significant figures, the sum of the first 8 terms of the series. [E]

69

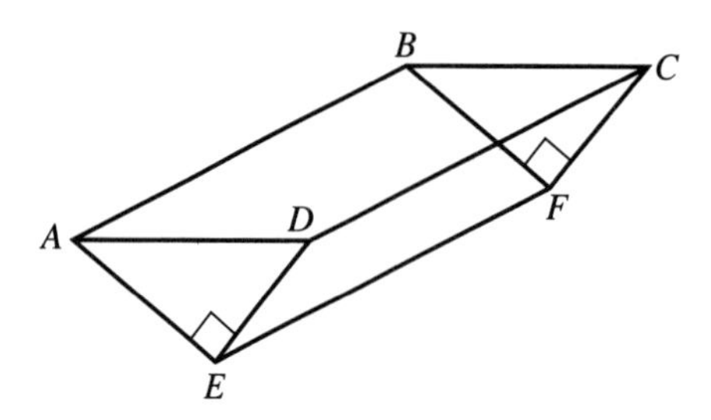

The diagram shows an open tank for storing water, $ABCDEF$. The sides $ABFE$ and $CDEF$ are rectangles. The triangular ends, ADE and BCF are isosceles and $\angle AED = \angle BFC = 90°$. The ends ADE and BCF are vertical and EF is horizontal. Given that $AD = x$ metres,

(a) show that the area of $\triangle ADE$ is $\frac{1}{4}x^2\,\text{m}^2$.

Given also that the capacity of the container is $4000\,\text{m}^3$ and that the total area of the two triangular and two rectangular sides of the container is $S\,\text{m}^2$,

(b) show that $S = \dfrac{x^2}{2} + \dfrac{16\,000\sqrt{2}}{x}$.

Given that x can vary,

(c) use calculus to find the minimum value of S

(d) justify that the value of S you have found is a minimum.

[E]

70 (i) The fifth term of an arithmetic series is 36 and the twenty-first term is -12.

Calculate

(a) the common difference of the series

(b) the first term of the series.

Given that the sum of the first n terms of this series is the same as the sum of the first $2n$ terms of this series,

(c) calculate the value of n.

(ii) The first term of a geometric series is a and the common ratio is r. The fifth term of this series is 16. The first term of another geometric series is $2a$ and the common ratio is $3r$. The second term of this series is 324.

Calculate

(a) the value of r (b) the value of a

(c) the sum of the first ten terms of the second series, giving your answer to 3 significant figures. [E]

71 A cylindrical biscuit tin has a close-fitting lid which overlaps the tin by 1 cm, as shown. The radii of the tin and the lid are both x cm. The tin and the lid are made from a thin sheet of metal of area $80\pi \text{ cm}^2$ and there is no wastage. The volume of the tin is $V \text{ cm}^3$.

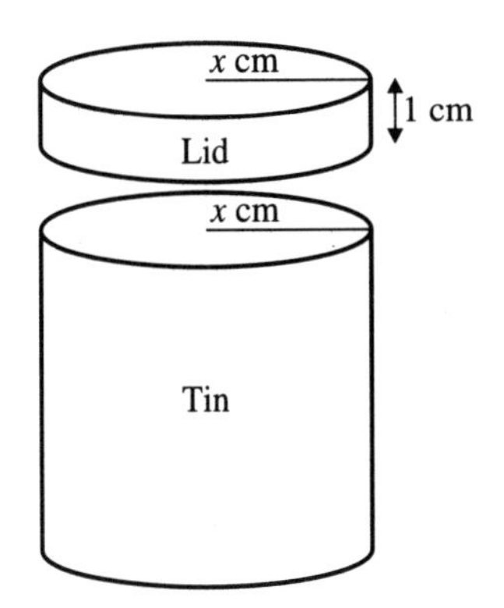

(a) Show that $V = \pi(40x - x^2 - x^3)$.

Given that x can vary,

(b) use differentiation to find the positive value of x for which V is stationary.

(c) Prove that this value of x gives a maximum value of V.

(d) Find this maximum value of V.

72

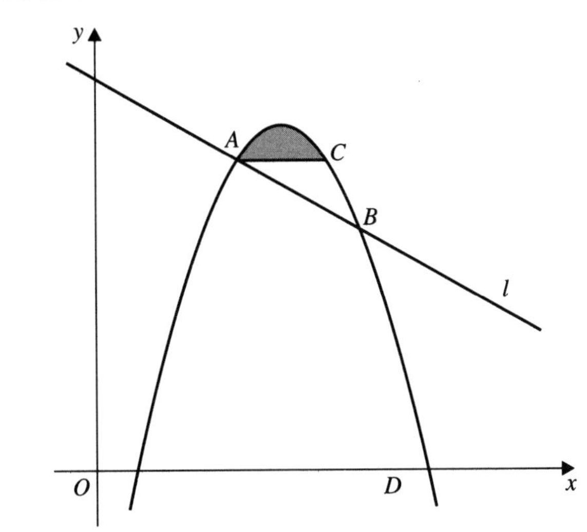

The diagram shows part of the curve with equation $y = p + 10x - x^2$, where p is a constant, and part of the line l with equation $y = qx + 25$, where q is a constant. The line l cuts the curve at the points A and B. The x-coordinates of A and B are 4 and 8 respectively. The line through A parallel to the x-axis intersects the curve again at the point C.

(a) Show that $p = -7$ and calculate the value of q.

(b) Calculate the coordinates of C.

The shaded region in the diagram is bounded by the curve and the line AC.

(c) Using algebraic integration and showing all your working, calculate the area of the shaded region. [E]

Essential skills

8

This chapter covers the essential skills required for the P1 course.

Most of the material in this chapter has been covered in the Higher tier of the GCSE maths course. If you have completed Intermediate tier GCSE maths you should look at this material before studying the rest of the book. If you have completed Higher tier GCSE maths you can use this chapter to revise those skills you feel you need to do extra work on.

8.1 Integers, fractions and real numbers

A **set** is a collection of objects, symbols or just things having some common property. The members of a set are called **elements**. You must be able to say whether an object is an element of a particular set and you must be able to distinguish elements individually. Already in mathematics, you will have met several different sets of numbers which are used frequently. These are:

$\mathbb{N}$ the set of natural numbers $\{1, 2, 3, \ldots\}$
$\mathbb{Z}$ the set of integers $\{0, \pm 1, \pm 2, \ldots\}$
$\mathbb{Q}$ the set of rational numbers (fractions)
$\left\{\frac{p}{q}, \text{ where } p \text{ and } q \text{ are integers and } q \text{ is positive}\right\}$
$\mathbb{R}$ the set of real numbers.

If a number x is an element of $\mathbb{R}$, then you write $x \in \mathbb{R}$ which read in words is 'the number x is an element of the set of real numbers', or more simply, 'x is a real number'.

These sets have evolved because of a need to solve practical problems. You should appreciate also that:

$$\mathbb{N} \text{ is a subset of } \mathbb{Z}$$

$$\mathbb{Z} \text{ is a subset of } \mathbb{Q}$$

$$\text{and } \mathbb{N}, \mathbb{Z} \text{ and } \mathbb{Q} \text{ are all subsets of } \mathbb{R}$$

because every element of $\mathbb{N}$ is a member of $\mathbb{Z}$, but not vice versa, and so on.

For years now, you should have been competent in processing members of these sets using operations like **addition**, **subtraction**, **multiplication** and **division**. You should be able to use a basic scientific calculator, but you should also be able to do simple sums without a calculator. In particular, you must be able to process both fractions and decimals. You should also understand the notation for writing numbers in **standard form**. That is, a number is in standard form when written as $A \times 10^n$, where $1 \leqslant A < 10$ and n is a positive or negative integer. This is a very convenient way of recording either very large or very small numbers.

Example 1

Express in standard form:

(a) 93 000 000 000 (b) 0.000 000 004 52

(a) $93\,000\,000\,000 = 9.3 \times 10^{10}$

(b) $0.000\,000\,004\,52 = 4.52 \times 10^{-9}$

8.2 Laws of indices for integer exponents

Index notation

3×3 can be written as 3^2.
$7 \times 7 \times 7 \times 7$ can be written as 7^4.
$y \times y \times y \times y \times y$ is usually written as y^5.
3^6 in words is 'three to the power of 6'.
a^2 in words is 'a squared' and a^3 is 'a cubed'.

In the expression a^n, the number a is usually called the **base** and the number n is called the **power** or the **index**. The plural of index is **indices**.

■ **In general** $\qquad \boldsymbol{x^a \times x^b = x^{a+b}}$

So when multiplying numbers *which have the same base* the indices are added. Remember that, in general, expressions involving indices can only be simplified when the numbers involved have the same base. The expression $5^2 \times 2^3$ can only be simplified by evaluating the expression; in this case, $25 \times 8 = 200$.

Example 2

Simplify $3^9 \times 3^{14}$.

$$3^9 \times 3^{14} = 3^{9+14} = 3^{23}$$

Example 3

Simplify $3x^4 \times 7x^6$.

$$\begin{aligned} & 3x^4 \times 7x^6 \\ &= 3 \times x^4 \times 7 \times x^6 \\ &= 3 \times 7 \times x^4 \times x^6 \\ &= 21 \times x^{4+6} \\ &= 21x^{10} \end{aligned}$$

Example 4

Simplify $3x^5 \times 4y^3 \times 2y^2 \times x^3$.

$$\begin{aligned} & 3x^5 \times 4y^3 \times 2y^2 \times x^3 \\ &= 3 \times 4 \times 2 \times x^5 \times x^3 \times y^3 \times y^2 \\ &= 24 \times x^{5+3} \times y^{3+2} \\ &= 24x^8y^5 \end{aligned}$$

Example 5

Simplify $2a^2 \times 3a^2b \times 2b^2 \times 3a^3$

$$\begin{aligned} & 2a^2 \times 3a^2b \times 2b^2 \times 3a^3 \\ &= 2 \times 3 \times 2 \times 3 \times a^2 \times a^2 \times a^3 \times b \times b^2 \\ &= 36 \times a^{2+2+3} \times b^{1+2} \\ &= 36a^7b^3 \end{aligned}$$

Evaluating $(x^a)^b$

x^3 means $x \times x \times x$.

$(x^3)^2$ means $(x \times x \times x)^2$ which is $(x \times x \times x) \times (x \times x \times x) = x^6$.

Similarly:

$$(a^2)^5 = (a \times a) \times (a \times a) \times (a \times a) \times (a \times a) \times (a \times a) = a^{10}$$

So $(x^3)^2 = x^6 = x^{3\times2}$ and $(a^2)^5 = a^{10} = a^{2\times5}$.

- **In general,** $(x^a)^b = x^{a\times b} = x^{ab}$

Example 6

Simplify (a) $(2^2)^3$ (b) $(8^5)^9$ (c) $(x^7)^4$.

(a) $(2^2)^3 = 2^{2\times3} = 2^6$

(b) $(8^5)^9 = 8^{5\times9} = 8^{45}$

(c) $(x^7)^4 = x^{7\times4} = x^{28}$

Dividing expressions involving indices

The expression $x^4 \div x^2$ means:

$$\frac{\not{x} \times \not{x} \times x \times x}{\not{x} \times \not{x}} = x^2$$

Similarly, $p^9 \div p^5$ means:

$$\frac{\not{p} \times \not{p} \times \not{p} \times \not{p} \times \not{p} \times p \times p \times p \times p}{\not{p} \times \not{p} \times \not{p} \times \not{p} \times \not{p}} = p^4$$

As $x^4 \div x^2 = x^2 = x^{4-2}$ and $p^9 \div p^5 = p^4 = p^{9-5}$ it should be clear that when dividing numbers with the same base, the indices are subtracted.

■ **In general, $x^a \div x^b = x^{a-b}$**

Example 7

Simplify (a) $x^9 \div x^3$ (b) $6x^5 \div 2x^3$ (c) $8a^7 \div 4a^2$ (d) $21x^3y^4 \div 3x^2y$.

(a) $x^9 \div x^3 = x^{9-3} = x^6$

(b) $6x^5 \div 2x^3 = \frac{6}{2}x^{5-3} = 3x^2$

(c) $8a^7 \div 4a^2 = \frac{8}{4}a^{7-2} = 2a^5$

(d) $21x^3y^4 \div 3x^2y = \frac{21}{3}x^{3-2}y^{4-1} = 7xy^3$

The meaning of x^0

Given that $x^a \div x^b = x^{a-b}$ then $5^4 \div 5^4 = 5^{4-4} = 5^0$.

But $5^4 \div 5^4 = \frac{5^4}{5^4} = 1$. So, $5^0 = 1$.

Here is another example: $9^7 \div 9^7 = 9^{7-7} = 9^0$.

But, $9^7 \div 9^7 = \frac{9^7}{9^7} = 1$. So, $9^0 = 1$.

In general $\qquad x^n \div x^n = x^{n-n} = x^0$

But $\qquad x^n \div x^n = \frac{x^n}{x^n} = 1$

■ **In general, $x^0 = 1$**

Negative indices

$x^a \div x^b = x^{a-b}$ so $2^3 \div 2^7 = 2^{-4}$.

But: $$2^3 \div 2^7 = \frac{\not{2} \times \not{2} \times \not{2}}{\not{2} \times \not{2} \times \not{2} \times 2 \times 2 \times 2 \times 2} = \frac{1}{2^4}$$

Similarly $$3^5 \div 3^7 = 3^{5-7} = 3^{-2}$$

Now:

$$3^5 \div 3^7 = \frac{\not{3} \times \not{3} \times \not{3} \times \not{3} \times \not{3}}{\not{3} \times \not{3} \times \not{3} \times \not{3} \times \not{3} \times 3 \times 3} = \frac{1}{3^2}$$

So $$2^{-4} = \frac{1}{2^4} \quad \text{and} \quad 3^{-2} = \frac{1}{3^2}$$

- **In general, $x^{-n} = \dfrac{1}{x^n}$**

Example 8

Simplify (a) $6^3 \div 6^7$ (b) $2^3 \times 4^{-2}$ (c) $3^{-2} \div 3^3$

(a) $6^3 \div 6^7 = \dfrac{6 \times 6 \times 6}{6 \times 6 \times 6 \times 6 \times 6 \times 6 \times 6} = \dfrac{1}{6^4} = 6^{-4}$

(b) $2^3 \times 4^{-2} = \dfrac{2^3}{4^2} = \dfrac{2 \times 2 \times 2}{4 \times 4} = \dfrac{1}{2} = 2^{-1}$

(c) $3^{-2} \div 3^3 = \dfrac{1}{3^2} \div 3^3 = \dfrac{1}{3^2} \times \dfrac{1}{3^3} = \dfrac{1}{3^5} = 3^{-5}$

Example 9

Evaluate $7.8 \times 10^{-3} + 6.9 \times 10^{-2}$, giving your answer in standard form.

$$7.8 \times 10^{-3} = 0.0078$$
$$6.9 \times 10^{-2} = 0.069$$

So: $$7.8 \times 10^{-3} + 6.9 \times 10^{-2} = 0.0768$$
$$= 7.68 \times 10^{-2}$$

Example 10

Given that $p = 2.5 \times 10^{-4}$, express $\dfrac{1}{p}$ in standard form

$$p = 2.5 \times 10^{-4} = 2.5 \times \frac{1}{10^4} = \frac{2.5}{10^4}$$

So: $$\frac{1}{p} = \frac{10^4}{2.5} = \frac{10 \times 10^3}{2.5} = 4 \times 10^3$$

8.3 Solving problems in ratio and proportion

Ratio

When preparing drawings of a new building, an architect needs to work to a **scale**. That is, the drawings need to be in a fixed *ratio* in terms of length to the dimensions of the building. Typical scales are 1 : 1000, 1 inch to 1 mile, 70 grams to 1 kilogram, etc. In mathematics, you frequently need to use the form $1 : n$.

Example 11

The costs of servicing three cars A, B and C are such that the cost for A to the cost for B is in the ratio 5 : 3 and the cost for B to the cost for C is in the ratio 5 : 4. The cost for servicing C is £240. Find the cost of the service for A.

You have:
$$\frac{\text{cost of servicing } B}{\text{cost of servicing } C} = \frac{5}{4}$$

and
$$\text{cost of servicing } C = £240$$
$$\text{cost of servicing } B = \tfrac{5}{4} \times £240 = £300$$

Also:
$$\frac{\text{cost of servicing } A}{\text{cost of servicing } B} = \frac{5}{3}$$
$$\text{cost of servicing } A = \tfrac{5}{3} \times £300$$
$$= £500$$

Notice also that you could write:

$$\text{cost of servicing } A : \text{cost of servicing } B : \text{cost of servicing } C$$
$$= 500 : 300 : 240$$
$$= 25 : 15 : 12$$

Written in the form, 25 : 15 : 12 are often called the **proportional parts** of the costs of servicing car A : car B : car C.

Variation

A quantity p **varies directly** as a quantity q if the ratio $\frac{p}{q}$ remains constant for all values of p and q. You can write $p = kq$, where k is a constant. The notation $y \propto x$ means y **varies directly** as x or y is **directly proportional** to x.

A quantity p **varies inversely** as a quantity q (or is **inversely proportional** to q) if the product pq remains constant for all values of p and q. You can write $pq = k$ or $p = \frac{k}{q}$, where k is a constant.

In this case $p \propto \frac{1}{q}$.

Example 12

Given that y is directly proportional to x^3 and that $y = 4$ when $x = 2$, find y in terms of x.

Here $y = kx^3$, where k is a constant. You know that $y = 4$ when $x = 2$, so:

$$4 = k(2)^3 \Rightarrow 4 = 8k \Rightarrow k = \tfrac{1}{2}$$

That is: $$y = \tfrac{1}{2}x^3$$

Example 13

Given that y is inversely proportional to $\sqrt{x}$ and that $y = 2$ when $x = 9$, find the value of y when $x = 25$.

Here $y = \dfrac{k}{\sqrt{x}}$, where k is a constant.

You know that $y = 2$ when $x = 9$.

So: $$2 = \frac{k}{3} \Rightarrow k = 6$$

That is: $$y = \frac{6}{\sqrt{x}}$$

When $x = 25$, $$y = \frac{6}{\sqrt{25}} = \frac{6}{5} = 1.2$$

Similarity relationships in plane figures and in solids

Two shapes or objects are mathematically similar if one is an **enlargement** of the other, regardless of orientation: corresponding angles are equal and the ratio of the lengths of corresponding sides is constant.

Consider two plane similar figures, in which each linear dimension of the larger is n times that of the smaller. Then you have:

$$\frac{\text{area of larger figure}}{\text{area of smaller figure}} = n^2$$

That is, the areas are proportional to the squares on corresponding sides.

For the similar triangles shown:

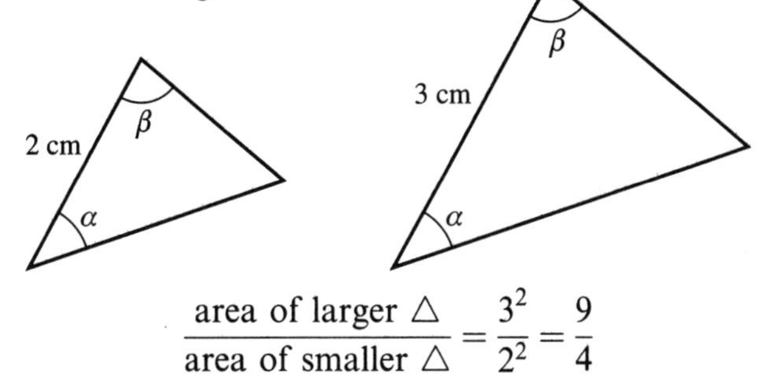

$$\frac{\text{area of larger } \triangle}{\text{area of smaller } \triangle} = \frac{3^2}{2^2} = \frac{9}{4}$$

Consider two similar solids, in which each linear dimension of the larger is n times that of the smaller. Then you have:

$$\frac{\text{volume of larger solid}}{\text{volume of smaller solid}} = n^3$$

For the similar cones shown:

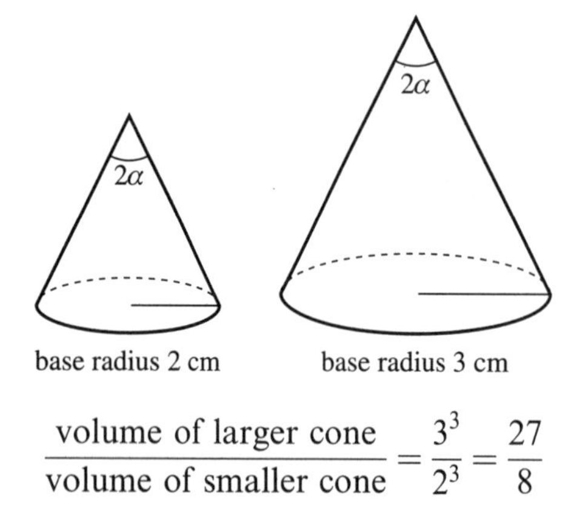

$$\frac{\text{volume of larger cone}}{\text{volume of smaller cone}} = \frac{3^3}{2^3} = \frac{27}{8}$$

Exercise 8A

Simplify:

1 $x \times x \times x \times x$

2 $a \times a \times a \times a \times a \times a$

3 $4 \times a \times a \times a$

4 $6 \times a \times a \times 5 \times b \times b$

5 $x^3 \times x^7$

6 $y^3 \times y^7 \times y^2$

7 $3y^2 \times 4y^5$

8 $2y^2 \times 3y^2 \times 4y^4$

9 $3a^3 \times 4b^2 \times 5b^2 \times a^2$

10 $3x^5 \times y \times 6x^2 \times 2y^3$

11 $6p^3 \times 2p^2 \times 4q^4 \times 2p^5$

12 $6a^3 \times 4b^2 \times 2a^2b^3$

13 $5p^3 \times 2p^2q^3 \times q^4 \times 3q^2$

14 $5x^3 \times 2y^2 \times 3z^3$

15 $2a^3 \times 4ab^2 \times 6bc^3 \times c^4$

16 $8a^3 \times 9b^2$

17 $2a^2 \times 3b^3 \times 6c \times 10a^3b^4c^5$

18 $6p^3 \times 2q^3 \times pq \times 7$

19 $a^3b \times ab^3 \times 2pq \times 3q^2$

20 $\frac{2}{3}ap^3 \times \frac{3}{4}pa^2 \times \frac{1}{2}aq^2 \times \frac{2}{5}qa^2$

21 $(a^5)^6$

22 $(x^2)^7$

23 $(a^2b^3)^4$

24 $x^0 \div x^{-4}$

25 $x^9 \div x^5$

26 $a^7 \div a^{-1}$

27 $p^2 \div p^5$

28 $15p^5 \div 3p^2$

29 $21a^3 \div 7a^4$

30 $12a^{15} \div (a^2)^4$

31 $4(a^3)^5 \div 2a^7$

32 $16(p^2)^2 \div 4p^4$

33 $8(x^5)^2 \div 2x^6$

34 $20(x^3)^9 \div 4(x^2)^{10}$

35 $15a^2b^7 \div 5ab^{-2}$

36 $14a^3b^2c \div 7abc^3$

37 $15(a^2b)^3 \div (3ab)^2$

38 $18a^2b^2c^2 \times 2(ab)^{-1}$

39 $3a^2bc^3 \times 4a^2bc \div 2a^2b^4c^2$

40 $9x^6y^4 \times 2x^5yz^3 \div 6(x^2yz)^2$

Evaluate, giving your final answer in standard form:

41 $2.9 \times 10^2 + 2.9 \times 10^3$ **42** $5.2 \times 10^3 - 6.1 \times 10^2$

43 $3.6 \times 10^2 \times 2.9 \times 10^3$ **44** $(6.3 \times 10^{-3}) \div (7.0 \times 10^{-4})$

45 $(4.5 \times 10^{-1})^2$

Evaluate:

46 $2^{-2} + 4^{-1}$ **47** $\left(\frac{3}{2}\right)^{-2}$ **48** $\left[\left(\frac{2}{5}\right)^{-1}\right]^3$

49 $2^{-1} - 3^{-1} - 4^{-1}$ **50** $\left(\frac{3}{4}\right)^{-3} \div 6^{-2}$

51 Given that y is inversely proportional to x^2 and that $y = 8$ when $x = 0.5$, find

(a) y in terms of x

(b) the possible values of x when $y = 50$. [E]

52 Given that $t = \dfrac{k}{v^2}$ and $t = 25$ when $v = 2$, find the value of k. Using this value of k, find the value of t when $v = 3$. [E]

53 Given that $\dfrac{x + 5y}{x - 3y} = 4$, find the ratio $x : y$.

54 The braking distance for a car varies as the square of the speed. When the car is moving at $40\,\text{km h}^{-1}$, the braking distance is 19.6 m. Find

(a) the braking distance when the speed is $80\,\text{km h}^{-1}$

(b) the speed when the braking distance is 122.5 m.

55 Given that y is directly proportional to $\sqrt{x}$ and that $y = 10$ when $x = 9$, find:

(a) the value of y when $x = 36$

(b) the value of x when $y = 24$.

56 Given that $y + 2$ is inversely proportional to x^3 and that $y = 1$ when $x = 2$, find y in terms of x.

57 Two similar miniature statues are made of the same uniform material; the larger is 6 cm tall and the smaller is 3.5 cm tall. Find the ratio of the areas of the surfaces of the larger to the smaller. Find also the mass of the larger if the smaller is of mass 0.6 kg, giving your answer in kg to 3 s.f.

58 Given that a carton of height 25 cm holds 1 litre calculate the height of a similar carton holding 2 litres, giving your answer to the nearest cm.

59 A cone of height 10 cm has a volume of $56\,\text{cm}^3$.

(a) Find the height of a similar cone whose volume is $7\,\text{cm}^3$.

(b) Find the volume of another similar cone whose height is 15 cm.

60 A scale model M is of a statue S. The scale factor for any length from M to the corresponding length of S is 2.5×10^4. Calculate in the same standard form:

(a) the scale factor for volume from M to S

(b) the scale factor for length from S to M. [E]

8.4 Basic algebra

The following examples cover work which you should be familiar with.

Example 14

Given that $x = -2$, $y = 3$ and $z = \frac{1}{2}$, evaluate:

(a) $x^2 - y^2$ (b) $x(2y - 3z)$ (c) $(z + x)(z - y)$.

(a) $x^2 - y^2 = (-2)^2 - (3)^2 = 4 - 9 = -5$

(b) $x(2y - 3z) = -2(6 - \frac{3}{2}) = -2(\frac{9}{2}) = -9$

(c) $(z + x)(z - y) = (\frac{1}{2} - 2)(\frac{1}{2} - 3) = (-\frac{3}{2})(-\frac{5}{2})$

$= \frac{15}{4} = 3\frac{3}{4}$

Example 15

Express as a single fraction in its lowest term:

(a) $\dfrac{2a - 3b}{3} - \dfrac{7a - 5b}{4}$ (b) $\dfrac{1}{2p} + \dfrac{1}{3p} - \dfrac{1}{6p}$.

(a) The lowest common denominator is 12, $\frac{1}{3} = \frac{4}{12}$ and $\frac{1}{4} = \frac{3}{12}$, so you can write:

$$\frac{2a - 3b}{3} - \frac{7a - 5b}{4} = \frac{4(2a - 3b)}{12} - \frac{3(7a - 5b)}{12}$$

$$= \frac{8a - 12b - 21a + 15b}{12}$$

That is: $$\frac{2a - 3b}{3} - \frac{7a - 5b}{4} = \frac{3b - 13a}{12}$$

(b) The common denominator is $6p$, $\dfrac{1}{2p} = \dfrac{3}{6p}$ and $\dfrac{1}{3p} = \dfrac{2}{6p}$.

So you have:

$$\frac{1}{2p} + \frac{1}{3p} - \frac{1}{6p} = \frac{3 + 2 - 1}{6p} = \frac{4}{6p} = \frac{2}{3p}$$

Example 16

Factorise (a) $x^2 - 9$ (b) $x^2 - 9x$ (c) $x^3 - 9x$.

(a) $x^2 - 9 = (x - 3)(x + 3)$, the difference of two squares
(b) $x^2 - 9x = x(x - 9)$, the factor x is common
(c) $x^3 - 9x = x(x^2 - 9) = x(x - 3)(x + 3)$, a two-stage process

You will find consolidation and extension of algebraic factorisation in chapter 1.

Example 17

Solve the equation $\frac{x - 2}{3} + \frac{2x - 5}{5} = 2$.

The lowest common denominator is 15 so you multiply each term in the equation by 15:

$$\frac{15(x - 2)}{3} + \frac{15(2x - 5)}{5} = 15 \times 2$$

That is:

$$5(x - 2) + 3(2x - 5) = 30$$

Removing the brackets:

$$5x - 10 + 6x - 15 = 30$$
$$11x - 25 = 30$$
$$11x = 30 + 25$$
$$11x = 55 \Rightarrow x = 5$$

Example 18

Solve for p and q the equations

$$8p + 3q = 11 \qquad (1)$$
$$6p - 2q = 21 \qquad (2)$$

Multiply (1) by 2 and (2) by 3:

$$16p + 6q = 22$$
$$18p - 6q = 63$$

Adding:

$$34p = 85$$
$$p = \tfrac{85}{34} = 2\tfrac{1}{2}$$

Substituting $p = 2\frac{1}{2}$ in (1) gives:

$$8\left(2\tfrac{1}{2}\right) + 3q = 11$$
$$20 + 3q = 11$$
$$3q = -9$$
$$q = -3$$

So the complete solution is $p = 2\frac{1}{2}$, $q = -3$.

You can also check your solution in equation (2) like this:

$$\begin{aligned}\text{left hand side} &= 6p - 2q\\ &= 6(2\tfrac{1}{2}) - 2(-3)\\ &= 15 + 6\\ &= 21 = \text{right hand side}\end{aligned}$$

8.5 Changing the subject of a formula

You encounter many formulae in mathematics and it is often necessary to rearrange a formula into an equivalent form. For example, $V = abc$ is the formula for the volume of a cuboid having sides of length a, b and c. This formula has V as its subject in this form, because V is given in terms of the other letters. If you divide each side of the formula by bc, you get

$$\frac{V}{bc} = \frac{abc}{bc}$$

That is:

$$a = \frac{V}{bc}$$

In this equivalent form, a is the subject of the formula because it is given in terms of the other letters.

Example 19

Make x the subject of the formula

$$x^2 + c^2 + d^2 = (x - c)(x - d)$$

Multiplying out the brackets:

$$x^2 + c^2 + d^2 = x^2 - cx - dx + cd$$

So: $$c^2 + d^2 = -cx - dx + cd$$

Rearranging: $$cx + dx = cd - c^2 - d^2$$

That is: $$x(c + d) = cd - c^2 - d^2$$

$$x = \frac{cd - c^2 - d^2}{c + d}$$

Example 20

Make b the subject of the formula

$$\frac{1}{a} = \frac{1}{b} + \frac{1}{c}$$

Multiply by the common denominator abc:

$$\frac{abc}{a} = \frac{abc}{b} + \frac{abc}{c}$$

That is: $$bc = ac + ab$$

Rearrange: $$bc - ab = ac$$

$$b(c - a) = ac$$

$$b = \frac{ac}{c - a}$$ is the form required.

Exercise 8B

Given that $p = 2$, $q = -3$, $r = \frac{1}{2}$ evaluate:

1 $pq - 6qr$ **2** $2q(3p + 4q)$ **3** $pq \div r$

4 $\frac{p}{q} + \frac{q}{r} + \frac{r}{p}$ **5** $(p + q + r)(p - q - r)$

Express as a single fraction:

6 $\frac{x+1}{2} - \frac{x-1}{3}$ **7** $\frac{2x-1}{5} + \frac{1-x}{7}$ **8** $\frac{6p}{5} - \frac{4p-3q}{3}$

9 $\frac{1}{x^2} - \frac{1}{x(x+1)}$ **10** $\frac{x}{2} - \frac{1-x}{3} + \frac{2-x}{4}$

Factorise completely:

11 $4 + 16y$ **12** $ab^2c - abc^2$ **13** $5x^2 - 15x$

14 $x^2 - 16$ **15** $x^2 + 20x$ **16** $2x^2 - 8$

17 $24 - 6y^2$ **18** $x^4 + 3x^3 + x^2$

19 $y + 1 + y(y + 1)$ **20** $4x^2 - (x - 1)^2$

Solve these equations:

21 $5x - 3 = 2x + 15$

22 $8x - 6 + 7x = 5x - 4 - 20$

23 $3(x + 4) + 2 = 2(4x + 3) - 2$

24 $5(x + 3) + 4(2x - 3) = 2(2x + 15)$

25 $2 + 7x - 4 - 2(3 + x) = 7 - 10x$

26 $3(2 + 3x) - 4(5 - 3x) = 28$

27 $\frac{x}{2} - \frac{x}{3} = 5$ **28** $\frac{x}{2} - \frac{1}{3} = \frac{1}{4} - \frac{x}{5}$

29 $\frac{2}{x} - \frac{5}{2x} = 2$ **30** $\frac{x+2}{3} + \frac{2x+1}{5} = 6$

31 $\frac{x+2}{3} - \frac{2x+1}{5} = 2$ **32** $\frac{3x-2}{4} - \frac{2x-3}{3} = \frac{x-1}{5}$

Solve these simultaneous linear equations:

33 $2x + y = 18$
$x - 2y = -1$

34 $x - 2y = 5$
$3x + y = 8$

35 $x + 2y = 4$
$3x + 5y = 9$

36 $5x + 2y = -30$
$3x + 4y = -32$

37 $14x + 4y = -1$
$-3x + 5y = 9$

38 $3x - 2y = 7$
$2x - 5y = 12$

39 $7p + 3q = 5$
$5p - q = 2$

40 $6x - 11y = 4$
$5x + 3y = -21$

41 $17x + 9y = 20$
$5x - 2y = -22$

42 $y = 7z$
$8y + 100 = 6z$

43 $2p + 3q = -5$
$3p - 5q = 21$

44 $5a - 6b = 7$
$8b - a = 2$

45 $2x - 7y = 57$
$3y - 11x = 6$

46 $3x - 2y = 2$
$9x - 4y = 1$

47 $24x + 12y + 7 = 0$
$6x + 12y = 5$

In questions 48–52 express x in terms of y given that:

48 $y(x + 1) = 5$

49 $x(y + 1) = y(2x + 1)$

50 $y = \frac{x - 2}{x + 3}$

51 $\frac{1}{x} + \frac{1}{y} = 3$

52 $y = \frac{2 + x}{3x}$

53 Make t the subject of the formula

$$\frac{t + a}{t + b} = \frac{t + c}{t + d}$$

54 Make a the subject of the formula

$$S = \frac{n}{2}[2a + d(n - 1)]$$

55 Make F the subject of the formula

$$9C = 5(F - 32)$$

8.6 The gradient of a straight line joining two points

Coordinate geometry is the study of the geometry of straight lines and curves using algebraic methods. When you graph a line or curve in two dimensions (2D) it exists in a plane – a two-dimensional flat surface. If two axes at right angles to each other are laid on the plane you can identify any point in the plane, using coordinates that show how far the point is from each axis. One axis is usually called the x-axis, and other is called the y-axis. The point where the axes meet is called the origin and is labelled O.

To define a point P in the plane using coordinate geometry you state the perpendicular distance of the point from the y-axis and its perpendicular distance from the x-axis. This defines the position of the point uniquely.

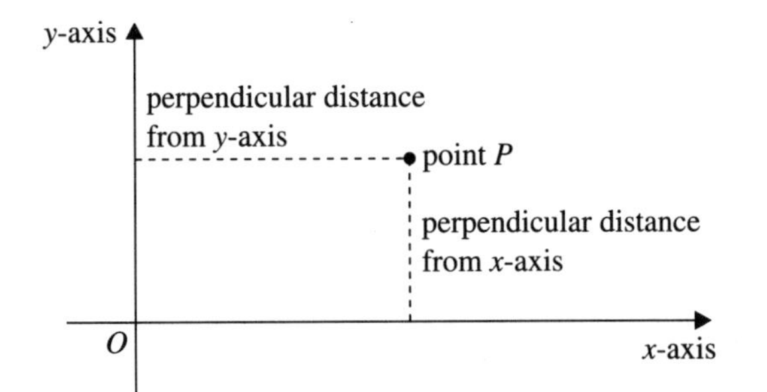

The part of the y-axis that lies above the x-axis is called the positive y-axis and is labelled with positive numbers. The part of the y-axis that lies below the x-axis is called the negative y-axis and is labelled with negative numbers. Similarly, the part of the x-axis to the left of the y-axis is called the negative x-axis and the part to the right is the positive x-axis.

The perpendicular distance of a point in the plane from the y-axis is called the **x-coordinate** (or **abscissa**) of the point. The perpendicular distance of the point from the x-axis is called the **y-coordinate** (or **ordinate**) of the point. These numbers must be labelled with a + or − sign to show where the point lies. The coordinates of a point are written (x, y), where the first number is the x-coordinate.

Example 21

The points A, B, C, D and E are shown with their coordinates on the diagram.

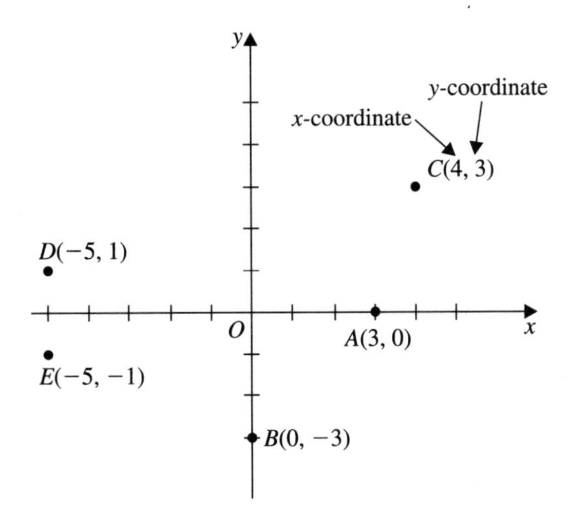

The **gradient** of a straight line is a measure of how steep that line is. The gradient of a line joining two points is defined as:

$$\frac{\text{change in } y\text{-coordinates}}{\text{change in } x\text{-coordinates}}$$

between the two points.

So the gradient joining the points $P(x_1, y_1)$ and $Q(x_2, y_2)$ is:

$$\frac{\text{change in } y\text{-coordinates}}{\text{change in } x\text{-coordinates}} = \frac{y_2 - y_1}{x_2 - x_1}$$

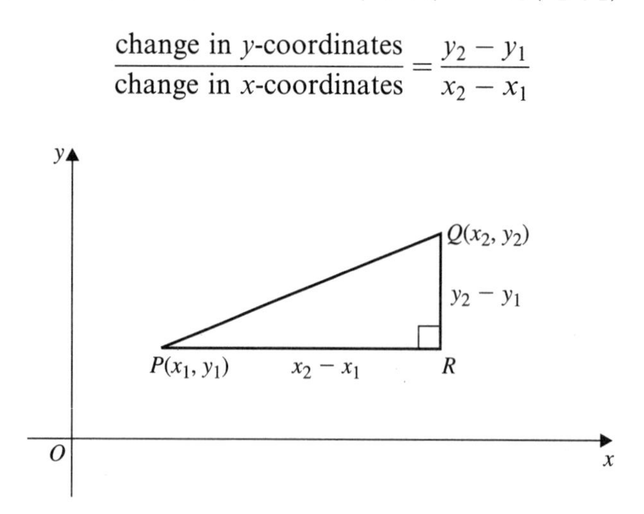

As $y_2 - y_1$ represents the distance QR and $x_2 - x_1$ represents the distance PR, the gradient of the line PQ is actually represented by the ratio

$$\frac{QR}{PR} = \tan \angle QPR$$ (see p. 222)

Example 22

Find the gradient of the line joining the points (2, 5) and (4, 8).

Gradient is:

$$\frac{y_2 - y_1}{x_2 - x_1} = \frac{8-5}{4-2}$$

$$= \frac{3}{2}$$

$$= 1.5$$

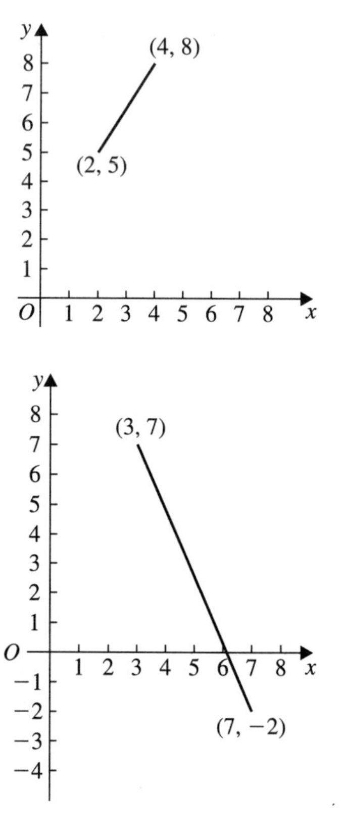

Example 23

Find the gradient of the line joining the points (3, 7) and (7, −2).

Gradient is:

$$\frac{y_2 - y_1}{x_2 - x_1} = \frac{-2-7}{7-3}$$

$$= \frac{-9}{4}$$

$$= -2.25$$

8.7 The equation of a straight line in the form $y = mx + c$

All straight lines are defined by an equation of the form $y = mx + c$, where m and c are constants. So $y = 2x + 6$, $y = -3x + 2$ and $y = 5x - 3$ each represent a straight line.

For the equation $y = mx + c$, when $x = 0$, $y = c$. Also, when $x = 1$, $y = m + c$.

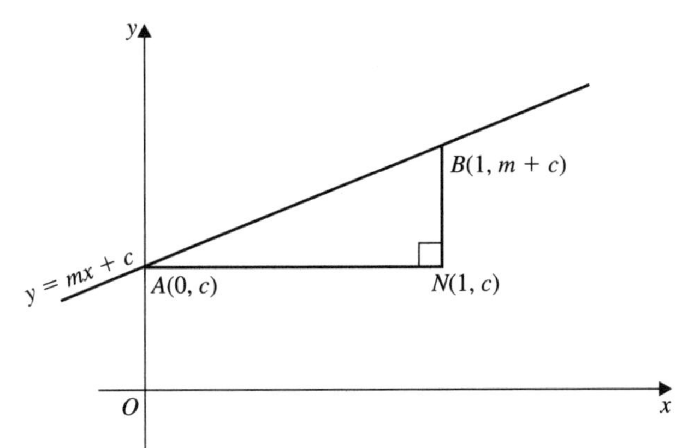

From the diagram you can see that $A(0, c)$ is the point where the line with equation $y = mx + c$ cuts the y-axis. So the distance OA is c. This distance is called the ***y*-intercept**. Also, since the coordinates of N are $(1, c)$ the gradient of the line AB is

$$\frac{(m + c) - c}{1 - 0} = \frac{m}{1} = m$$

So m is the gradient of the straight line.

- **If a straight line has equation $y = mx + c$ then m is the gradient of the line and c is the y-intercept.**

Example 24

Given that an equation of a straight line is $y = -5x + 4$, find

(a) the gradient of the line
(b) the coordinates of the point where the line cuts the y-axis.

Compare $y = -5x + 4$ with $y = mx + c$. Notice that $m = -5$ and $c = 4$. So the gradient of the line is -5 and it cuts the y-axis at $(0, 4)$.

Example 25

A straight line has equation $3y + 4x + 6 = 0$. Find

(a) the gradient of the line (b) the y-intercept.

The equation $3y + 4x + 6 = 0$ can be written as $3y = -4x - 6$ or as $y = -\frac{4}{3}x - 2$.
If you compare this with $y = mx + c$, then $m = -\frac{4}{3}$ and $c = -2$.

So the gradient of the line is $-\frac{4}{3}$ and the y-intercept is -2.

You know that if an equation of a straight line (linear) graph is given in the form $y = mx + c$ it has gradient m and y-intercept c. So it is very easy to find an equation of a given straight line if you are told its gradient and its y-intercept. You reverse the process shown in the previous examples to find an equation.

Example 26
A straight line has gradient 3 and cuts the y-axis at the point with coordinates (0, 2). Find an equation of the line.

As the gradient is 3 and the y-intercept is 2, an equation of the line is $y = 3x + 2$.

Example 27
A straight line has gradient $-\frac{1}{2}$ and cuts the y-axis at the point with coordinates $(0, -\frac{2}{3})$. Find an equation of the line.

An equation of the line with gradient m and y-intercept c is $y = mx + c$, so an equation of the line with gradient $-\frac{1}{2}$ and y-intercept $-\frac{2}{3}$ is $y = -\frac{1}{2}x - \frac{2}{3}$ or $6y = -3x - 4$ or indeed $6y + 3x + 4 = 0$.

8.8 Finding the distance between two points in two dimensions

Suppose you want to find the distance PQ between the points $P(2, 3)$ and $Q(7, 5)$. First draw the right-angled triangle PQR. Notice that the x-coordinate of R is 7 as it is the same perpendicular distance from the y-axis as is Q. The y-coordinate of R is 3. That is, R is the point (7, 3). The distance PR is $7 - 2 = 5$ units in length. The distance QR is $5 - 3 = 2$ units in length.

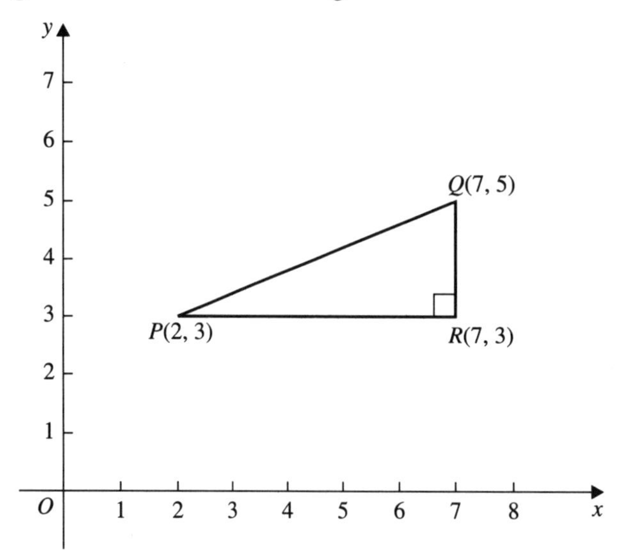

Now that you know the lengths of PR and QR it is easy to calculate the length of PQ using Pythagoras' theorem.

So:
$$\begin{aligned} PQ^2 &= 5^2 + 2^2 \\ &= 25 + 4 \\ &= 29 \end{aligned}$$

and so $PQ = \sqrt{29} = 5.39$ (3 s.f.).

To find a formula that can always be used to find the distance between two points, you can generalise this process using two points with coordinates (x_1, y_1) and (x_2, y_2).

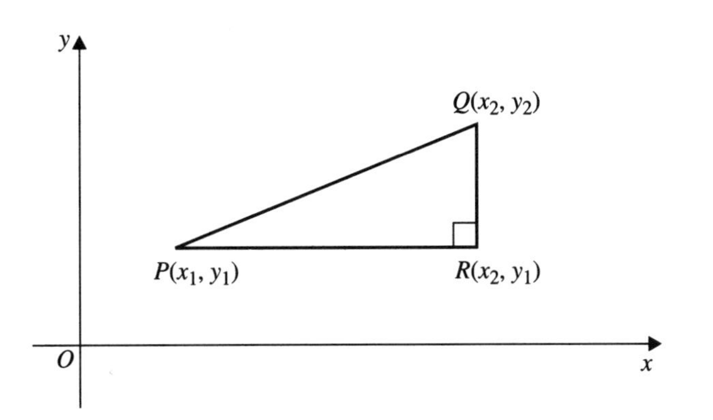

P is the point (x_1, y_1) and Q is the point (x_2, y_2). Drawing the right-angled triangle PQR gives the point R as (x_2, y_1).

The distance PR is $x_2 - x_1$, and the distance QR is $y_2 - y_1$. Using Pythagoras' theorem gives:

$$PQ^2 = (x_2 - x_1)^2 + (y_2 - y_1)^2$$

- **The distance between the points $P(x_1, y_1)$ and $Q(x_2, y_2)$ is:**
$$PQ = \sqrt{[(x_2 - x_1)^2 + (y_2 - y_1)^2]}$$

Example 28

Find the distance between the points $P(2, 7)$ and $Q(4, 10)$.

Using the formula:

$$\begin{aligned} PQ &= \sqrt{[(x_2 - x_1)^2 + (y_2 - y_1)^2]} \\ &= \sqrt{[(4 - 2)^2 + (10 - 7)^2]} \\ &= \sqrt{(2^2 + 3^2)} \\ &= \sqrt{(4 + 9)} \\ &= \sqrt{13} \\ &= 3.61 \quad \text{(3 s.f.)} \end{aligned}$$

Example 29

Find the distance between the points $P(-3, -5)$ and $Q(2, -4)$.

Using the formula:

$$\begin{aligned}PQ &= \sqrt{[(x_2 - x_1)^2 + (y_2 - y_1)^2]}\\ &= \sqrt{[(2+3)^2 + (-4+5)^2]}\\ &= \sqrt{(5^2 + 1^2)}\\ &= \sqrt{(25+1)}\\ &= \sqrt{26}\\ &= 5.10 \text{ (3 s.f.)}\end{aligned}$$

Exercise 8C

Find the gradients of the lines joining the following points:

1 (4, 3) and (8, 11)

2 (3, 7) and (1, 9)

3 (6, 7) and (3, −2)

4 (5, 3) and (10, −5)

5 (−8, −7) and (−4, 5)

6 (3, 4) and (0, 13)

7 (6, 0) and (8, 15)

8 (5, −3) and (2, −9)

9 (3, 9) and (−2, −4)

10 (1, 7) and (−6, −14)

11 (−2, 4) and (7, −3)

12 (−3, −6) and (2, −7)

13 (−5, −2) and (−7, −8)

14 (2, −7) and (−7, 2)

15 $A(2, -3)$, $B(7, 5)$ and $C(-2, 9)$ are the vertices of $\triangle ABC$. Find the gradient of each of the sides of the triangle.

16 $A(3, 2)$, $B(6, 4)$ and $C(4, 0)$ are three points. Find the gradients of the lines AB, BC and CA.

17 $A(2, 5)$, $B(4, -2)$ and $C(-7, 9)$ are the vertices of $\triangle ABC$. Calculate the gradients of AB, BC and CA.

18 $A(1, 2)$, $B(3, 1)$ and $C(9, -2)$ are three points. Find the gradient of AB and of AC. What deduction do you make?

19 Three points are $P(-1, -5)$, $Q(1, -2)$ and $R(5, 4)$. Find the gradient of PQ and QR. What deduction is possible from your result?

20 Show that the line passing through the points $A(0, -2)$ and $B(3, 1\frac{1}{2})$ also passes through the point $C(-6, -9)$.

21 Find the gradient of the straight line with equation:

(a) $y = 5x - 7$ (b) $y = -7x + 2$ (c) $y = 4 - 2x$

(d) $y = -7 - x$ (e) $2y = 3x + 4$ (f) $3y = 7x + 1$

(g) $-3y = 4 - 2x$ (h) $-4y = 8 - 3x$ (i) $2y + 3x = 6$

(j) $3y + 4x = 5$ (k) $2y - 5x - 2 = 0$ (l) $5y + 6x - 4 = 0$

22 Find the y-intercept of the straight line with equation:

(a) $y = x - 4$
(b) $y = 2x + 6$
(c) $y = 3 - 5x$
(d) $y = -2 - 4x$
(e) $2y = 3x + 7$
(f) $3y = 7x + 5$
(g) $2y = 6 - 5x$
(h) $5y = 7 - 6x$
(i) $-2y = 3x - 7$
(j) $-5y = 10x - 8$
(k) $3x - 2y - 7 = 0$
(l) $5y - 6x + 20 = 0$

23 Find the coordinates of the point where the following straight lines cut the y-axis:

(a) $y = -3x + 4$
(b) $y = -2x - 7$
(c) $2y = 5x + 3$
(d) $3y = 6x - 2$
(e) $2y + 3x = 5$
(f) $3y - 5x = 9$
(g) $4x + 3y - 7 = 0$
(h) $5x + 4y - 6 = 0$

24 Find an equation of the straight line with gradient 2 and y-intercept 3.

25 Find an equation of the straight line with gradient 5 and y-intercept -4.

26 Find an equation of the straight line with gradient -1 and y-intercept 2.

27 Find an equation of the straight line with gradient $-\frac{2}{3}$ that passes through the point (0, 1).

28 Find an equation of the straight line with gradient $-\frac{1}{2}$ that passes through the point (0, -7).

29 Find an equation of the straight line with gradient 2 that passes through (0, 1).

30 Find an equation of the straight line with gradient -3 that passes through (0, 3).

31 Find an equation of the straight line with gradient $\frac{1}{2}$ that passes through (0, 5).

32 Find an equation of the straight line with gradient $-\frac{4}{5}$ that passes through (0, -2).

33 Find an equation of the straight line with gradient -7 that passes through the point (0, 6).

34 Find an equation of the straight line with gradient 3 that passes through the point (0, $-\frac{1}{2}$).

Find the lengths of the lines joining:

35 (2, 9) and (3, 12)

36 (4, 5) and (7, 10)

37 (2, −3) and (3, 8)

38 (−5, 7) and (6, 9)

39 (2, 10) and (−3, −2)

40 (−3, −5) and (−6, −3)

41 (8, 7) and (2, 3)

42 (9, −4) and (−3, −6)

43 (−2, −6) and (2, 8)

44 (−8, −2) and (−3, −4)

45 (−2, 7) and (3, −8)

46 (5, −9) and (2, −7)

47 Show that the triangle ABC is isosceles, where A is the point (−5, 0), B is the point (−1, 3) and C is the point (2, 7).

48 The vertices of a triangle are $A(5, 12)$, $B(-12, 5)$ and $C(-7, 17)$. Show that $\angle ACB$ is 90°.

49 Calculate the lengths of the sides of the triangle whose vertices are $P(-2, 3)$, $Q(4, 1)$ and $R(-1, -1)$.

50 Two opposite vertices of a square are $P(3, 2)$ and $Q(-5, -10)$. Find the length of:
(a) a diagonal of the square
(b) a side of the square.

51 Show that the point $C(9, 3)$ is at the same distance from $A(2, 2)$ and $B(4, 8)$.

52 Show that the points (1, −1) (−1, 1) and ($\sqrt{3}$, $\sqrt{3}$) are the vertices of an equilateral triangle.

53 Using the fact that the diagonals of a rectangle are of equal length, show that the points (6, 0), (2, 4), (3, −3) and (−1, 1) are the vertices of a rectangle.

8.9 Basic trigonometry

For any right-angled triangle, the three basic trigonometric functions are defined as:

$$\sin\theta = \frac{BC}{AC} = \frac{\textbf{side opposite } \theta}{\textbf{hypotenuse}} \quad \textbf{or} \quad \frac{\textbf{opposite}}{\textbf{hypotenuse}} \quad \textbf{for short}$$

$$\cos\theta = \frac{AB}{AC} = \frac{\textbf{side adjacent to } \theta}{\textbf{hypotenuse}} \quad \textbf{or} \quad \frac{\textbf{adjacent}}{\textbf{hypotenuse}} \quad \textbf{for short}$$

$$\tan\theta = \frac{BC}{AB} = \frac{\textbf{side opposite } \theta}{\textbf{side adjacent to } \theta} \quad \textbf{or} \quad \frac{\textbf{opposite}}{\textbf{adjacent}} \quad \textbf{for short}$$

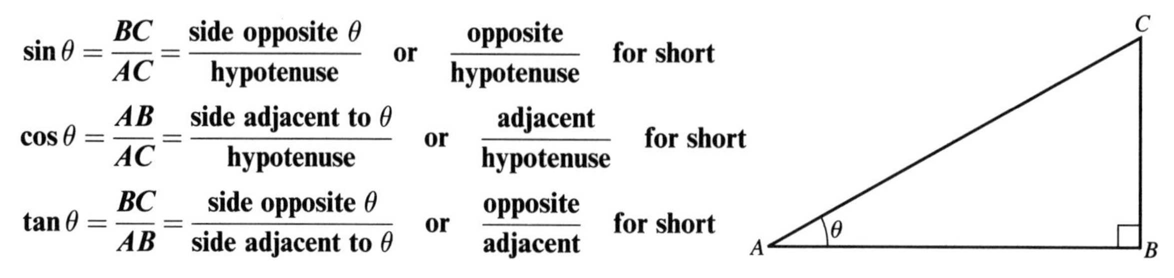

These definitions must be remembered.

Example 30

From your calculator write down to 3 significant figures the value of:

(a) $\sin 70.1°$ (b) $\cos 47.3°$ (c) $\tan 37.8°$ (d) $\cos 81.7°$ (e) $\tan 52.9°$

(a) $\sin 70.1° = 0.940\,288\,1 = 0.940$ (3 s.f.)

(b) $\cos 47.3° = 0.678\,159\,6 = 0.678$ (3 s.f.)

(c) $\tan 37.8° = 0.775\,679\,5 = 0.776$ (3 s.f.)

(d) $\cos 81.7° = 0.144\,356\,2 = 0.144$ (3 s.f.)

(e) $\tan 52.9° = 1.322\,237 \quad = 1.32$ (3 s.f.)

If you can use a calculator to find the value of $\sin\theta$, $\cos\theta$ or $\tan\theta$, where θ is acute, then you can also reverse the process and find θ, if it is acute, from the value of $\sin\theta$, $\cos\theta$ or $\tan\theta$. The inverse functions to sin, cos and tan are called **arcsin**, **arccos** and **arctan**. There should be inverse buttons that allow you to use these functions on your calculator. Notice that on some calculators $\arcsin\theta$ appears as $\sin^{-1}\theta$, $\arccos\theta$ as $\cos^{-1}\theta$ and $\arctan\theta$ as $\tan^{-1}\theta$. However this book uses the notation $\arcsin\theta$ and so on.

For example, if $\sin\theta = 0.7$, write $\theta = \arcsin 0.7$ and find this using your calculator. Similarly, if $\cos\theta = 0.7$ write $\theta = \arccos 0.7$ and find this using your calculator. The same applies to $\tan\theta = 0.7$ which we can write as $\theta = \arctan 0.7$.

Example 31

Find, in degrees to one decimal place, the acute angle such that:

(a) $\sin\theta = 0.7976$ (b) $\cos\theta = 0.3391$ (c) $\tan\theta = 1.5923$

(d) $\sin\theta = 0.8824$ (e) $\cos\theta = 0.1125$.

(a) $\sin\theta = 0.7976$

So:

$$\begin{aligned}\theta &= \arcsin 0.7976\\ &= 52.901\,52\ldots°\\ &= 52.9° \text{ (1 d.p.)}\end{aligned}$$

(b) $\cos\theta = 0.3391$

So:

$$\begin{aligned}\theta &= \arccos 0.3391\\ &= 70.177\,94\ldots°\\ &= 70.2° \text{ (1 d.p.)}\end{aligned}$$

(c) $\tan\theta = 1.5923$

So:

$$\begin{aligned}\theta &= \arctan 1.5923\\ &= 57.870\,260\ldots°\\ &= 57.9° \text{ (1 d.p.)}\end{aligned}$$

(d) $\sin\theta = 0.8824$

So: $$\begin{aligned}\theta &= \arcsin 0.8824\\ &= 61.933\,243\,..^\circ\\ &= 61.9^\circ\ (1\,\text{d.p.})\end{aligned}$$

(e) $\cos\theta = 0.1125$

So: $$\begin{aligned}\theta &= \arccos 0.1125\\ &= 83.540\,550\,..^\circ\\ &= 83.5^\circ\ (1\,\text{d.p.})\end{aligned}$$

Example 32

Given that θ is acute and $\tan\theta = 1.1714$, work in degrees to find, to 3 significant figures, the value of $\sin\theta$.

$\tan\theta = 1.1714$

So: $$\begin{aligned}\theta &= \arctan 1.1714\\ &= 49.513\,298\,..^\circ\end{aligned}$$

and: $$\begin{aligned}\sin\theta &= \sin 49.513\,298\,..^\circ\\ &= 0.760\,556\,68\,..\\ &= 0.761\ (3\,\text{s.f.})\end{aligned}$$

Notice that in questions such as this you should leave the full display, 49.513 298 . . . , in the calculator before pressing the sin button. Do not approximate the angle to 3 s.f. first, or you will introduce approximation errors.

8.10 Using basic trigonometry to solve problems

The definitions of sine, cosine and tangent for acute angles can be used to help solve problems in two dimensions. So far our definitions of sine, cosine and tangent are limited to acute angles which lie in right-angled triangles, so you can only use them to solve problems in which the triangle contains a right angle.

Example 33

Find to 3 significant figures the length of the side AB in the triangle ABC.

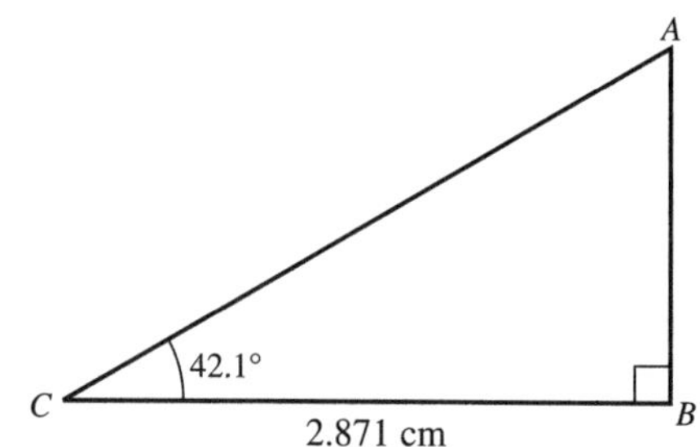

$$\begin{aligned}\frac{AB}{2.871} &= \tan 42.1^\circ\\ AB &= 2.871 \times \tan 42.1^\circ\\ &= 2.871 \times 0.903\,569\,..\\ &= 2.594\,14\,..\\ AB &= 2.59\,\text{cm}\ (3\,\text{s.f.})\end{aligned}$$

Example 34

Find to 3 significant figures the length of the side AB in the triangle ABC.

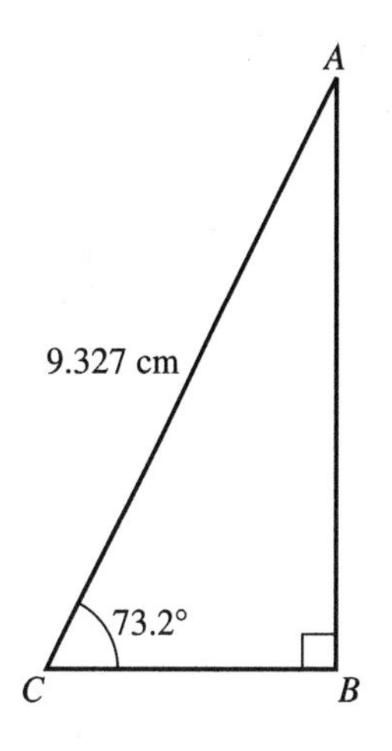

$$\frac{AB}{9.327} = \sin 73.2°$$

$$AB = 9.327 \times \sin 73.2°$$

$$= 9.327 \times 0.957\,319\ldots$$

$$= 8.928\,918\,9\ldots$$

$$AB = 8.93\,\text{cm (3 s.f.)}$$

Example 35

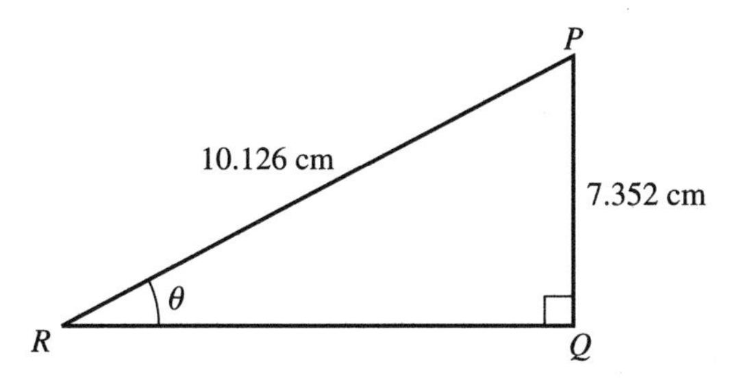

Find in degrees to 1 decimal place the size of the angle θ.

$$\sin\theta = \frac{7.352}{10.126}$$

$$= 0.726\,051\,7\ldots$$

$$\theta = \arcsin 0.726\,051\,7\ldots$$

$$= 46.556\,41\ldots°$$

$$\theta = 46.6°\ \text{(1 d.p.)}$$

Example 36

Calculate to 3 significant figures the length of the side PR.

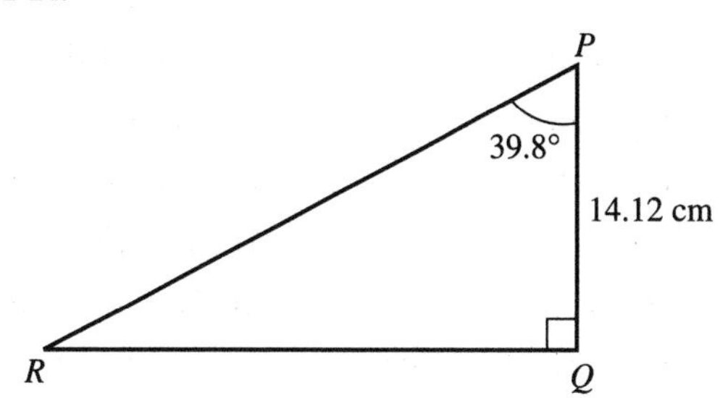

$$\frac{14.12}{PR} = \cos 39.8°$$

$$\frac{PR}{14.12} = \frac{1}{\cos 39.8°}$$

$$PR = \frac{14.12}{\cos 39.8°}$$

$$= \frac{14.12}{0.768\,283\ldots}$$

$$= 18.378\,63\ldots$$

$$PR = 18.4\,\text{cm (3 s.f.)}$$

Example 37

Calculate to 3 significant figures the length of CD.

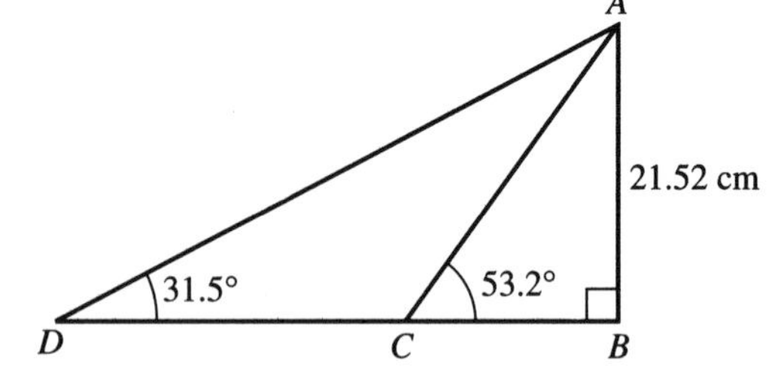

$$\angle BAC = 90^\circ - 53.2^\circ = 36.8^\circ$$

So: $$\frac{BC}{21.52} = \tan 36.8^\circ$$

$$BC = 21.52 \times \tan 36.8^\circ$$

$$= 21.52 \times 0.748\,095$$

$$BC = 16.099\,01\,..\text{ cm}$$

$$= 16.10\text{ cm } (4\text{ s.f.})$$

$$\angle BAD = 90^\circ - 31.5^\circ = 58.5^\circ$$

So: $$\frac{BD}{21.52} = \tan 58.5^\circ$$

$$BD = 21.52 \times \tan 58.5^\circ$$

$$= 21.52 \times 1.631\,851\,6\,..$$

$$BD = 35.117\,448\,..\text{ cm}$$

$$= 35.12\text{ cm}(4\text{ s.f.})$$

$$CD = BD - BC = 35.12 - 16.10$$

$$CD = 19.0\text{ cm } (3\text{ s.f.})$$

Notice that in questions such as this each length must be stored to at least 4 significant figures if you want to give the final answer to 3 significant figures. If you approximate too soon you will usually get an inaccurate final answer.

Exercise 8D

1 Find to 3 significant figures the value of:

(a) $\sin 28.1^\circ$ (b) $\cos 63.2^\circ$

(c) $\tan 57.3^\circ$ (d) $\cos 31.4^\circ$

(e) $\tan 48.4^\circ$ (f) $\sin 15.7^\circ$

(g) $\tan 23.8^\circ$ (h) $\cos 44.4^\circ$

(i) $\cos 77.3^\circ$ (j) $\tan 43.4^\circ$

2 Find in degrees to 1 decimal place the acute angle such that:

(a) $\sin\theta = 0.341\,72$ (b) $\cos\theta = 0.112\,16$

(c) $\tan\theta = 1.521\,34$ (d) $\cos\theta = 0.331\,32$

(e) $\cos\theta = 0.471\,26$ (f) $\sin\theta = 0.131\,47$

(g) $\tan\theta = 0.661\,67$ (h) $\sin\theta = 0.781\,23$

(i) $\tan\theta = 0.531\,32$ (j) $\cos\theta = 0.616\,14$

3 Given that θ is acute, find to 3 significant figures the value of:

(a) $\sin\theta$, given that $\cos\theta = 0.271\,36$

(b) $\cos\theta$, given that $\tan\theta = 1.0123$

(c) $\tan\theta$, given that $\sin\theta = 0.8021$

(d) $\cos\theta$, given that $\sin\theta = 0.741\,59$

(e) $\sin\theta$, given that $\tan\theta = 0.982\,81$

(f) $\tan\theta$, given that $\cos\theta = 0.772\,16$

(g) $\sin\theta$, given that $\cos\theta = 0.210\,12$

(h) $\cos\theta$, given that $\tan\theta = 1.9527$

(i) $\tan\theta$, given that $\sin\theta = 0.334\,51$

(j) $\sin\theta$, given that $\tan\theta = 1.142\,41$

In questions 4–15, find the sizes of the angles and the lengths of the sides marked. Give the angles in degrees to 1 decimal place and the sides in cm to 3 significant figures.

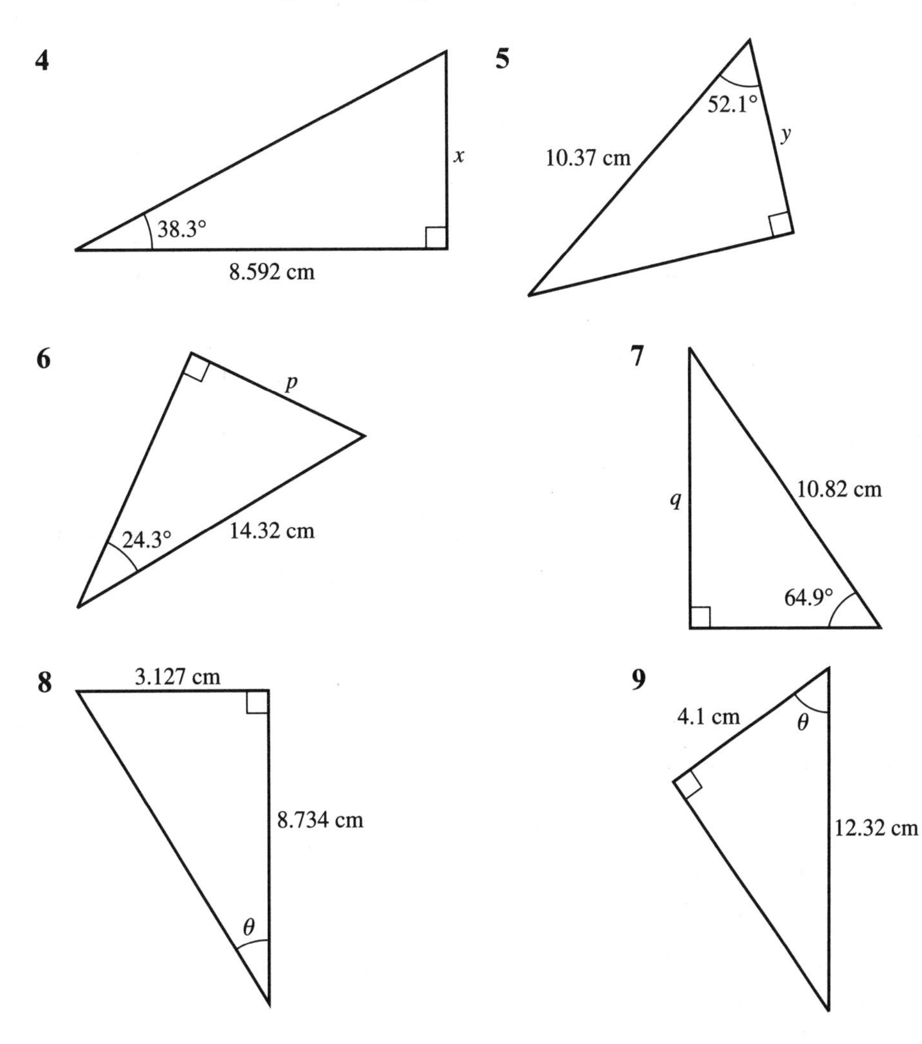

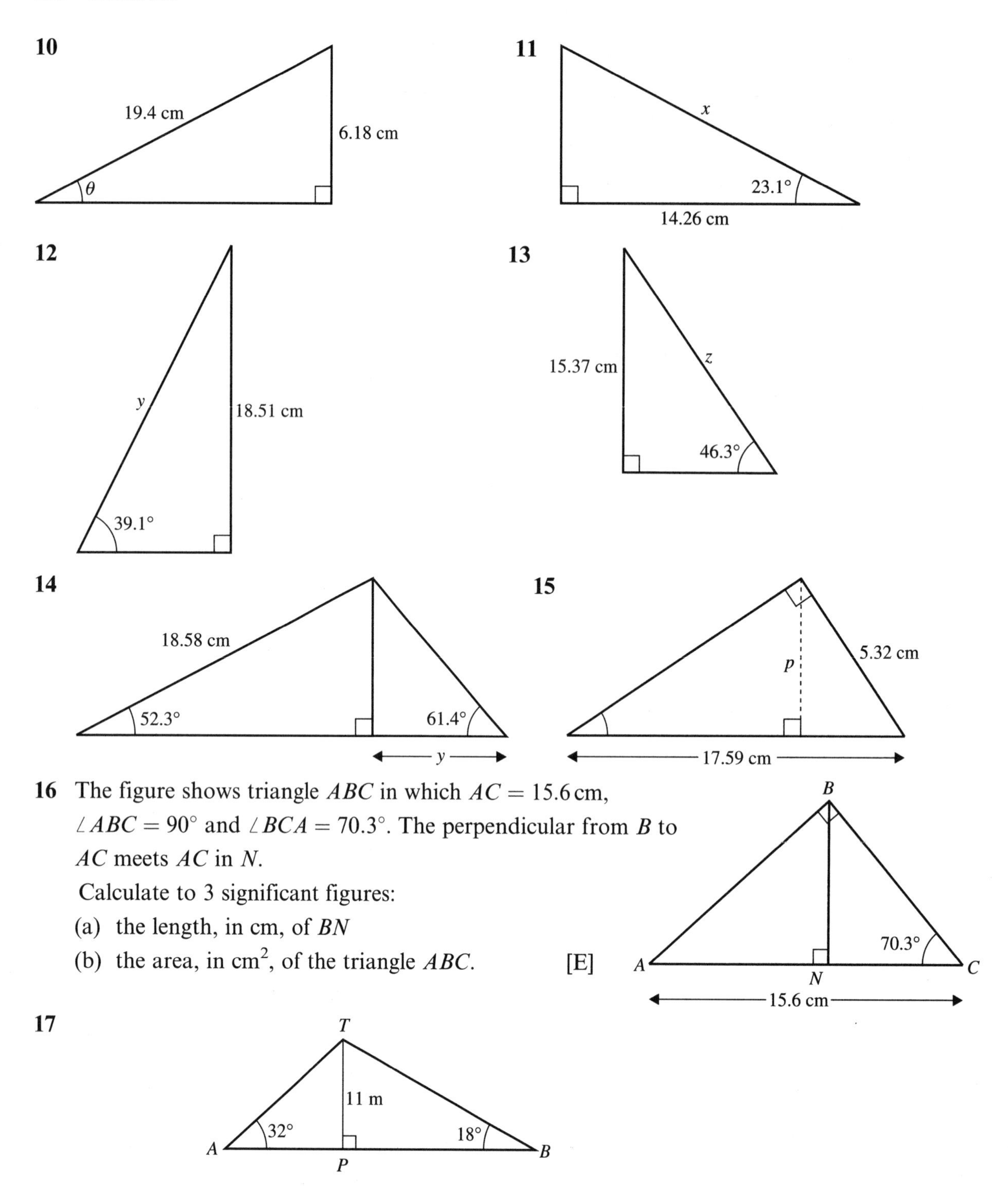

16 The figure shows triangle ABC in which $AC = 15.6$ cm, $\angle ABC = 90°$ and $\angle BCA = 70.3°$. The perpendicular from B to AC meets AC in N.

Calculate to 3 significant figures:

(a) the length, in cm, of BN

(b) the area, in cm^2, of the triangle ABC. [E]

17

The vertical flagpole PT is of height 11 m and stands on level ground at P. The points A and B are on the same horizontal level as P and $\angle TAP = 32°$, $\angle TBP = 18°$.

(a) Calculate, in m to one decimal place, the length AP.

(b) Calculate, in m to one decimal place, the length BT. [E]

18

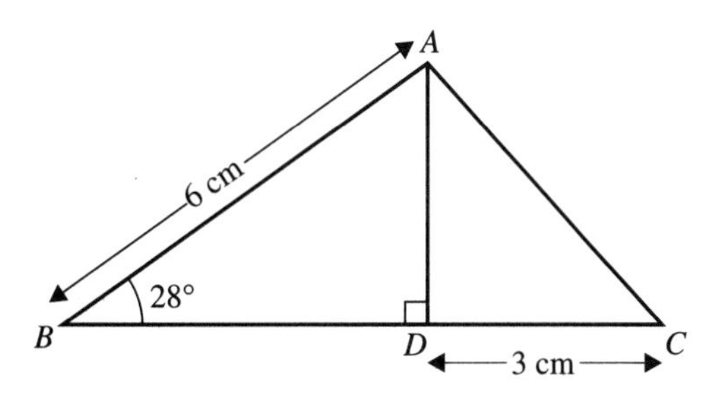

In the figure D is the point on BC such that $\angle ADB$ is a right angle. Given that $AB = 6\,\text{cm}$, $DC = 3\,\text{cm}$ and $\angle ABD = 28°$, calculate:

(a) AD (b) BD (c) $\angle ACD$. [E]

19 In the figure, $ABCD$ represents the cross-section through a building. The perpendicular from D meets AB in E. CF is perpendicular to DE. Angle $DAE = 41°$.
$DF = FE = CB = 10$ m and $EB = 8$ m.
Calculate to 3 significant figures:

(a) angle DCF
(b) the length, in metres, of DC
(c) the length, in metres, of AB
(d) the area, in square metres, of $ABCD$. [E]

20

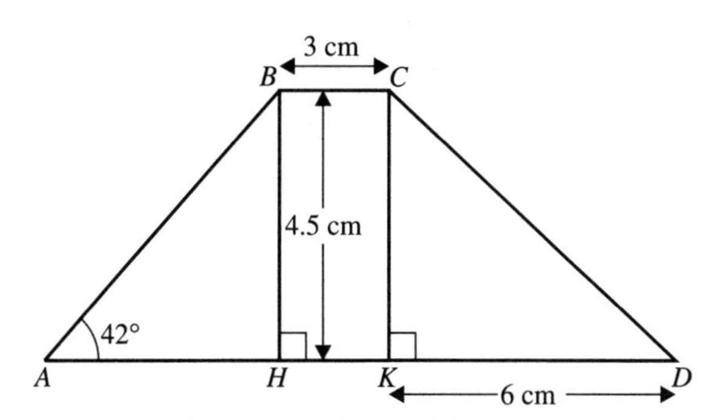

In the figure $ABCD$ is a trapezium with BC parallel to AD. The lines BH and CK are each perpendicular to AD and are 4.5 cm in length. Given that $BC = 3\,\text{cm}$, $KD = 6\,\text{cm}$ and $\angle BAH = 42°$, calculate:

(a) CD
(b) AB
(c) AD
(d) the perimeter of the trapezium $ABCD$
(e) the area of the trapezium $ABCD$. [E]

21

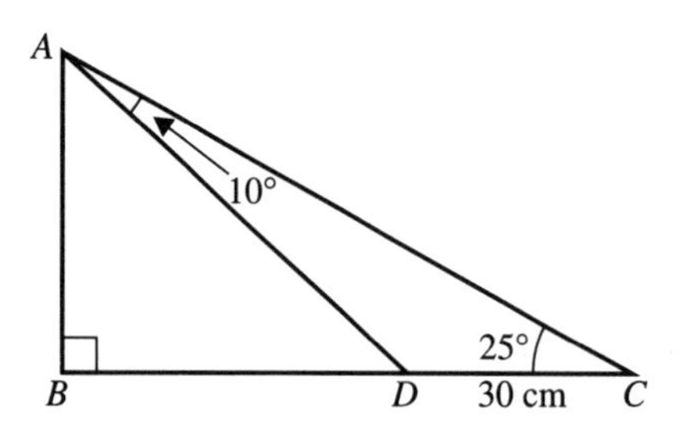

In the figure the triangle ABC is right-angled at B and the point D, on BC, is such that $DC = 30\,\text{cm}$. Given that $\angle ACD = 25°$ and $\angle DAC = 10°$, calculate the area of triangle ABC, giving your answer in square centimetres to three significant figures. [E]

22

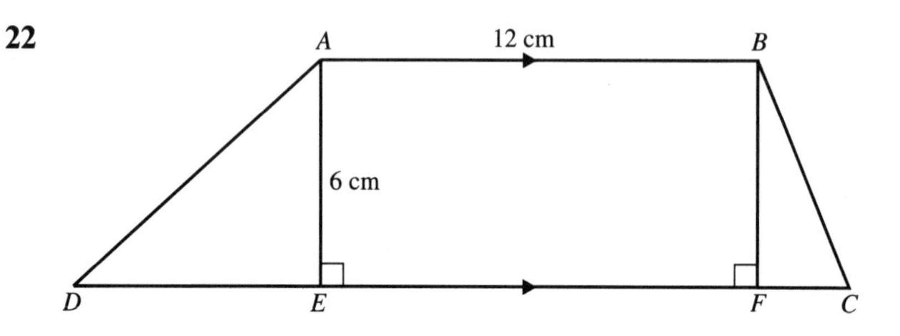

The figure shows the trapezium $ABCD$ in which AB is parallel to DC, $AB = 12$ cm, $\angle ABC = 117°$ and $\angle ADC = 46°$. The perpendicular from A to DC meets DC in E and $AE = 6$ cm. The perpendicular from B to DC meets DC in F. Calculate, in centimetres, giving your answers to 3 significant figures:

(a) the length of DE

(b) the length of AD

(c) the length of FC.

Find also, in square centimetres, giving your answer to 3 significant figures, the area of the trapezium. [E]

23 The point O is at the top of a vertical cliff. From a boat which is at a point A, due south of O, the angle of elevation of O is 28°. The boat now sails 50 metres due north to a point B from which the angle of elevation of O is 70°. Calculate, in metres to 3 significant figures:

(a) the height of the cliff

(b) the distance of B from the foot of the cliff. [E]

24 A ship sails from port A to port B, a distance of 5 km, on a bearing of 036°.

(a) Calculate, in km to 2 decimal places, the distance by which B is

(i) east of A

(ii) north of A.

The ship then sails to a point C, a further distance of 8 km, on a bearing of 138°.

(b) Calculate, in km to 2 decimals places, the distance by which C is:

(i) east of A

(ii) south of A.

(c) Calculate:

(i) the bearing, to the nearest degree, of A from C

(ii) the distance, in km to 2 decimal places, of A from C. [E]

8.11 The sine rule and the cosine rule

You should use the following notation when dealing with the general triangle ABC denoted by $\triangle ABC$:

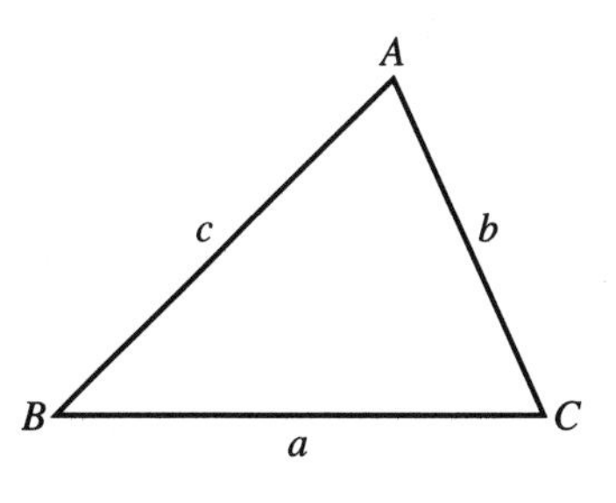

The angles BAC, ABC and ACB are called A, B and C respectively. The lengths of the sides BC, CA and AB are called a, b and c respectively, and it is assumed that the unit of length is the same for each in any given triangle.

In triangles where some data are given, the sine and the cosine rules are used to find the length of an unknown side or the size of an unknown angle.

The sine rule

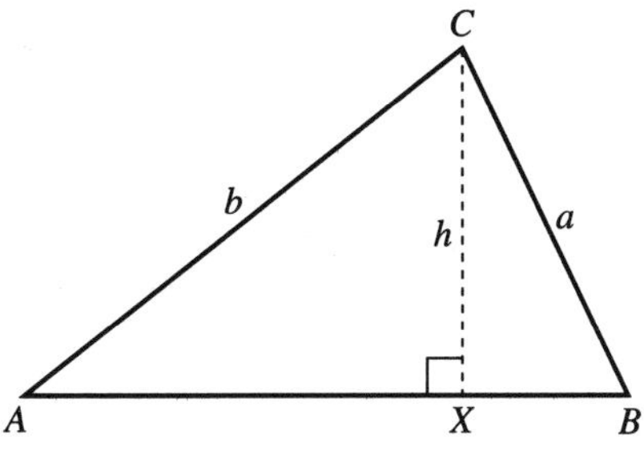

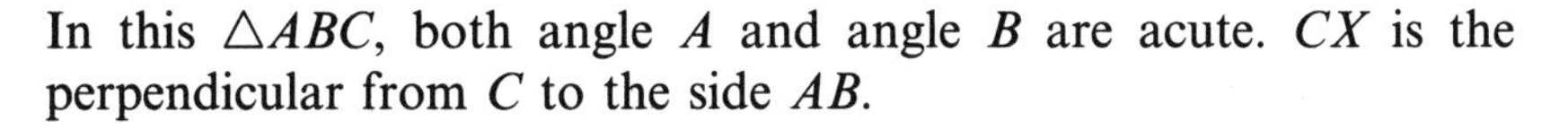

In this $\triangle ABC$, both angle A and angle B are acute. CX is the perpendicular from C to the side AB.

Let $CX = h$

In $\triangle AXC$,

$$\frac{h}{b} = \sin A$$

So: $$h = b \sin A$$

In $\triangle BXC$,

$$\frac{h}{a} = \sin B$$

So: $$h = a \sin B$$

Thus $$a \sin B = b \sin A$$

$\div \sin A$: $$\frac{a \sin B}{\sin A} = \frac{b \cancel{\sin A}}{\cancel{\sin A}}$$

$\div \sin B$: $$\frac{a \cancel{\sin B}}{\sin A \cancel{\sin B}} = \frac{b}{\sin B}$$

So: $$\frac{a}{\sin A} = \frac{b}{\sin B}$$

In a similar way, by drawing the perpendicular from B to the line AC you can show that

$$\frac{a}{\sin A} = \frac{c}{\sin C}$$

Putting these two results together, you get:

$$\frac{a}{\sin A} = \frac{b}{\sin B} = \frac{c}{\sin C}$$

Now suppose that $\angle B$ is obtuse:

CX is the perpendicular from C to the side AB. (This time the side AB has to be extended.)

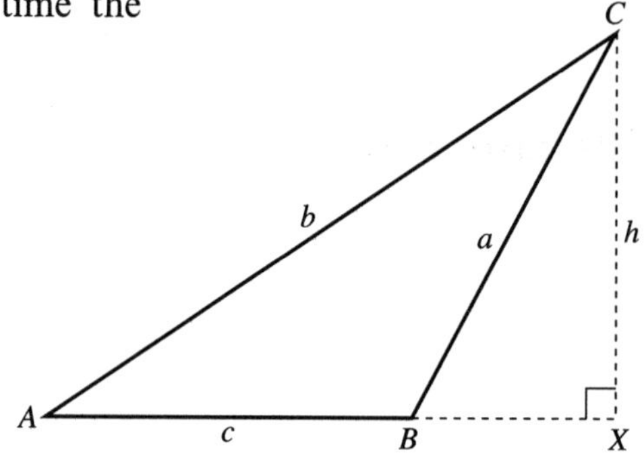

Let $CX = h$, as before.

In $\triangle AXC$,

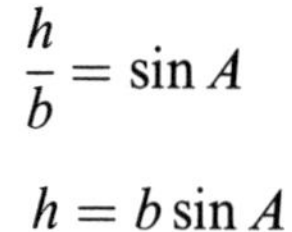

$$\frac{h}{b} = \sin A$$

So: $$h = b \sin A$$

In $\triangle BXC$,

$$\frac{h}{a} = \sin(180° - B)$$

But $$\sin(180° - B) = \sin B$$

So: $$\frac{h}{a} = \sin B$$

or $$h = a\sin B$$

Thus, as before: $$a\sin B = b\sin A$$

and so: $$\frac{a}{\sin A} = \frac{b}{\sin B}$$

Again, if you draw the perpendicular from B to the side AC then you can show that

$$\frac{a}{\sin A} = \frac{c}{\sin C}$$

Consequently, once again you have

■ $$\boldsymbol{\frac{a}{\sin A} = \frac{b}{\sin B} = \frac{c}{\sin C}}$$

This is known as the **sine rule**.

Remember, although these results have been proved by using a construction line that is perpendicular to one of the sides of the triangle, in neither case was the triangle ABC right-angled. The sine rule applies to *any* triangle.

You can use the sine rule, in general, if the given triangle contains two known angles and one known side and you need to find another side. It can also be used where the given triangle contains two known sides and one known angle, *which is not the angle between the two sides* (often called the 'non-included' angle), and where you need to find another angle.

Example 38

Calculate, in cm to 3 significant figures, the length of the side AB of the triangle ABC in which $\angle ACB = 62°$, $\angle ABC = 47°$ and $AC = 7$ cm.

By the sine rule:

$$\frac{c}{\sin 62°} = \frac{7}{\sin 47°}$$

So: $$\frac{c}{\sin 62°} \times \sin 62° = \frac{7}{\sin 47°} \times \sin 62°$$

$$c = \frac{7\sin 62°}{\sin 47°}$$

$$= 8.4509$$

$$= 8.45 \text{ cm (3 s.f.)}$$

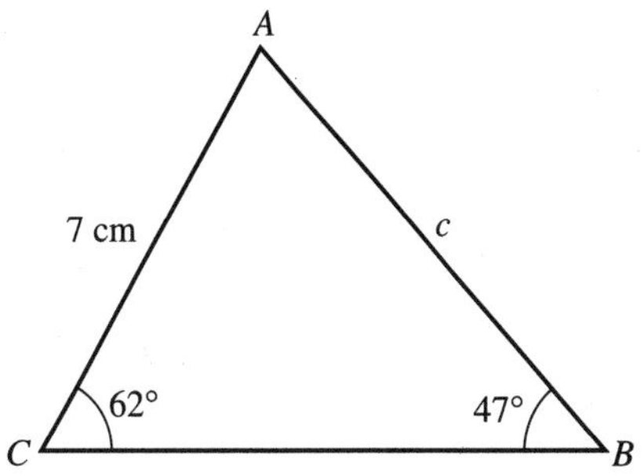

Example 39

Calculate, in cm to 3 significant figures, the length of the side AC of the triangle ABC in which $BC = 6.4\,\text{cm}$, $\angle ACB = 43^\circ$ and $\angle BAC = 71^\circ$.

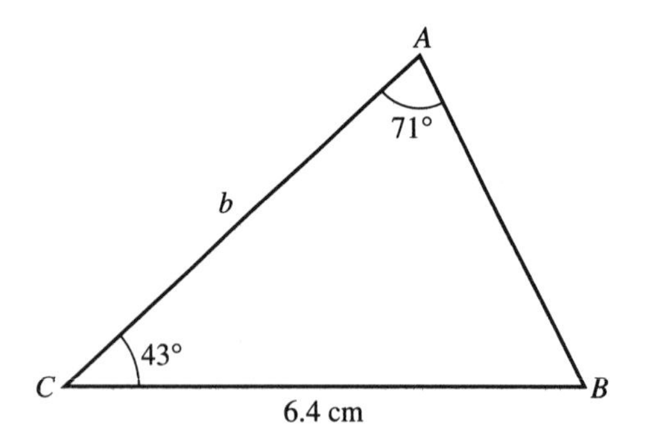

Since the sum of the three angles in any triangle is 180°,

$$71^\circ + 43^\circ + \angle ABC = 180^\circ$$

So: $$\angle ABC = 180^\circ - 114^\circ = 66^\circ$$

By the sine rule:

$$\frac{b}{\sin 66^\circ} = \frac{6.4}{\sin 71^\circ}$$

$$\frac{b}{\cancel{\sin 66^\circ}} \times \cancel{\sin 66^\circ} = \frac{6.4}{\sin 71^\circ} \times \sin 66^\circ$$

$$b = \frac{6.4 \sin 66^\circ}{\sin 71^\circ}$$

$$= 6.183\text{ cm}$$

$$= 6.18\text{ cm (3 s.f.)}$$

When the information given about the triangle is two sides and the non-included angle, you must be careful. It is sometimes possible from such information to obtain *two* solutions. That is, it is sometimes possible to find two *different* triangles with the same given data, as the next example demonstrates.

Example 40

Calculate, in degrees to 1 decimal place, the size of the angles CAB and ACB in the triangle ABC, where $AC = 4\,\text{cm}$, $BC = 5\,\text{cm}$ and $\angle ABC = 42^\circ$.

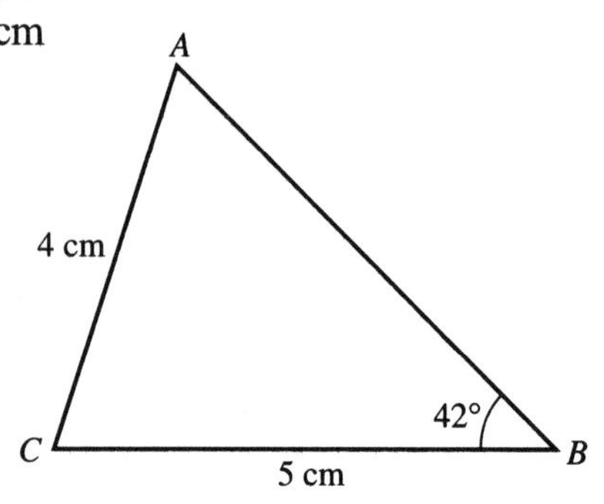

By the sine rule, $$\frac{5}{\sin A} = \frac{4}{\sin 42^\circ}$$

That is: $$\frac{\sin A}{5} = \frac{\sin 42^\circ}{4}$$

So: $$\frac{\sin A}{\cancel{5}} \times \cancel{5} = \frac{\sin 42^\circ}{4} \times 5$$

$$\sin A = \frac{5 \sin 42^\circ}{4}$$

$$\sin A = 0.8364$$

So: $A = 56.76^\circ$ *or* $A = 180^\circ - 56.76^\circ$
$= 123.24^\circ$

Since the angles of a triangle add up to 180°,

when $A = 56.76^\circ$, $C = 180^\circ - 56.76^\circ - 42^\circ$

$\Rightarrow$ $A = 56.8^\circ$, $C = 81.2^\circ$ (1 d.p.)

when $A = 123.24^\circ$, $C = 180^\circ - 123.24^\circ - 42^\circ$

$\Rightarrow$ $A = 123.2^\circ$, $C = 14.8^\circ$ (1 d.p.)

Thus there are two possible triangles that can be drawn in this case with the given information, as shown below.

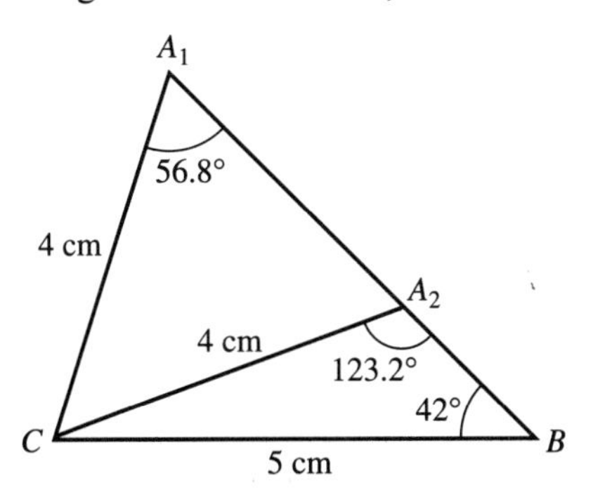

Example 41

Find the length of AB and the sizes of $\angle ACB$ and $\angle BAC$ for the triangle ABC in which $AC = 6.3$ cm, $BC = 4.8$ cm and $\angle ABC = 53^\circ$.

$$\frac{4.8}{\sin A} = \frac{6.3}{\sin 53^\circ}$$

So: $$\frac{\sin A}{4.8} = \frac{\sin 53^\circ}{6.3}$$

$$\frac{\sin A}{4.8} \times 4.8 = \frac{\sin 53^\circ}{6.3} \times 4.8$$

$$\sin A = \frac{4.8 \sin 53^\circ}{6.3}$$

$$= \frac{4.8 \times 0.7986}{6.3}$$

$$\sin A = 0.6084$$

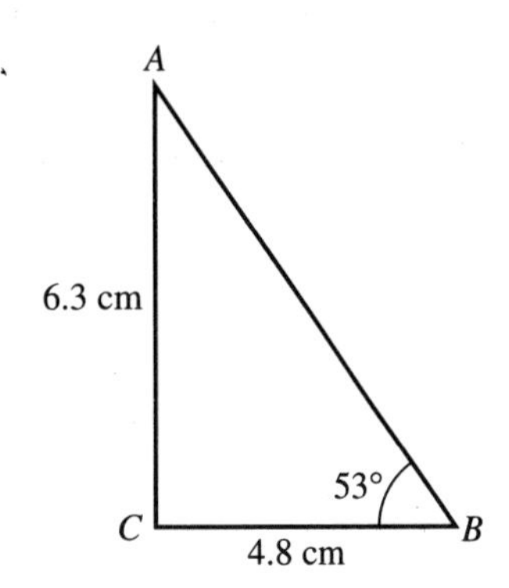

So: $A = 37.47^\circ$
$= 37.5^\circ$ (1 d.p.)

and: $C = 180^\circ - 37.47^\circ - 53^\circ$
$= 89.5^\circ$ (1 d.p.)

The cosine rule

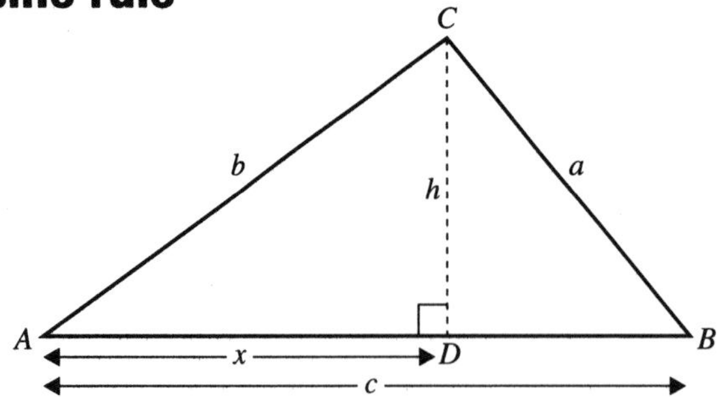

In $\triangle ABC$, CD is the perpendicular from C to the side AB, $CD = h$ and $AD = x$. Angle B is acute. In $\triangle ADC$ you can use Pythagoras' theorem to write

$$h^2 + x^2 = b^2 \quad (1)$$

In $\triangle BCD$, you can use Pythagoras' theorem to write

$$h^2 + DB^2 = a^2$$

But: $DB = c - x$

So: $h^2 + (c - x)^2 = a^2 \quad (2)$

From (1): $h^2 = b^2 - x^2$

Substitute for h^2 in (2):

$$b^2 - x^2 + (c - x)^2 = a^2$$

That is: $b^2 - x^2 + c^2 - 2cx + x^2 = a^2$

or $a^2 = b^2 + c^2 - 2cx \quad (3)$

But in $\triangle ADC$, $\frac{x}{b} = \cos A$

or: $x = b \cos A$

Substitute this into equation (3):

$$a^2 = b^2 + c^2 - 2c(b \cos A)$$

or $a^2 = b^2 + c^2 - 2bc \cos A$

Now consider the case when A is obtuse:

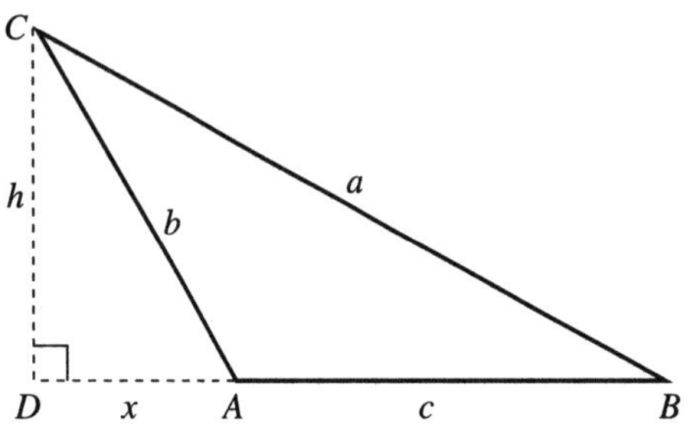

In $\triangle ABC$, $AB = c$, $BC = a$, $AC = b$, CD is the perpendicular from C to the line BA (which is extended or **produced**), $DA = x$ and $CD = h$.

In $\triangle ADC$, you can use Pythagoras' theorem, to write:

$$h^2 + x^2 = b^2 \quad (1)$$

In $\triangle BCD$, you can use Pythagoras' theorem, to write

$$h^2 + (x + c)^2 = a^2 \quad (2)$$

From (1): $h^2 = b^2 - x^2$

Substitute this into equation (2):

$$b^2 - x^2 + (x + c)^2 = a^2$$

So: $b^2 - x^2 + x^2 + 2cx + c^2 = a^2$

or $a^2 = b^2 + c^2 + 2cx \quad (3)$

Now in $\triangle ADC$, $\dfrac{x}{b} = \cos \angle CAD$

So: $x = b \cos \angle CAD$

$= b \cos(180° - A)$

$= -b \cos A$ (since $\cos(180° - A) = -\cos A$)

By substituting this into equation (3) you get:

$$a^2 = b^2 + c^2 + 2c(-b \cos A)$$

or $a^2 = b^2 + c^2 - 2bc \cos A$

as before.

This is the **cosine rule**. It can written as:

■ $\boldsymbol{a^2 = b^2 + c^2 - 2bc \cos A}$

OR $\boldsymbol{b^2 = a^2 + c^2 - 2ac \cos B}$

OR $\boldsymbol{c^2 = a^2 + b^2 - 2ab \cos C}$

Now if

$$a^2 = b^2 + c^2 - 2bc \cos A$$

then: $a^2 - b^2 - c^2 = -2bc \cos A$

So: $-a^2 + b^2 + c^2 = 2bc \cos A$

or $\dfrac{-a^2 + b^2 + c^2}{2bc} = \cos A$

So: $\cos A = \dfrac{b^2 + c^2 - a^2}{2bc}$

This is another form of the cosine rule. You can get two other similar formulae if you rearrange the other two forms of the cosine rule. So you can get:

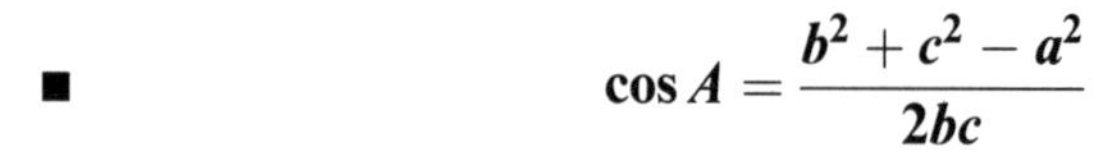

$$\cos A = \frac{b^2 + c^2 - a^2}{2bc}$$

OR

$$\cos B = \frac{a^2 + c^2 - b^2}{2ac}$$

OR

$$\cos C = \frac{a^2 + b^2 - c^2}{2ab}$$

You can use the cosine rule when you are given the lengths of two sides in a triangle and the size of the angle between them. You can then find the length of the third side of the triangle, using either

$$a^2 = b^2 + c^2 - 2bc \cos A$$

or

$$b^2 = a^2 + c^2 - 2ac \cos B$$

or

$$c^2 = a^2 + b^2 - 2ab \cos C$$

You can also use the cosine rule when you are given the lengths of all three sides of a triangle and want to find the sizes of its angles. In this case you can find the angles using

$$\cos A = \frac{b^2 + c^2 - a^2}{2bc}$$

or

$$\cos B = \frac{a^2 + c^2 - b^2}{2ac}$$

or

$$\cos C = \frac{a^2 + b^2 - c^2}{2ab}$$

Example 42

Find the length of the side BC of the triangle ABC in which $AB = 7\,\text{cm}$, $AC = 9\,\text{cm}$ and $\angle BAC = 71°$.

By the cosine rule:

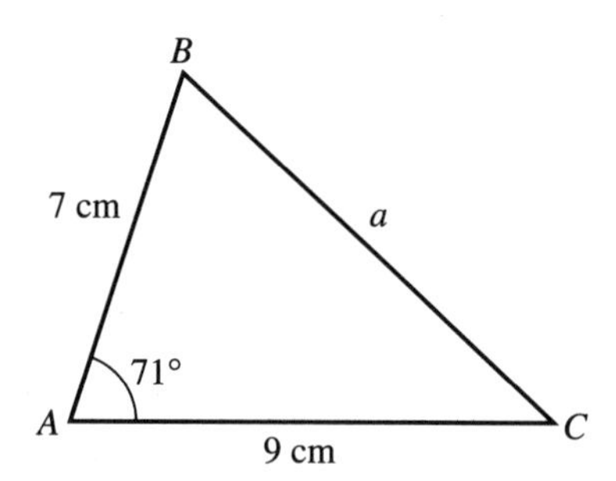

$$\begin{aligned} a^2 &= b^2 + c^2 - 2bc \cos A \\ &= 9^2 + 7^2 - 2 \times 9 \times 7 \cos 71° \\ &= 81 + 49 - 126 \cos 71° \\ &= 130 - 126 \times 0.3255 \\ &= 130 - 41.02 \end{aligned}$$

$$a^2 = 88.98$$

So: $a = \sqrt{88.98} = 9.43\,\text{cm}$ (3 s.f.)

Example 43

Find the length of the side AB of the triangle ABC in which $BC = 15.3\,\text{cm}$, $AC = 9.4\,\text{cm}$ and $\angle ACB = 121°$.

By the cosine rule:

$$c^2 = a^2 + b^2 - 2ab\cos C$$

$$= 15.3^2 + 9.4^2 - 2 \times 15.3 \times 9.4\cos 121°$$

$$= 234.09 + 88.36 - 287.64\cos 121°$$

$$= 322.45 - 287.64(-\cos 59°)$$

$$= 322.45 + 287.64\cos 59°$$

$$= 322.45 + 287.64 \times 0.5150$$

$$= 322.45 + 148.14$$

$$c^2 = 470.59$$

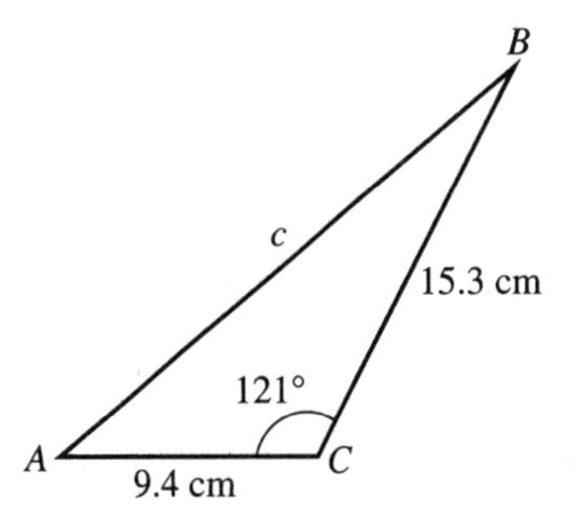

So: $c = 21.7\,\text{cm}$ (3 s.f.)

Example 44

Calculate the size of $\angle ABC$ of the triangle ABC in which $AB = 3.6\,\text{cm}$, $BC = 5.2\,\text{cm}$ and $CA = 4.3\,\text{cm}$.

By the cosine rule:

$$\cos B = \frac{a^2 + c^2 - b^2}{2ac}$$

$$= \frac{5.2^2 + 3.6^2 - 4.3^2}{2 \times 5.2 \times 3.6}$$

$$= \frac{27.04 + 12.96 - 18.49}{37.44}$$

$$\cos B = \frac{21.51}{37.44} = 0.5745$$

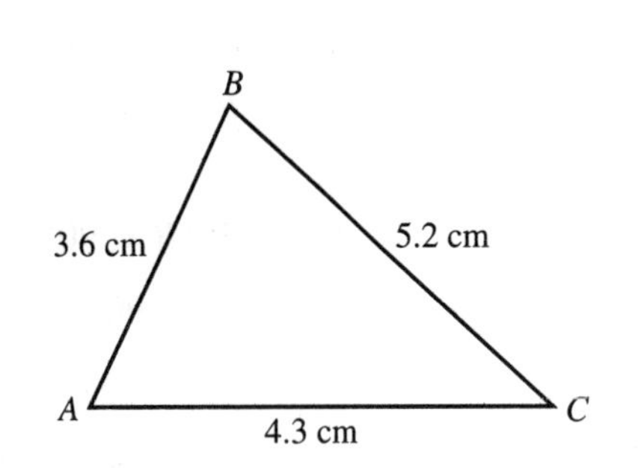

So: $B = 54.9°$ (1 d.p.)

Example 45

Find the size of $\angle ABC$ of the triangle ABC in which $AB = 13.7\,\text{cm}$, $BC = 12.1\,\text{cm}$ and $AC = 19.3\,\text{cm}$.

By the cosine rule:

$$\cos B = \frac{a^2 + c^2 - b^2}{2ac}$$

$$= \frac{12.1^2 + 13.7^2 - 19.3^2}{2 \times 12.1 \times 13.7}$$

$$= \frac{146.41 + 187.69 - 372.49}{331.54}$$

$$= \frac{-38.39}{331.54}$$

$$\cos B = -0.1157$$

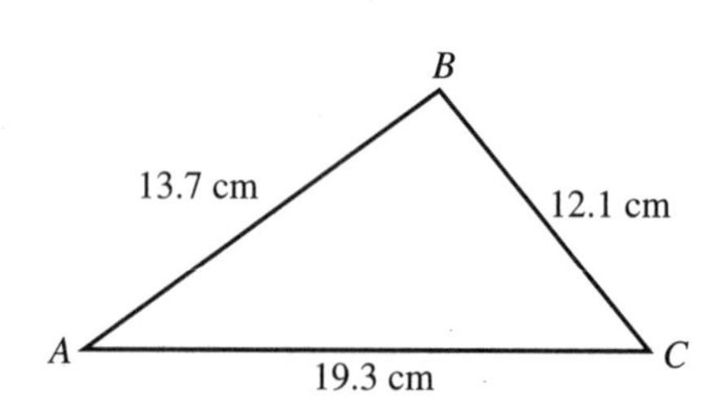

So: $B = 96.6°$ (1 d.p.)

Sometimes two facts are required from a given triangle, requiring both the sine rule and the cosine rule to be used. Here is an example set in a practical context.

Example 46

Two coastguard stations A and B are 5 km apart and B is due east of A. From A the bearings of two ships P and Q are 025° and 061° respectively and from B the bearings are 290° and 338° respectively. Find the distance between the ships.

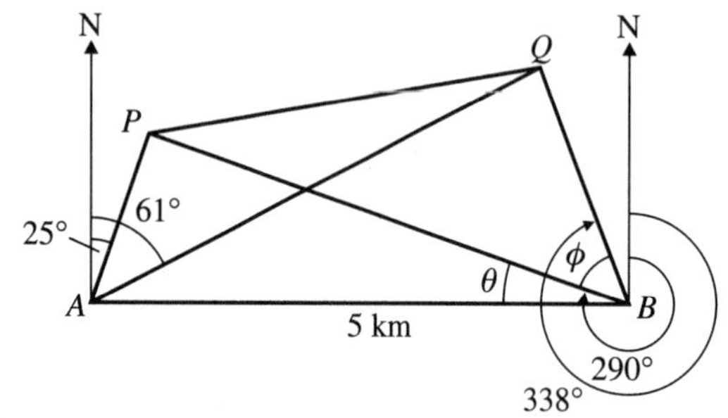

From the diagram you can see that

$$\angle PAQ = 61° - 25° = 36°$$

and $\angle PAB = 90° - 25° = 65°$

also: $\theta = 290° - 270° = 20°$

and $\theta + \phi = 338° - 270° = 68°$

So: $\phi = 68° - 20° = 48°$

In $\triangle PAB$, $\angle PAB = 65°$, $\angle PBA = 20°$

So: $\angle APB = 180° - 65° - 20° = 95°$

Use the sine rule in $\triangle APB$:

$$\frac{AP}{\sin 20^\circ} = \frac{5}{\sin 95^\circ}$$

$$AP = \frac{5\sin 20^\circ}{\sin 95^\circ}$$

$$= \frac{5\sin 20^\circ}{\sin 85^\circ}$$

$$= 1.716 \text{ km}$$

In $\triangle AQB$, $\angle ABQ = 68^\circ$ and $\angle QAB = 90^\circ - 61^\circ = 29^\circ$.

So: $\angle AQB = 180^\circ - 68^\circ - 29^\circ = 83^\circ$

Use the sine rule in $\triangle AQB$:

$$\frac{AQ}{\sin 68^\circ} = \frac{5}{\sin 83^\circ}$$

$$AQ = \frac{5\sin 68^\circ}{\sin 83^\circ}$$

$$= 4.670 \text{ km}$$

Use the cosine rule in $\triangle APQ$:

$$PQ^2 = 1.716^2 + 4.670^2 - 2 \times 1.716 \times 4.670 \cos 36^\circ$$
$$= 2.944 + 21.80 - 12.966$$

$$PQ^2 = 11.78$$

So: $PQ = 3.43 \text{ km (3 s.f.)}$

The area of a triangle

You should know that one formula for the area of $\triangle ABC$ is $\frac{1}{2}ah$, where h is the height of the triangle. This formula is not always the most useful one.

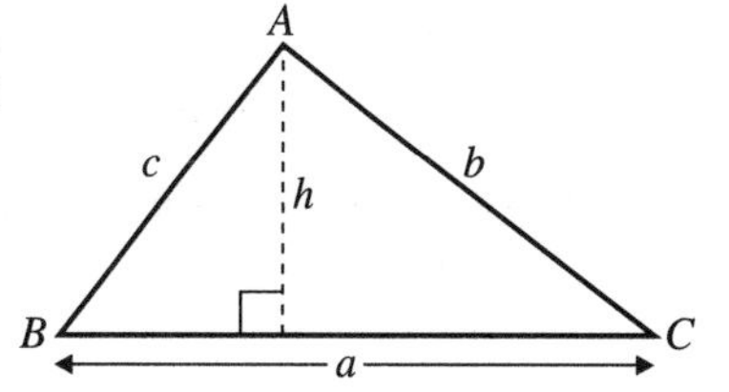

In the diagram,

$$\frac{h}{b} = \sin C$$

So: $h = b\sin C$

Thus another useful formula for the area of a triangle is:

$$\tfrac{1}{2}a(b\sin C) = \tfrac{1}{2}ab\sin C$$

This has two other forms:

$$\tfrac{1}{2}bc\sin A$$

and $\tfrac{1}{2}ac\sin B$

Example 47

Find the area of the triangle shown.

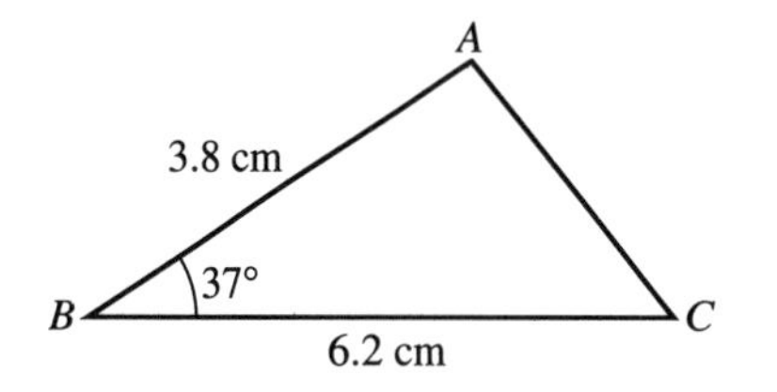

Here, in the usual notation,

$a = 6.2$, $c = 3.8$ and $B = 37°$.

Using the formula:

$$\begin{aligned}\text{Area } \triangle ABC &= \tfrac{1}{2}ac\sin B \\ &= \tfrac{1}{2}(6.2)(3.8)\sin 37°\ \text{cm}^2 \\ &= 7.09\ \text{cm}^2\ (2\ \text{d.p.})\end{aligned}$$

8.12 Mensuration of a right circular cone and a sphere

The right circular cone

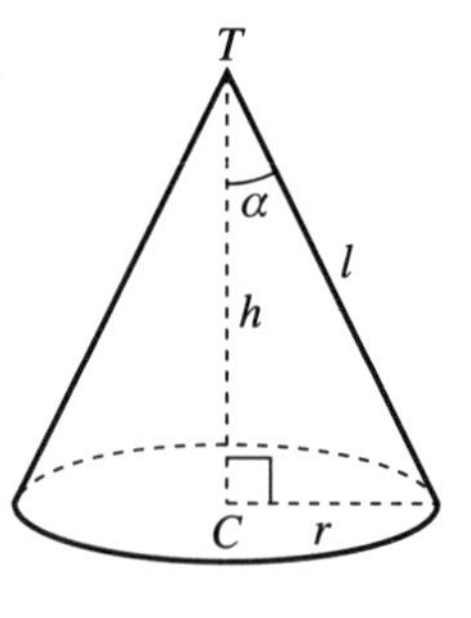

The right circular cone shown has **base radius r, height h, slant-height l** and **semi-vertical angle α. Its name comes from the fact that the vertex T is directly above the centre** C of its horizontal circular base.

By Pythagoras, $h^2 + r^2 = l^2$.

From trigonometry, $r = h\tan\alpha$, $h = l\cos\alpha$, $r = l\sin\alpha$.

The area of the base $= \pi r^2$

The curved surface area $= \pi rl$

The volume of the cone $= \frac{1}{3}\pi r^2 h$

Notice also that the **total surface area** $= \pi r^2 + \pi rl$

$$= \pi r(r + l)$$

The sphere

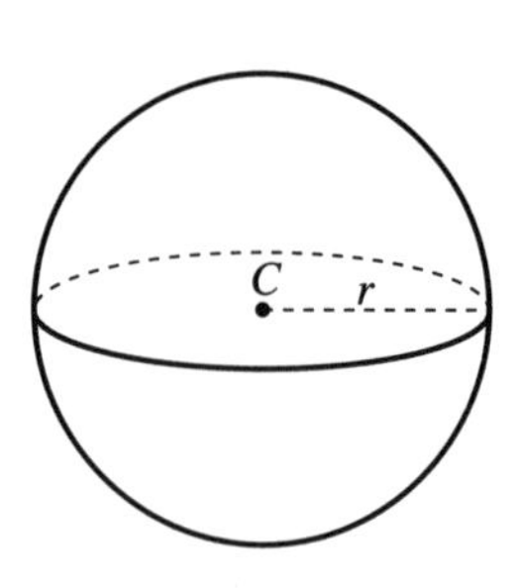

The sphere shown has centre C and radius r.

Surface area $= 4\pi r^2$

Volume $= \frac{4}{3}\pi r^3$

You should memorise these results because you will often need to use them in practical problems, such as finding maximum and minimum values in calculus (see Chapter 5).

Example 48

A toy consists of a solid cone of height 14 cm and base diameter 21 cm, stuck along its base to a hemisphere having the same circular base (see the diagram).

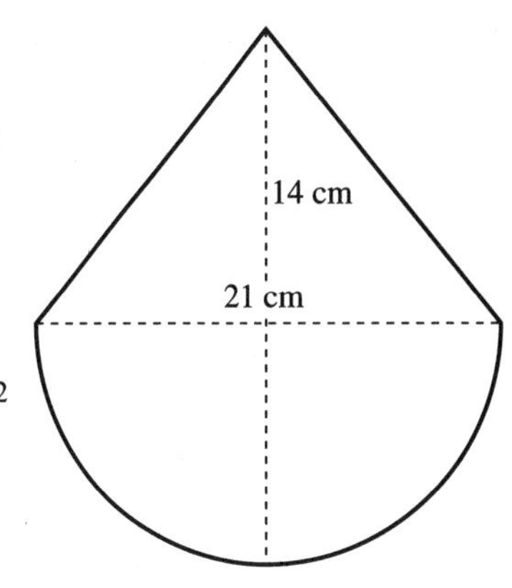

Calculate (a) the surface area (b) the volume of the toy.

(a) Surface area $= \pi rl$ (cone) $+ 2\pi r^2$ (hemisphere)

$$= \left[\pi \times 10.5 \times \sqrt{(10.5^2 + 14^2)} + 2 \times \pi \times 10.5^2\right] \text{cm}^2$$

$$= 1270 \text{ cm}^2 \text{ (3 s.f.)}$$

(b) Volume $= \frac{1}{3}\pi r^2 h$ (cone) $+ \frac{2}{3}\pi r^3$ (hemisphere)

$$= \left[\tfrac{1}{3} \times \pi \times 10.5^2 \times 14 + \tfrac{2}{3} \times \pi \times 10.5^3\right] \text{cm}^3$$

$$= 4040 \text{ cm}^3 \text{ (3 s.f.)}$$

8.13 Simple geometrical properties of a circle

You need to know and to memorise the following properties.

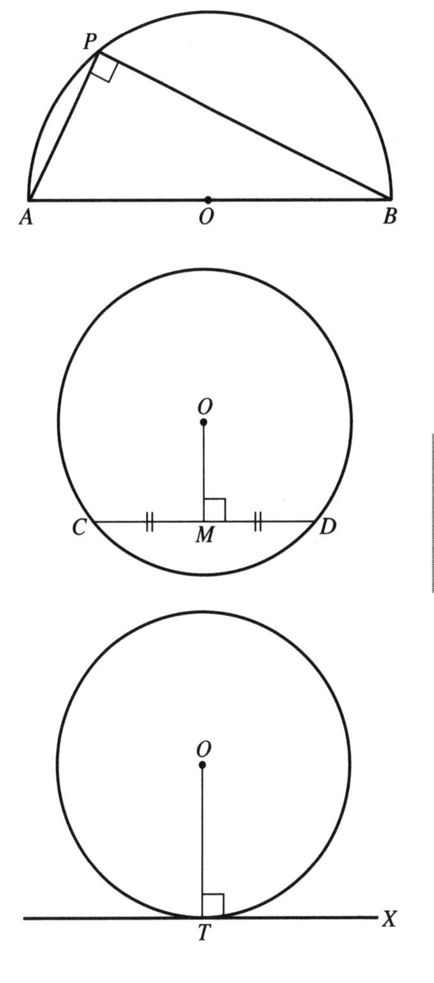

- **The angle in a semicircle is a right angle.**

That is, for a circle centre O with AB as a diameter, the angle APB is 90° for all positions of P on the semicircle.

- **The perpendicular drawn from the centre of a circle to a chord bisects the chord.**

That is, for a circle centre O with CD any chord, the line OM, perpendicular to CD, is such that $CM = MD$.

- **The radius of a circle drawn to the point of contact of a tangent to the circle is at right angles to the tangent.**

That is, for the tangent TX, contact point T, to the circle, centre O, the radius OT is perpendicular to TX. That is, $\angle OTX = 90°$.

Exercise 8E

Use the sine rule to calculate the lengths of the unknown sides and the sizes of the unknown angles in these triangles:

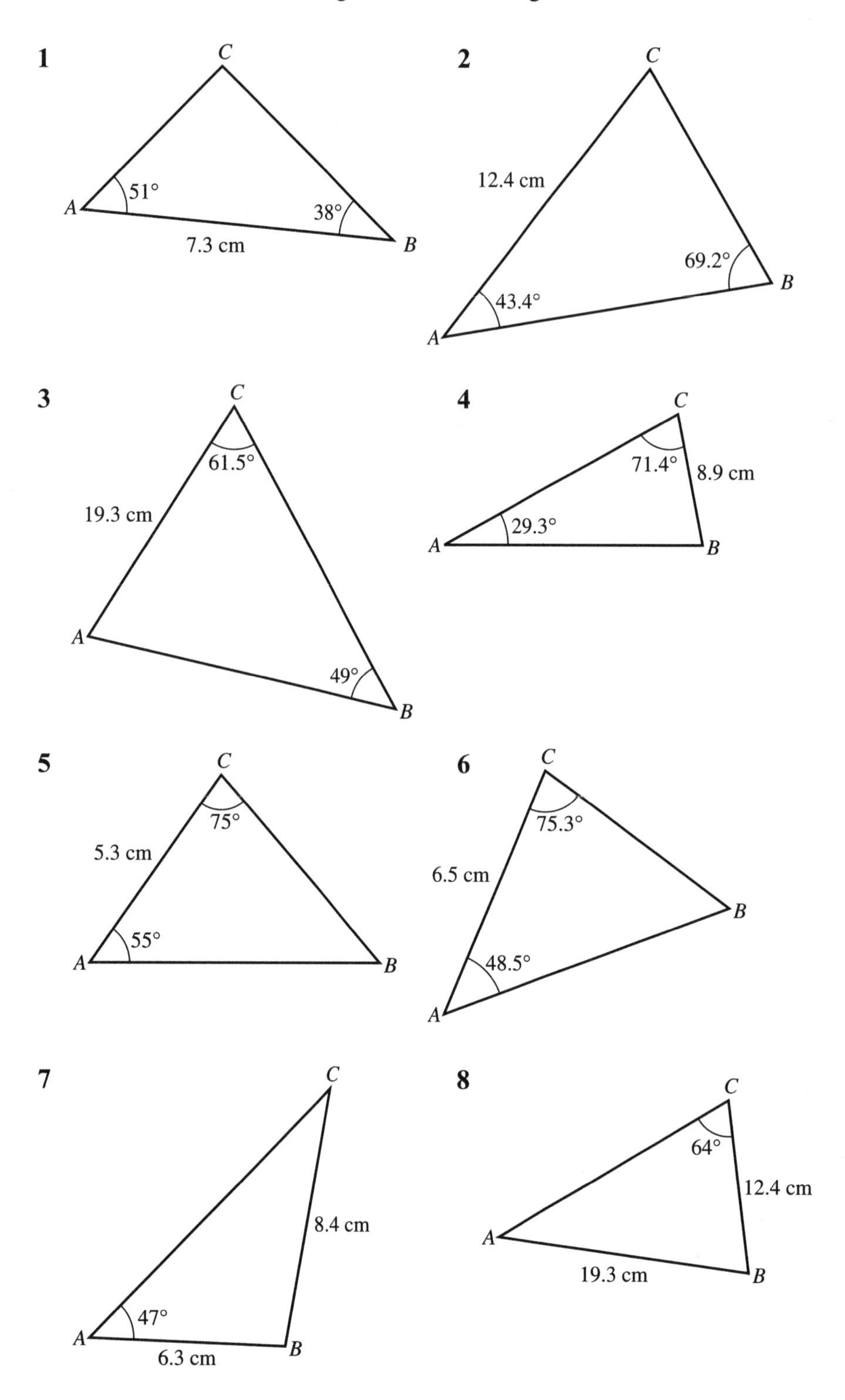

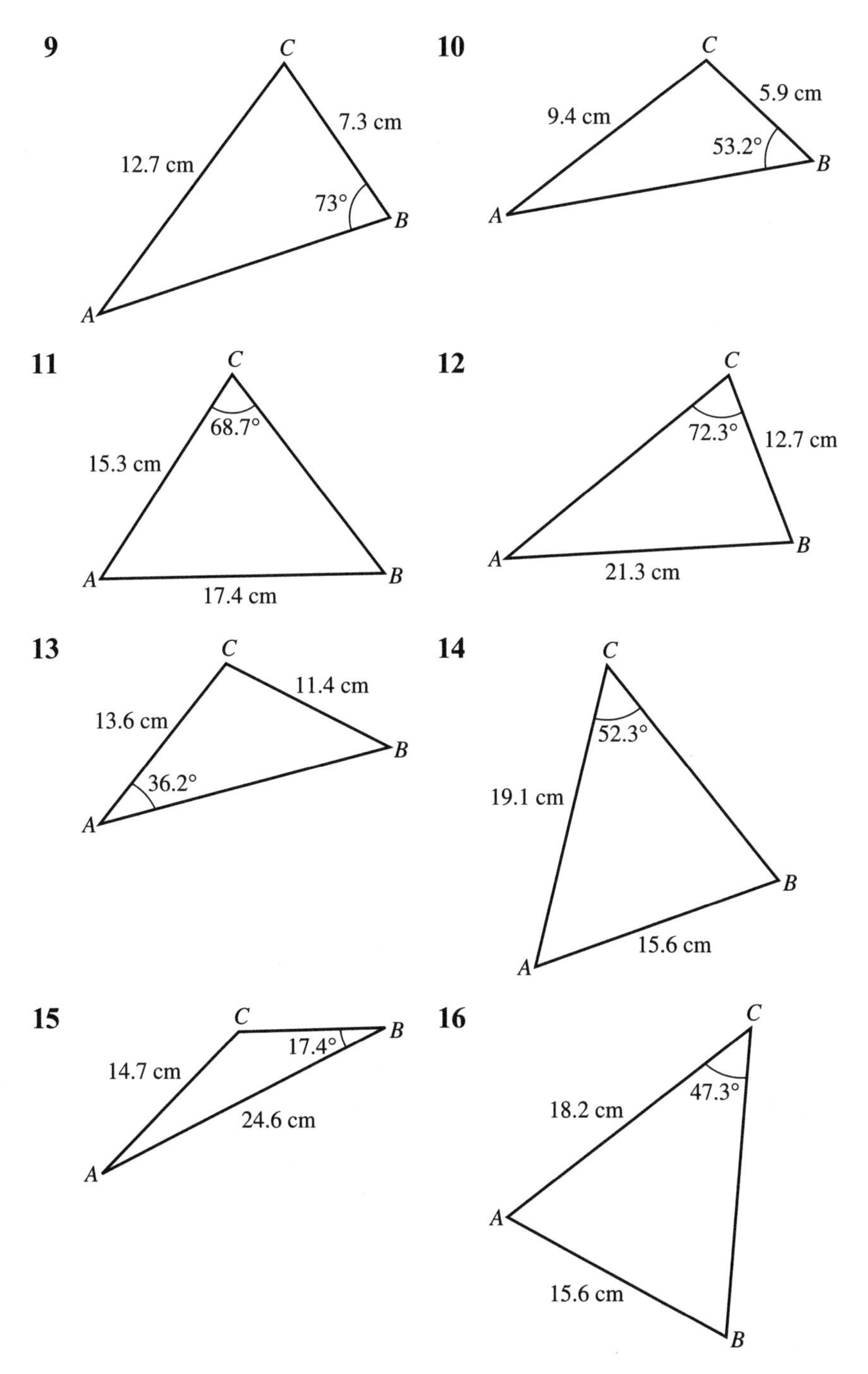

Use the cosine rule to calculate x in cm to 3 s.f. in the following triangles:

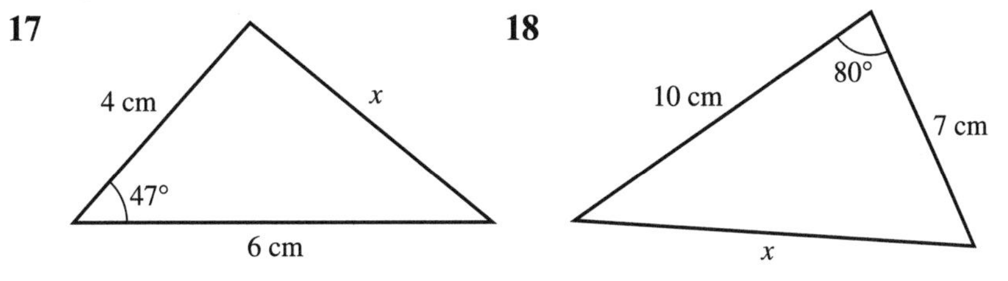

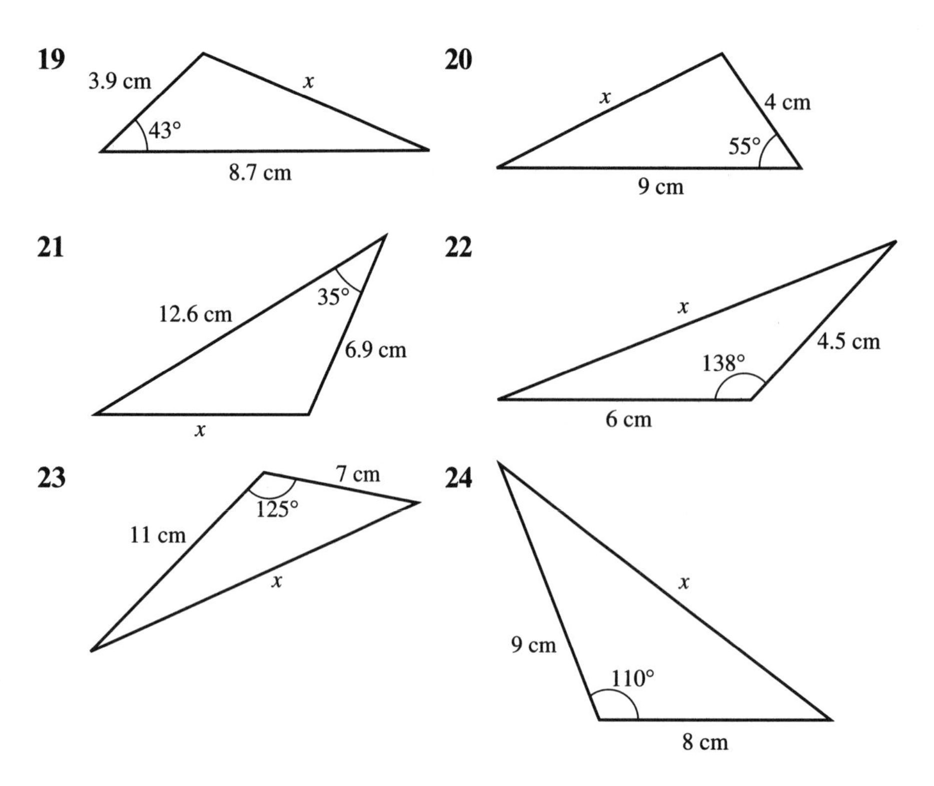

Use the cosine rule to calculate θ, to 0.1°, in the following triangles:

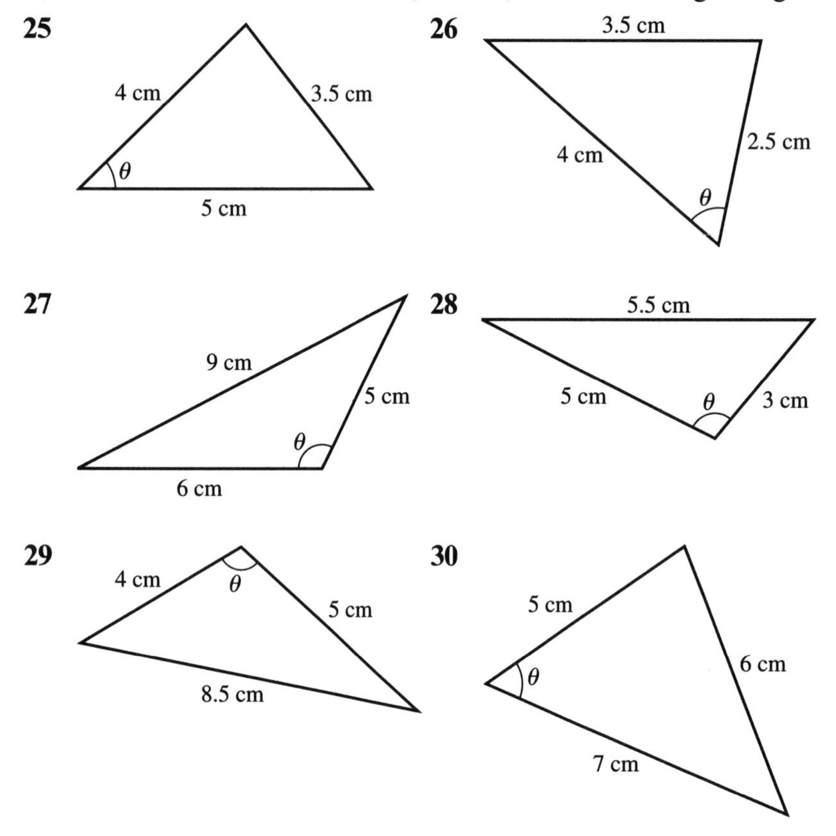

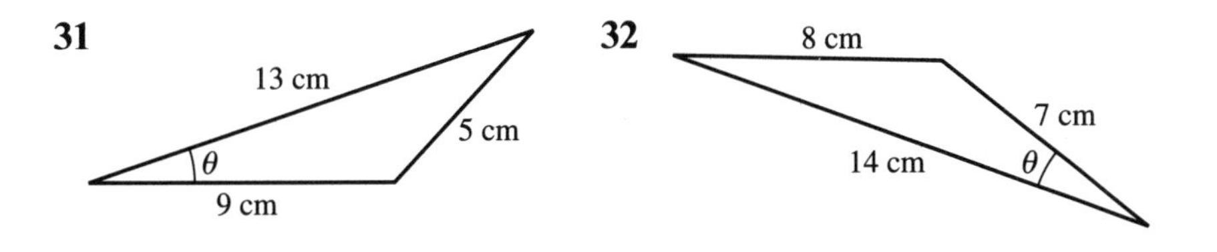

Solve the following triangles, giving lengths in cm to 3 significant figures and angles to 0.1°. Find also the area, in cm^2 to 3 significant figures, of each triangle.

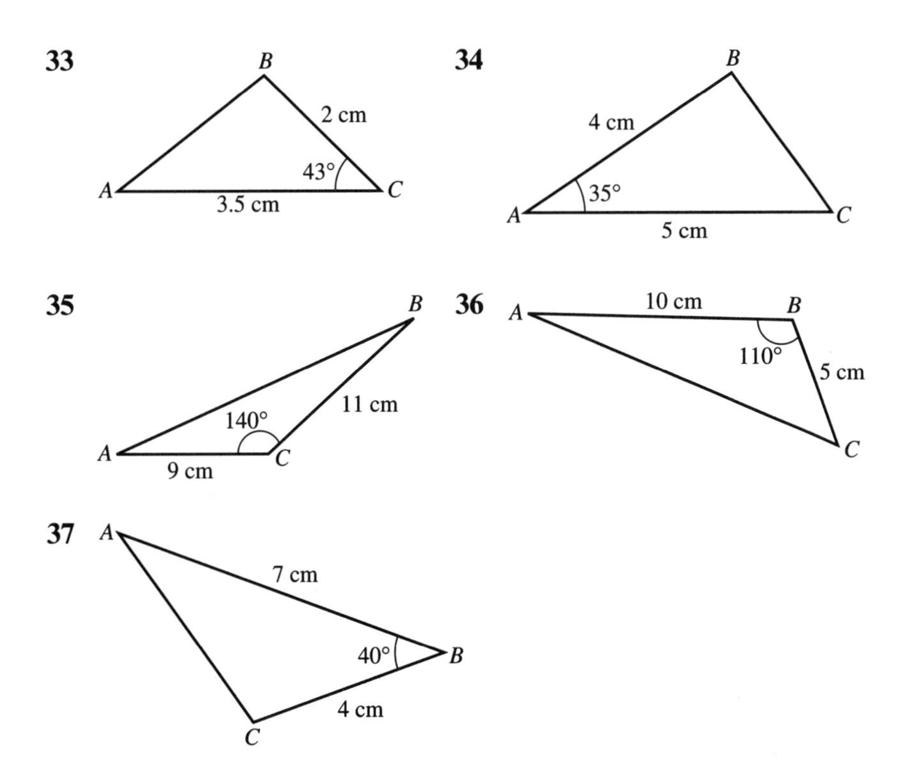

38 A ship sails 3 nautical miles from P to Q on a bearing 071° and then 4 nautical miles from Q to R on a bearing 292°. Calculate the distance PR and the bearing of R from P.

39 From a lighthouse A at a given time, two ships P and Q lie on bearings 300° and 020° respectively. Given that P is 6 km from the lighthouse and Q is 3.5 km from the lighthouse, calculate the distance between the ships at the given time.

40 Three points A, B and C lie on level ground and B is due south of A. The point C lies 350 m from A on a bearing 065° and the distance BC is 450 m. Calculate the bearing of B from C.

In each of the following C is the centre of the circle shown.

41 Find the lengths of

(a) PB (b) the radius of the circle (c) BQ

giving your answers in cm to 1 d.p.

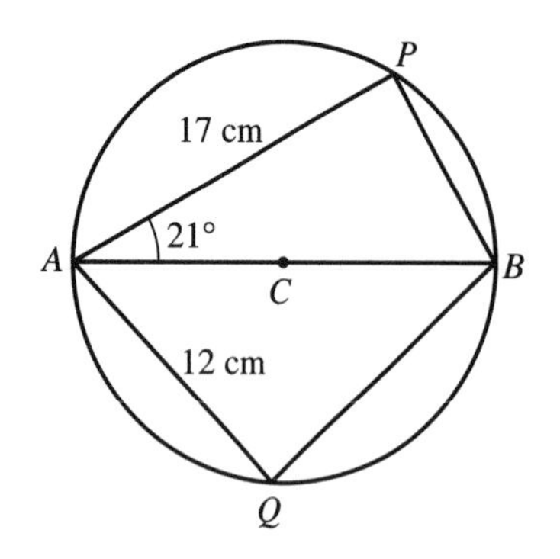

42 Calculate the area of the circle giving your answer in cm^2 to 1 d.p.

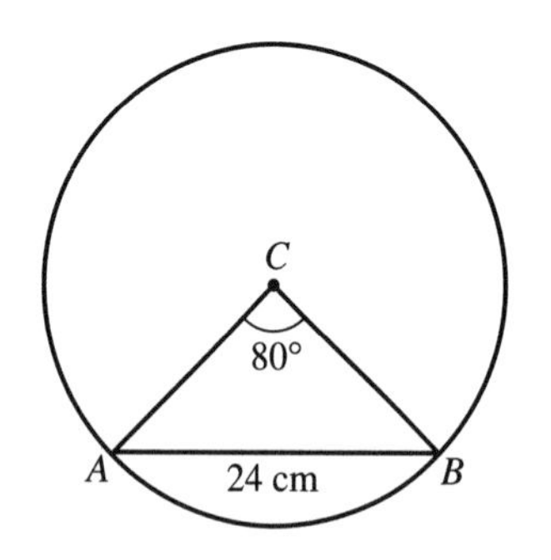

43 TB is a tangent to the circle. Calculate the lengths of

(a) TB (b) AB

giving your answers in cm to 1 d.p.

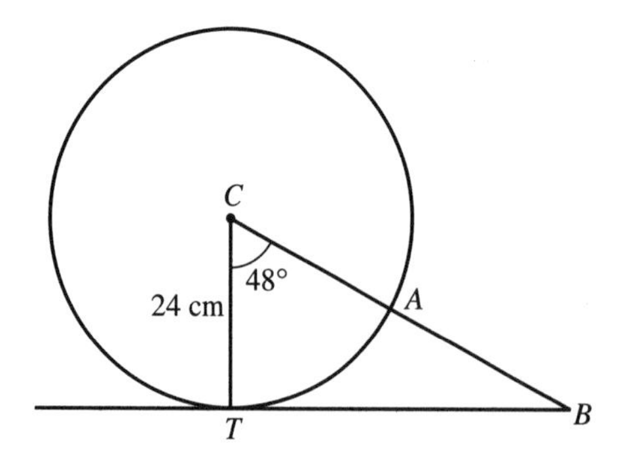

Examination style paper

P1

Use only a basic scientific calculator when answering this paper.

Answer all questions **Time 90 minutes**

Full marks (75) may be obtained by correct solutions to **all** questions.

1. Prove that the roots of the equation

$$x^2 + (k + 2)x + 2k = 0$$

are real for all values of k. **(4 marks)**

2. Find the pairs of values (x, y) which satisfy the simultaneous equations

$$3x + 2y = 0$$
$$x^2 + xy + y^2 = 7$$
(5 marks)

3.

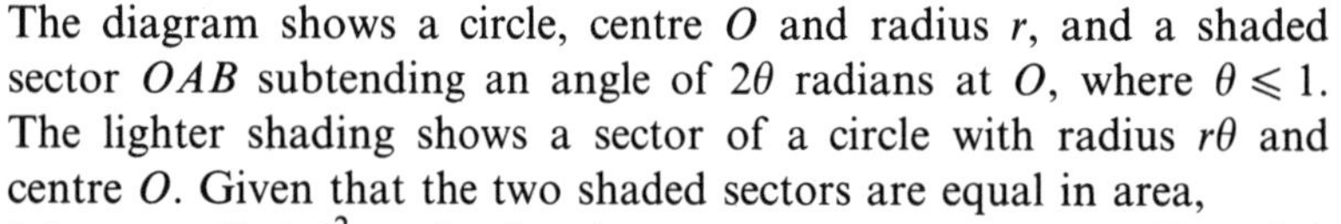

The diagram shows a circle, centre O and radius r, and a shaded sector OAB subtending an angle of 2θ radians at O, where $\theta \leqslant 1$. The lighter shading shows a sector of a circle with radius $r\theta$ and centre O. Given that the two shaded sectors are equal in area,

(*a*) prove that $\theta^2 - \pi\theta + 1 = 0$, **(4 marks)**

(*b*) find θ to 2 decimal places. **(4 marks)**

4. (*a*) Express the following angles in radians in terms of π:

(i) $60°$, (ii) $300°$ **(2 marks)**

(*b*) Find, in radians to 2 decimal places, the value of x for which

$$7 - 2\cos x = 8\sin^2 x, \quad 0 \leqslant x \leqslant 6$$
(8 marks)

5. Given that $f(x) = 2x^{-\frac{1}{2}} + 10x^{\frac{3}{2}}$, find

(*a*) $f'(9)$, **(5 marks)**

(*b*) $\int_1^4 f(x)\,dx$. **(6 marks)**

6. The points $A(-3, 3)$, $B(2, -1)$ and $C(5, 1)$ are given and O is the origin. Find, in the form $ax + by + c = 0$, an equation of the line

(*a*) l_1 passing through A and C, **(3 marks)**

(*b*) l_2 through B perpendicular to AC, **(3 marks)**

(*c*) l_3, through M, the mid-point of AB, and parallel to AC. **(3 marks)**

(*d*) Find the coordinates of the point where the lines l_3 and BC meet. **(3 marks)**

7. (*a*) An arithmetic series has first term 13 and second term 11. Find the sum of the first 35 terms. **(3 marks)**

(*b*) A geometric series has first term 13 and second term 11.

(i) Find the sum to infinity. **(3 marks)**

(ii) Find the least number of terms, n, such that the sum to n terms exceeds 84. **(6 marks)**

8. The volume of a closed right circular cylinder is $250\pi\,\text{cm}^3$. The radius of the cylinder is r cm and the height is h cm. The total surface area is $A\,\text{cm}^2$.

(*a*) Prove that $A = 2\pi\left(r^2 + \frac{250}{r}\right)$. **(4 marks)**

Given that r and h may vary, find

(*b*) the minimum value of A, verifying that the value you have found is, in fact, a minimum. **(9 marks)**

Answers

Exercise 1A

1	2	**2**	4	**3**	2	**4**	3
5	27	**6**	$\frac{1}{4}$	**7**	1	**8**	$\frac{1}{100}$
9	$-\frac{1}{343}$	**10**	$\frac{5}{2}$	**11**	$1\frac{1}{3}$	**12**	$\frac{3}{2}$
13	$\frac{4}{9}$	**14**	9	**15**	1	**16**	$\frac{6}{5}$
17	$1\frac{11}{25}$	**18**	$\frac{25}{4}$	**19**	$\frac{5}{4}$	**20**	$\frac{13}{9}$
21	$\frac{1}{25}$	**22**	6	**23**	$\frac{49}{64}$	**24**	$\frac{343}{64}$

Exercise 1B

1	$3\sqrt{3}$	**2**	$3\sqrt{5}$	**3**	$9\sqrt{2}$	**4**	$4\sqrt{3}$
5	$5\sqrt{3}$	**6**	$7\sqrt{3}$	**7**	$9\sqrt{7}$	**8**	$4\sqrt{7}$
9	$\sqrt{3}$	**10**	$\sqrt{2}$	**11**	3	**12**	3
13	$17\sqrt{3}$	**14**	$\sqrt{2}$	**15**	$-4\sqrt{5}$	**16**	$10\sqrt{6}$
17	$4\sqrt{7}$	**18**	$\frac{\sqrt{2}}{2}$	**19**	$\frac{\sqrt{7}}{7}$	**20**	$\frac{7\sqrt{5}}{5}$
21	$\frac{\sqrt{6}}{9}$	**22**	$\frac{1}{2}$	**23**	$\frac{1}{3}$	**24**	$\frac{\sqrt{7}}{7}$

25 $\frac{\sqrt{3}}{6}$ **26** $-\sqrt{2}-1$ **27** $\frac{\sqrt{3}+1}{2}$

28 $\frac{\sqrt{5}+1}{2}$ **29** $\frac{\sqrt{7}+4}{3}$ **30** $\frac{3+\sqrt{7}}{2}$

31 $-\frac{4+\sqrt{11}}{5}$ **32** $\frac{\sqrt{7}+\sqrt{3}}{4}$

33 $\frac{1}{3}(\sqrt{11}+\sqrt{5})$ **34** $\frac{7}{10}(\sqrt{13}+\sqrt{3})$

35 $\frac{7}{3}(\sqrt{7}-2)$

36 $\frac{4\sqrt{3}-8-2\sqrt{11}+\sqrt{33}}{5}$

37 $28+7\sqrt{14}$ **38** $4-\sqrt{15}$

39 $\frac{1}{3}(10-\sqrt{91})$ **40** $\frac{1}{10}(33-\sqrt{989})$

Exercise 1C

1 $11x^7+2x^6+5x^5+3x^2+2x$

2 $3x^5+4x^4-2x^3+x^2+6x+2$

3 $2x^3+2x^2-8$

4 $6x^4+5x^3-7x^2+10x$

5 $5x^6+3x^5-2x^4+x^3+7x+7$

6 $-3x^4+8x^3-x^2+9x-3$

7 $5x^7-6x^6+5x^4-4x^3-2x-5$

8 $-2x^3+3x^2+10x-7$

9 x^5+2x^3+7x-9 **10** $3x^3-7x^2+7x+1$

11 $-2x^5+9x^3-9x^2+3x-1$

12 $-3x^5+4x^3+9x-9$

13 $6x^4-3x^3+10x^2+4x+2$

14 $3x^3+5x-2$ **15** $6x^5-3x^3-7x+3$

16 $7x^6+3x^3+7x-10$

17 $8x^4-6x^3-2x^2+8x+15$

18 $12x^5+8x^4+37x^3+32x^2+40x+24$

19 $10x^7-26x^5-17x^4-12x^3+3x$

20 $24x^4+8x^3+20x^2+15x-7$

21 $6x^5-9x^4-x^3-6x^2+16x-6$

22 $16x^8-28x^7-66x^6+66x^5-6x^4-23x^3$
$-6x^2+81x-42$

23 $2x^7-4x^6+14x^5-13x^4+20x^3-61x^2+63x$
-18

24 $24x^7-16x^6-41x^5+6x^4-9x^3+23x^2+22x$
-12

25 $6x^7-2x^6-29x^5+24x^4-16x^3+4x^2+35x$

Exercise 1D

1	$x^2(2-3x)$	**2**	$y(5y^4+2y+7)$
3	$2x(x+9)$	**4**	$8(5y-1)$
5	$6x(y+3)$	**6**	$5xy(y^2+3x)$
7	$2(x-3y+5xy)$	**8**	$x(x^2+3x-6)$
9	$3pq(2q+3p)$	**10**	$2c(c+3d)$
11	$(x-6)(x+6)$	**12**	$(y-9)(y+9)$
13	$(10-b)(10+b)$	**14**	$2(2a-3b)(2a+3b)$
15	$2(3-5b)(3+5b)$	**16**	$a^2(1-b)(1+b)$

17 $2(x-3)(x+3)$ **18** $3(x-3)(x+3)$
19 $17(a-2)(a+2)$ **20** $3(4-7c)(4+7c)$
21 $(x+3)^2$ **22** $(x-4)^2$
23 $(x+7)^2$ **24** $(x-9)^2$
25 $3(x-6)^2$ **26** $(3x+5)^2$
27 $(2x-7)^2$ **28** $5(3x-1)^2$
29 $2(3x^2+6x+1)$ **30** $(5x-3)^2$
31 $(5x+2)^2$ **32** $(3x-7)^2$ **33** $(x+2)^3$
34 $(x-3)^3$ **35** $(2x+1)^3$ **36** $(3x-1)^3$
37 $(2x+3)^3$ **38** $(2x-5)^3$ **39** $y(x+2)^3$
40 $3(x-3)^3$ **41** $(x+2)(x+7)$
42 $(x-5)(x-3)$ **43** $(x-4)(x+3)$
44 $(x+12)(x+1)$ **45** $(x-10)(x+1)$
46 $(x-2)(x-1)$ **47** $(x+7)(x+8)$
48 $(x+6)(x-2)$ **49** $(x+5)(x+2)$
50 $(x-8)(x+5)$ **51** $(x-4)(x-3)$
52 $(x-9)(x-5)$ **53** $(x+9)(x+2)$
54 $(x-3)(x-7)$ **55** $(x+5)(x+4)$
56 $(x-2)(x-15)$ **57** $(x-10)(x-12)$
58 $(x+15)(x-12)$ **59** $(x-11)(x+10)$
60 $(x+31)(x-20)$ **61** $(3x-1)(x-1)$
62 $(2x+1)(x-1)$ **63** $(4x+1)(2x+1)$
64 $(4x-1)(x+1)$ **65** $(7x-1)(2x+1)$
66 $(7x+1)(2x-1)$ **67** $(6x+1)(3x-1)$
68 $(6x+1)(3x+1)$ **69** $2(x-5)(x-7)$
70 $3(x-8)(x+6)$ **71** $(2x+3)(x+2)$
72 $(2x-3)(x+3)$ **73** $(3x+1)(x+2)$
74 $(5x+1)(x-3)$ **75** $(3y+7)(y+2)$
76 $(2y+1)(y-5)$ **77** $(7y-2)(y-4)$
78 $(5y-2)(y-5)$ **79** $(3y+5)(y-5)$
80 $(3y-2)(y+2)$ **81** $(2x+3)(2x+1)$
82 $(3x-2)(2x+1)$ **83** $(4x-3)(2x+5)$
84 $(5x-4)(2x-3)$ **85** $(4x-3)(3x+1)$
86 $(5x-7)(2x+5)$ **87** $(4x-1)(3x-2)$
88 $(3x+2)(3x+5)$ **89** $(2x+7)(2x-3)$
90 $(7x+3)(5x-2)$ **91** $x(x-7)(x+3)$
92 $x(2-x)(3-2x)$ **93** $x^3(5x+2)(2x+1)$
94 $4(3+5x)(1-2x)$ **95** $y(3x-1)(2x+7)$
96 $(3x^2+1)(x^2+1)$ **97** $(3y+2)(2y-3)$
98 $(3y+2)(2y+3)$ **99** $(7x+4)(4x-5)$
100 $(2x-1)(2x+1)(x-1)(x+1)$

Exercise 1E

1 (i) b (ii) b (iii) a (iv) c (v) c
2 $A=1, B=3$ **3** $A=2, B=-1$
4 $A=B=3$ **5** $A=5, B=-4$
6 $A=10, B=-2$ **7** $A=1, B=2, C=3$
8 $A=C=1, B=-2$
9 $A=1, B=2, C=-3$
10 $A=2, B=4, C=-3$
11 $A=10, B=3, C=-2$
12 $A=1, B=0, C=-1$
13 $A=2, B=12, C=-3$
14 $A=1, B=0, C=-1$
15 $A=\frac{1}{3}, B=-\frac{1}{3}, C=1$
16 $A=3, B=1, C=-5$
17 $A=2, B=-3, C=-4$
18 $A=3, B=2, C=0$
19 $A=4, B=2, C=-1$
20 $A=5, B=-2, C=-1$
21 $A=1, B=-2, C=-2$
22 $A=4, B=-\frac{3}{2}, C=16$; Minimum value $=16$
23 $A=3, B=3, C=-32$; Minimum value $=-32$
24 $A=20, B=2, C=1$; Maximum value $=20$
25 $A=25, B=5$

Exercise 1F

1 $2x^2+x-1$ **2** x^2+2x+1
3 $2x^2-x+7$ **4** $-3x^3+2x^2-7x+5$
5 $6x^2-7x+5$ **6** $2x^3+4x^2-2x$
7 $2x^2+x-1$ remainder 2
8 $5x^3-2x^2+3x-1$ remainder 6
9 $-3x^3+2x^2+11x-20$ remainder -10
10 $4x^3-2x^2+x-6$ remainder 4
11 $3x^3-2x^2+x+1$ **12** $-x^2+6x+5$
13 $-2x^2-3x+1$ **14** x^3+3x^2+5x-7
15 $2x^3+4x^2-7x+6$
16 $3x^2-4x+6$ remainder $2x-1$
17 x^3-x^2+2x+6 remainder $-2x+7$
18 $-2x^3+x^2+6$ remainder $-3x-2$

19 $2x^2 - 4x + 3$ remainder $-4x - 4$

20 $3x^3 - 2x + 1$ remainder $2x - 2$

Exercise 1G

1 1, 2 or 3; $(x-1)(x-2)(x-3)$

2 $-2, -1$ or 1; $(x-1)(x+1)(x+2)$

3 $-2, -1$ or 2; $(x+2)(x+1)(x-2)$

4 $-3, -2$ or 2; $(x+3)(x+2)(x-2)$

5 $-2, -1$ or 3; $(x+2)(x+1)(x-3)$

6 $(2x-1)(x^2+x+1)$

7 x^2+1

8 $(x-3)(2x-1)(x+2)$

9 $(2x-3)(2x+1)(x-4)$

10 $(3x-2)(x^2-2x+7)$

Exercise 1H

1 16, 4 **2** 25, 5

3 $\frac{25}{4}, \frac{5}{2}$ **4** 1, 1

5 12, 2 **6** 7, 1

7 $4(x-2)^2 - 15$ **8** $-3(x-\frac{5}{6})^2 + \frac{25}{12}$

9 $2(x+\frac{3}{4})^2 + \frac{23}{8}$ **10** $-(x+\frac{1}{2})^2 + \frac{5}{4}$

11 $4(x-\frac{3}{8})^2 - \frac{89}{16}$ **12** $5(x+\frac{9}{10})^2 - \frac{141}{20}$

21 $\frac{16}{3}; -\frac{1}{3}$

22 Other factors are 3 and $x+1$.

Exercise 1I

1 ± 7 **2** ± 2.5 **3** ± 12 **4** $0, -5$

5 0, 7 **6** $2, -1$ **7** $-3, 2$ **8** $-2, -5$

9 4, 5 **10** $-3, -3$ **11** $-4, -8$

12 $10, -1$ **13** $4, -3$ **14** $-6, 2$

15 $12, -1$ **16** $-1, -12$ **17** $2, -\frac{1}{2}$

18 $1, -\frac{1}{2}$ **19** $-2, -\frac{1}{2}$ **20** $2, \frac{1}{3}$

21 $\frac{2}{3}, -7$ **22** $5, -\frac{2}{3}$ **23** $\frac{1}{4}, \frac{1}{3}$

24 $\frac{3}{2}, -\frac{2}{3}$ **25** $\frac{5}{4}, \frac{3}{2}$ **26** $\frac{9}{4}, -\frac{3}{7}$

27 $-\frac{2}{3}, -\frac{4}{5}$ **28** $\frac{9}{5}, -\frac{1}{4}$

29 $2.41, -0.41$ **30** $-7.16, -0.84$

31 $4.19, -1.19$ **32** $0.66, -0.46$

33 $-0.23, -0.63$ **34** $1.18, -0.43$

35 1.43, 0.23 **36** $0.53, -1.13$

37 $1.72, -0.39$ **38** 2.65, 0.85

Exercise 1J

1 $(2, -1), (-\frac{32}{17}, \frac{49}{17})$ **2** $(0, 1), (\frac{4}{5}, -\frac{3}{5})$

3 $(-2, 0), (\frac{1}{3}, \frac{7}{3})$ **4** $(4, -2), (-2, -4)$

5 $(6, 3), (4\frac{1}{2}, 4\frac{1}{2})$ **6** $(2, 1), (-2, -1)$

7 $(3, 12), (1, 4)$ **8** $(8, +5), (-5, -8)$

9 $(1, -2), (-2\frac{3}{7}, -\frac{2}{7})$

10 $(-1, 3), (-2, 2)$

Exercise 1K

1 $x > 1$ **2** $x < 3$ **3** $x < -2$ **4** $x < -2$

5 $x \geqslant 2$ **6** $x \leqslant 3$ **7** $x < 6\frac{1}{2}$

8 $x \leqslant -2\frac{1}{3}$ **9** $x < \frac{1}{4}$ **10** $x \geqslant 3$

11 $x > 3$ **12** $x > 5$ **13** $x \geqslant 4$ **14** $x < 7$

15 $x \leqslant 5$ **16** $x \geqslant 3\frac{1}{2}$ **17** $x < -3$

18 $x \leqslant -\frac{29}{7}$ **19** $x < \frac{5}{3}$ **20** $x \geqslant \frac{5}{8}$

21 $x \geqslant -2, x \leqslant -5$ **22** $2 \leqslant x \leqslant 3$

23 $x > 3, x < -4$ **24** $3 < x < 6$

25 $-2 \leqslant x \leqslant 9$ **26** $x < -7, x > -4$

27 $x < \frac{3}{2}, x > 4$ **28** $x < -2, x > \frac{4}{3}$

29 $-\frac{2}{3} < x < 7$ **30** $x < 3, x > \frac{7}{2}$

31 $-4 \leqslant x \leqslant -\frac{5}{2}$ **32** $x < -1, x > \frac{9}{5}$

33 $x \leqslant -9, x \geqslant -\frac{7}{3}$ **34** $\frac{3}{4} < x < 5$

35 $x < \frac{5}{3}, x > 6$ **36** $-\frac{2}{3} \leqslant x \leqslant 2$

37 $-2 < x < \frac{3}{2}$ **38** $-\frac{1}{4} \leqslant x \leqslant \frac{2}{3}$

39 $x < -\frac{2}{3}, x > \frac{3}{7}$ **40** $-\frac{4}{3} < x < \frac{5}{2}$

Exercise 2A

1 (a) 180° (b) 90° (c) 30°
(d) 45° (e) 270° (f) 135°
(g) 720° (h) 540° (i) 225°
(j) 150°

2 (a) $\frac{\pi}{8}$ (b) $\frac{\pi}{12}$ (c) π (d) $\frac{7\pi}{6}$
(e) $\frac{2\pi}{3}$ (f) $\frac{3\pi}{4}$ (g) $\frac{5\pi}{4}$ (h) $\frac{11\pi}{6}$
(i) $\frac{5\pi}{3}$ (j) $\frac{7\pi}{4}$

3 (a) 57° (b) 138° (c) 286°
(d) 201° (e) 23° (f) 97°
(g) 246° (h) 315° (i) 229°
(j) 344°

4 (a) 0.35^c (b) 0.87^c (c) 1.22^c (d) 2.27^c
(e) 2.97^c (f) 4.01^c (g) 4.36^c (h) 1.48^c
(i) 0.66^c (j) 2.65^c

5 (a) 3 cm, 4.5 cm^2 (b) 14 cm, 49 cm^2
(c) 12 cm, 48 cm^2 (d) 28.5 cm, 135.4 cm^2
(e) 2.75 cm, 7.56 cm^2
(f) 25.2 cm, 113.4 cm^2
(g) 49.5 cm, 272.25 cm^2
(h) 47.5 cm, 225.6 cm^2
(i) 52.64 cm, 294.8 cm^2
(j) 50.63 cm, 210.1 cm^2

6 2.5^c **7** 3.6 cm **8** 1.25^c **9** 8.45 cm

10 (a) 3 cm^2 (b) 0.9375 cm^2
(c) 2.1875 cm^2 (d) 5.6875 cm^2
(e) 6.336 cm^2

Exercise 2B

1 (a) $-\sin 80°$ (b) $-\cos 40°$
(c) $\tan 5°$ (d) $-\tan 5°$
(e) $-\cos 43°$ (f) $\sin 54°$
(g) $\sin 14°$ (h) $\cos 24°$
(i) $\tan 36°$ (j) $-\tan 32°$
(k) $-\sin 83°$ (l) $-\cos 68°$
(m) $-\sin 29°$ (n) $\cos 67°$
(o) $\tan 84°$ (p) $\tan 36°$
(q) $-\sin 21°$ (r) $-\cos 59°$
(s) $-\tan 53°$ (t) $\sin 18°$

2 (a) $\frac{\sqrt{3}}{2}$ (b) $-\frac{\sqrt{3}}{2}$ (c) 1
(d) $\frac{1}{2}$ (e) $-\frac{1}{2}$ (f) $-\frac{1}{2}$
(g) $-\frac{\sqrt{3}}{2}$ (h) $\frac{\sqrt{3}}{2}$ (i) $\frac{1}{\sqrt{2}}$
(j) $-\frac{1}{\sqrt{2}}$ (k) $\frac{1}{\sqrt{3}}$ (l) $-\frac{1}{\sqrt{3}}$
(m) $-\frac{1}{2}$ (n) $-\frac{\sqrt{3}}{2}$ (o) $\sqrt{3}$
(p) $\frac{1}{\sqrt{3}}$ (q) $\frac{1}{2}$ (r) $\frac{1}{2}$
(s) $\frac{1}{\sqrt{2}}$ (t) $-\frac{1}{\sqrt{3}}$

3 (a) 66°, 204° (b) 53°, 307°
(c) 72.5°, 287.5°

4 missing values 2.87, 2.83; $38 < x < 68$

5 −0.62, −1, −0.62, 0.25, 1; 38.2°, 141.8°

8 $P(\frac{\pi}{2}, 1)$, $Q(-\frac{\pi}{2}, -1)$

9 1.05, 4.19, 7.33; 2.09, 5.24, 8.38

10 max (149°, 5.83); min (−31°, −5.83)

Exercise 2C

1 (a) 22.8°, 157.2°, 202.8°, 337.2°
(b) 67.2°, 292.8°, 112.8°, 247.2°
(c) 36.3°, 323.7°, 143.8°, 216.3°

2 (a) 91° (b) 215°

3 19.4°, 160.6° **4** 132.2°, 312.2°

5 138.5°, 221.5° **6** 21.0°, 201.0°

7 20.4°, 299.6° **8** 76.4°, 256.4°

9 14.2°, 75.8°, 194.2°, 255.8°

10 19.6°, 100.4°, 139.6°, 220.4°, 259.6°, 340.4°

11 29.7°, 80.3°, 260.3°, 209.7°

12 50.7°, 269.3° **13** 31.4°, 127.0°

14 23.8°, 83.8°, 143.8°, 203.8°, 263.8°, 323.8°

15 30°, 150°

16 90°, 270°, 48.6°, 131.4°

17 14.5°, 165.5°, 30°, 150°

18 116.6°, 135°, 296.6°, 315°

19 90°

20 50.2°, 230.2°

21 26.6°, 206.6°

22 30°, 60°, 150°, 300°

23 19.5°, 123.7°, 160.5°, 303.7°

24 23.9°, 83.9°, 143.9°, 203.9°, 263.9°, 323.9°

25 14.5°, 30°, 150°, 165.5°

26 0.99, 2.15

27 (a) 135° (b) 105°, 165°

28 $\frac{\pi}{6}, \frac{5\pi}{6}, \frac{7\pi}{6}, \frac{11\pi}{6}$

29 $\frac{\pi}{3}, \frac{2\pi}{3}, \frac{4\pi}{3}, \frac{5\pi}{3}$

30 0, 1.11, 3.14, 4.25, 6.28

31 1.57, 3.48, 5.94

32 0.84, 1.57, 4.71, 5.44

33 $\frac{\pi}{4}, \frac{3\pi}{4}, \frac{5\pi}{4}, \frac{7\pi}{4}$

34 $\frac{3\pi}{2}$

35 0.52, 5.76

Exercise 3A

1 $2x - y - 9 = 0$ **2** $3x + y - 27 = 0$

3 $2x - 4y + 21 = 0$ **4** $4x + 5y - 2 = 0$

5 $6x - 2y - 7 = 0$ **6** $7x + y - 27 = 0$
7 $2x - 3y - 1 = 0$ **8** $3x - y - 2 = 0$
9 $x + y - 1 = 0$ **10** $x + 3y - 11 = 0$
11 $13x + y + 20 = 0$ **12** $6x - y - 26 = 0$
13 $3x - 2y - 18 = 0$ **14** $2x + y + 7 = 0$
15 (3, 0); (0, 2) **16** (−5, 0); (0, 3)
17 $(-\frac{7}{2}, 0); (0, -7)$ **18** $(\frac{9}{4}, 0); (0, -3)$
19 $(-\frac{22}{5}, 0); (0, 2)$ **20** $(-\frac{8}{3}, 0); (0, 4)$
21 2 **22** $\frac{1}{3}$ **23** $\frac{4}{3}$
24 $-\frac{5}{9}$ **25** $\frac{1}{3}$ **26** $-\frac{17}{13}$
27 $x - 3y - 14 = 0$ **28** $6x - 3y - 20 = 0$
30 (a) $x - y - 1 = 0, x + y + 1 = 0$
(b) 4 units^2

Exercise 3B

1 $(6, 5\frac{1}{2})$ **2** $(4\frac{1}{2}, -4)$ **3** (−6, −4)
4 (1, 4) **5** (2, 3) **6** (−3.1, −1.5)
7 $4x - 2y = 13$ **8** $6x + 4y = 11$
9 $x + 2y + 14 = 0$ **10** $x - 1 = 0$
11 $55x - 17y = 59$ **12** $70x - 30y + 172 = 0$
13 $x - y - 8 = 0$ **14** $-\frac{20}{3}$
15 50
16 (a) $4x - y - 5 = 0$ (b) $6x + y + 3 = 0$
(c) $3x + 2y + 4 = 0$
17 (a) $x + 3y - 22 = 0$ (b) $x - 5y - 28 = 0$
(c) $3x - 2y - 4 = 0$
19 $\frac{3}{2}$ or −5
20 (0, 2), (7, 1); 25 units^2
21 (a) $(3\frac{1}{2}, 2\frac{1}{2})$ (b) (3, 6) (c) 20 units^2
22 (a) $2x + y = 6$ (b) $(\frac{19}{5}, -\frac{8}{5})$

Review exercise 1

1 (1, −1), (2, −4) **2** $\frac{1}{4}$
3 19.5, 160.5, 270
4 $5\sqrt{10}$
5 (a) 3.89, −0.39 (b) 14
7 (3, 11)
8 $p = -38$, $(x + 4)(3x - 2)(x - 3)$
9 (a) $x \leqslant 14$ (b) $-4 < x < 4$
10 3, −3 **11** (3, −2)
12 (a) (1, 3), (−4, −2); $5\sqrt{2}$
13 (a) $3(x - 1)(x + 1)$; $\dfrac{x - 4}{3(x + 1)}$
(b) 3.08, −1.08
14 (a) 20π cm (b) $100(\pi - 1)$ cm
(c) 45.5 cm^2
15 $\frac{3}{4}$
16 (a) $y = 3x - 7$
(b) $y = 13 - 2x$
(c) (4, 5)
17 (a) 1.7, −0.34
(b) $-12 + 14x + 19x^2 - 15x^3$
18 $\dfrac{\sqrt{5}(\sqrt{19} - \sqrt{3})}{80}$
19 $\frac{1}{3}$
20 (a) 1.90, −1.23
(b) −13, −43, 27
21 (a) 44.4°, 135.6°
(b) ±45.6°
(c) 35°, −145°
22 $4x + 3y = 29$
23 $x < -3$ or $x > 4$
24 (a) 5.85, −0.85
(b) $x < -1$ or $x > 6$
25 $k = -4$, max $y = 6$
26 5 units^2
27 (a) $\dfrac{\pi}{18}, \dfrac{5\pi}{18}, \dfrac{13\pi}{18}, \dfrac{17\pi}{18}$
(b) $\dfrac{3\pi}{4}$
28 $\dfrac{q^2 + pq}{p^2 - p^4 q}$
29 (a) 3, 2, −7 (b) −7
(c) −0.5, −3.5
30 $\left(135°, \frac{\sqrt{2}}{2}\right), \left(315°, -\frac{\sqrt{2}}{2}\right)$
31 (a) 0.719, 2.78
(b) $-6x^3 + 13x^2 - 9x + 2$
32 $k = \pm 3$
$3(x + 1)^2(x - 3); -3(x - 1)^2(x + 3)$
33 (a) $\sqrt{3}$ (b) $y = x\sqrt{3} + 2\sqrt{3}$
(d) 6 (e) 60°
34 (a) 15.6 cm^2 (b) 25.4 cm^2

35 $A(0, 3)$, $B(45°, 4.1)$, $C(135°, 2.1)$
(b) 75°, 195°, 255°

36 $2 + 2\sqrt{2}$, $2 - \sqrt{2}$

37 $-\frac{1}{2} < x < \frac{3}{2}$

38 (0, 1), (−2, −3)

39 (a) $1 \pm \sqrt{13}$
$x > 1 + \sqrt{13}$ or $x < 1 - \sqrt{13}$

40 $\left(-\frac{\pi}{6}, 1\right)$ or $\left(\frac{\pi}{2}, -\frac{1}{2}\right)$

41 −145°, 35°, 215°

42 (a) $\frac{4}{3}x + \frac{2}{3}$ (b) $(-\frac{5}{3}, -\frac{14}{9})$

43 (a) 45.5 cm (b) 106 cm^2

44 $k = -13$, $p = -35$, $q = 27$

45 (a) −5, 4 (b) $x < -5$ or $x > 4$

46 (a) $2y = 3x - 3$ (b) (3, 3)

47 0, 120, 300, 360

48 (a) $y = \frac{9}{2}x + \frac{3}{2}$ (b) $-\frac{1}{9}$

49 (a) $4x - 3y = 6$ (b) 6

50 (a) 10°, 50° (b) 0, 45°

51 $p = -5$, $q = 8$; $(x + 1)(x - 2)(x - 4)$

52 (a) $6y^{-1}$ (b) −216, 1

53 (2, 1)

54 (a) (i) $2 - \sqrt{3}$ (ii) $-2 - \sqrt{3}$
(b) 0.73, 2.41, 4.71

55 (a) $9x + 13y + 19 = 0$

56 (a) 8.62 cm^2 (b) 7.90 cm

57 0, 2

58 $a = 3$, $b = 1$, $(x + 3)(x - 1)(x + 1)$

59 9, 4, −16; $-\frac{8}{3} > x > -\frac{16}{3}$

60 (a) 3.67, 5.76 (b) 0.64, 5.64
(c) 1.25, 4.39

61 (a) $7x + 5y - 18 = 0$
(b) $\frac{162}{35}$

62 (a) 15°, 75°, −105°, −165°
(b) ±30°, ±90°, ±150°

63 $\dfrac{\sqrt{17} + \sqrt{13}}{4}$

64 $(\frac{1}{3}, 12)$, $(8, \frac{1}{2})$

65 $\frac{1}{16}$, 64

66 $3 < x < 4$

67 $2x^3 + 9x^2 + x - 12$

68 (a) $2 - 5x + 2x^2 + x^3$
(b) 1, −17

69 (a) $\frac{11}{12}\pi$, $\frac{23}{12}\pi$
(b) $\frac{2}{3}\pi$, $\frac{5}{6}\pi$, $\frac{5}{3}\pi$, $\frac{11}{6}\pi$

70 (a) 90 (b) 120, 300 (c) 1, 5
(d) Turning points at (30, 5), (120, 1) (210, 5), (300, 1)

71 $\dfrac{66 - 15\sqrt{11}}{19}$

72 $-\frac{1}{2} \pm \sqrt{2}$

73 (i) (a) 3 (b) $\frac{1}{9}$ (c) 1 (ii) −1

74 $(-\frac{1}{2}, 0)$, (−1, −1)

75 (a) $-\frac{3}{8}\pi$, $\frac{1}{8}\pi$ (b) 0.730
(c) −0.137

76 $-1 < x < 4$

77 $\dfrac{2\sqrt{7} - 5}{3}$

78 (a) $-\frac{1}{2}$, $-2\frac{1}{2}$ (b) 1.2, −4.7

79 (a) 8.75 cm^2 (b) 0.70 cm^2

80 (a) $\sqrt{884}$
(b) $11x - 10y + 19 = 0$

81 (a) (0, 1)
(b) $\dfrac{17\pi}{24}, \dfrac{23\pi}{24}, \dfrac{41\pi}{24}, \dfrac{47\pi}{24}$

82 (7, −3), $(5, -\frac{5}{3})$

83 {2, 3, 4}

84 (a) −151.8, −61.8, 28.2, 118.2
(b) 30, 150, 289.5, 340.5

85 (a) (i) 3 (ii) $(x + 3)(3x + 2)(x - 2)$
(b) $(4, \frac{1}{3})$, (−4, −5)

86 $-5 < x < \frac{7}{4}$

87 $\dfrac{7\sqrt{5} - 32}{19}$

88 $\frac{1}{2}$, $\frac{3}{2}$

89 (a) 2.54, −3.54 (b) $x^3 - 3x - 2$

90 (7, 3), (−7, −4)

91 (a) $2(p + 3)(p - q)$ (b) 0.88, −1.88

92 $-1 < x < \frac{1}{2}$

93 (a) (4, 3) (c) $x - y - 1 = 0$
(d) $(x - 4)^2 + (y - 3)^2 = 8$
(e) (6, 5), (2, 1) (f) $x + y = 11$

94 (b) $A(0, 1)$, $B\left(45, \frac{\sqrt{2}}{2}\right)$, $C\left(225, -\frac{\sqrt{2}}{2}\right)$
(d) 3; 30°, 150°, 270°
(e) $y = \cos x$

95 -1

96 -1; $(x+1)(2x+1)(x-2)$

97 $(\frac{1}{5}, \frac{13}{10})$, $(\frac{1}{3}, \frac{3}{2})$

98 (a) $x > 10.5$ (b) $x(x-5) < 104$
(c) $10.5 < x < 13$

99 -2

100 (b) 80.9 m (c) 267 m (d) 847 m^2

101 $-\frac{1}{2} < x < 3$

102 (a) $x + 2y = 6$ (b) $4y = x + 9$
(c) 1.12 (d) 18.75

103 (c) $y^2 - \frac{10\,001}{10}y + 100 = 0$
(d) -1, 3

104 (a) 60, 150, 240, 330
(b) (i) (340, 0) (ii) $p = 1.5$, $q = 60$

105 1.42, -0.422

107 (a) 0.78, -1.28 (b) -2

108 (a) (ii) $p = -3$, $q = 12$
(b) (2, 0), (4, 6)

109 $A = 1$, $B = -4$
(b) 37.76, 142.24, 217.76

110 (a) 6.71 (b) $y = -\frac{1}{2}x + 18$
(d) $(1\frac{1}{2}, 17\frac{1}{2})$

111 (b) $(3x+2)(x+2)(x-1)$
(c) 90°, 222°, 318°

Exercise 4A

1 (a) 3, 6, 9, 12 (b) 1, 6, 11, 16
(c) -1, 5, 15, 29 (d) 1, 2, 4, 8

2 (a) 11, 13; $2n+1$ (b) 243, 729; 3^n
(c) $\frac{5}{6}, \frac{6}{7}; \frac{n}{n+1}$ (d) 65, 129; $1 + 2^{n+1}$
(e) 30, 42; $n^2 + n$

3 (a) 17, 26 (b) 41, 122
(c) 677, 458 330

4 $1, \frac{1}{4}, \frac{1}{9}, \frac{1}{16}$: converges to zero

5 (a) oscillates between $-\frac{1}{2}$ and $\frac{1}{2}$
(b) periodic with period 8
(c) converges to zero

6 $0, \frac{1}{3}, \frac{1}{2}, \frac{3}{5}$; the nth term approaches 1 as n approaches ∞

7 (a) converges to 3; (c) and (d) both converge to zero; (b) and (e) are non-convergent

8 (a) 15, 22 : $(7n + 1)$
(b) -13, -19 : $(-6n + 5)$
(c) -3, -19 : $(13 - 8n)$

9 30°, 29.7°, 29.403°, 29.108 97°
20th term 24.785°, 100th term 11.092°

10 7, 19, 43; $n = 9$

Exercise 4B

1 3, 425 **2** 4, 777 **3** -7, -536

4 0.25, 116.25 **5** £25, £56 925

6 276 **7** 432 **8** 196 **9** 9900

10 37 500 **11** 7, 1530 **12** 314, 4940

13 10, -1, 55

14 31 terms, 50 terms

15 (a) £7600 (b) £30 200

16 1, 55, $(6n - 5)$ **17** -2.5, $\frac{1}{2}$, 735

18 $x = 3$; -60 **19** $\frac{r-4}{6}$, $16\frac{1}{2}$

20 52 **21** 9 **22** 7

23 -1, 9 **24** 15 000 **25** $\frac{y-x}{2(n-2)}$

Exercise 4C

1 (a) 5120, 10 230 (b) -5120, -3410
(c) $\frac{5}{256}$, $19\frac{251}{256}$

2 (a) 12 285 (b) -4095 (c) 8.93
(d) 1.79

3 (a) $\frac{1}{32}$, $7\frac{31}{32}$ (b) -2048, -4092
(c) $-\frac{16}{729}$, 36.0 (d) -28.24, -745.8

4 series (b) has no sum to infinity
(a) 8 (c) 36 (d) -1000

5 $\frac{1}{9} - \frac{1}{3} + 1 - \ldots$; $-14\,762.\dot{2}$

6 (a) $\frac{1}{3}[1 - (\frac{1}{10})^n]$; $\frac{1}{3}$
(b) $\frac{32}{3}[1 - (-\frac{1}{2})^n]$; $\frac{32}{3}$
(c) $\frac{16}{5}[1 - (-\frac{3}{2})^n]$; diverges

7 $-\frac{2}{5}$, 5.00 **8** 3, $\frac{2}{3}$, 8.93

9 2, -3; no sum to infinity because $|r| > 1$

10 35 years

11 (a) 36 (b) 40

12 $\frac{64}{27}$, $\frac{2}{3}$; 81

13 (a) £55 300 (b) £20 441 700

14 (a) 105 (b) 11.1

15 −14.1

Exercise 4D

1 $p = 18$, $q = 2$; 27

2 $15\frac{15}{16}$

3 90 m

4 7, 8

5 $243\sqrt{6}$; $364(\sqrt{6} + \sqrt{2})$

6 6, 2

7 $-\frac{8}{9}$

8 (a) −10.1 (b) 49

9 $\frac{10^8}{9}$, $-\frac{10^8}{11}$

10 76.21

11 $8(4n + 3)$, $n > 1$

12 (a) 52 (b) −70

13 $r = -3$ $(a = 2)$, 4th term $= -54$

14 34, 1717

15 93 750

Exercise 5A

1 y-coordinate:

(a) 8 (b) 3.375 (c) 1.331

(d) 1.030 301 (e) 1.0003

gradient PQ:

(a) 7 (b) 4.75 (c) 3.31

(d) 3.0301 (e) 3.0000

Gradient of tangent to curve at P is 3

2 (a) −2 (b) $\frac{1}{2}$ (c) $-\frac{1}{3}$

3 (a) $4x^3$ (b) $-3x^{-4}$ (c) 0

(d) 3 (e) $12x^2$ (f) $-10x^{-6}$

(g) $-\frac{1}{2x^2}$ (h) $-\frac{3}{x^3}$ (i) $2x + 2x^{-3}$

(j) $-3x^{-2} + 4x^{-3}$ (k) $2x + 1$

(l) $3x^2 - 3$ (m) 1

(n) $16x^3 - 24x$ (o) $x + \frac{1}{2x^2}$

4 (a) −6 (b) −12 (c) −3

(d) $-\frac{1}{16}$ (e) $3\frac{1}{4}$ (f) 8

(g) $-1\frac{1}{4}$

5 2, 0, −2

6 −12, −3, −12, −3

7 $y = 1$, $\frac{dy}{dx} = -5$

8 $A(100, 0)$, 100, −100. $H(50, 2500)$. H is the highest point reached by the arrow in its flight.

9 4, $11\frac{3}{4}$, $17\frac{25}{27}$

10 2, $2\frac{2}{3}$

11 (2, −13), (−2, 15)

12 $a = 1$, $b = -4$, $c = 5$

Exercise 5B

1 $-3x^2$; f is decreasing because $f'(x) < 0$

2 (a) (i) $x > 2$ (ii) $x < 2$

(b) (i) $x > 0$ (ii) $x < 0$

(c) (i) $x < 0$ (ii) $x > 0$

(d) (i) $x < 1$ (ii) $x > 1$

3 $\frac{3}{4}$; $-\frac{3}{2}$

4 greatest value is 4 at $x = 2$; $y \leqslant 4$

5 (a) (0, 0) min, $(-\frac{2}{3}, \frac{4}{27})$ max

(b) (1, 0) min, (−1, 4) max

(c) (1, 2) min, (−1, −2) max

(d) (2.38, 17.0) min

(e) $(\frac{2}{3}, -\frac{25}{27})$ min, $(-\frac{1}{2}, \frac{9}{4})$ max

(f) (2, 2) min, (0, 6) max

6 (1, −1) min, (−2, 26) max

7 Cuts x-axis where $x = -\sqrt{3}$, 0 and $\sqrt{3}$. Max. at (1, 2); min. at (−1, −2)

8 $\frac{1}{3}$ and −2

9 (a) (1, 4) and $(1\frac{2}{3}, 3\frac{5}{27})$

(b) $\frac{2}{3} < x < 2$

10 (a) $y - 2 = 4(x - 1)$, $y - 2 = -\frac{1}{4}(x - 1)$

(b) $y + 4 = -4(x + 1)$, $y + 4 = \frac{1}{4}(x + 1)$

(c) $y - 2 = \frac{1}{4}(x - 4)$, $y - 2 = -4(x - 4)$

(d) $y - \frac{1}{2} = -\frac{1}{48}(x - 8)$, $y - \frac{1}{2} = 48(x - 8)$

11 (a) $y - 18 = -8(x + 1)$, $y - 18 = \frac{1}{8}(x + 1)$

12 minimum 0, maximum −4

13 500 m, 10

14 a minimum value of 12 at $x = 2$

15 2500

16 20

17 125 000

18 3.30, 65.4

19 4

20 16π cm^3, 2 cm

21 18 000 cm^3

22 $2(x + \frac{12}{x})$; 13.86 cm ($4\sqrt{12}$ cm)

23 0.1 cm s^{-1}; 0.3 cm s^{-1}; $A = \frac{\pi t^4}{100}$; 0.016 cm^2 s^{-1}

0.424 cm^2 s^{-1}

24 $V = \frac{\pi t^3}{6}$; $A = \pi t^2$; $\frac{9\pi}{2}\,\text{cm}^3\,\text{s}^{-1}$; $6\pi\,\text{cm}^2\,\text{s}^{-1}$

26 $-3 < x < 4$ **27** $1.96\,\text{m}^3\,\text{s}^{-1}$

28 1.762, −0.1144

29 $(\frac{1}{2}, \frac{4}{3})$, $(-\frac{1}{2}, -\frac{4}{3})$

30 At $x = 2$, y takes maximum value of 7.
tangent $y + 13 = -24(x - 4)$
normal $y + 13 = \frac{1}{24}(x - 4)$

Exercise 6A

1 (a) $\frac{x^2}{2} + C$ (b) $\frac{x^5}{5} + C$
(c) $\frac{3x^4}{2} + C$ (d) $-x^{-1} + C$
(e) $-\frac{2}{3}x^{-3} + C$ (f) $3x^3 + C$
(g) $2x^2 + 5x + C$ (h) $\frac{x^3}{3} - \frac{x^2}{2} + C$
(i) $x - \frac{x^2}{2} + C$ (j) $\frac{x^3}{3} + x^2 + x + C$
(k) $4x - 2x^2 + \frac{x^3}{3} + C$
(l) $\frac{x^3}{3} - 2x - \frac{1}{x} + C$
(m) $-x^{-1} - \frac{x^3}{3} + C$
(n) $x - 2x^{-1} - \frac{1}{3}x^{-3} + C$
(o) $\frac{x^6}{2} + x^{-2} + C$
(p) $-5x^{-3} + C$
(q) $3x^3 - 12x - 4x^{-1} + C$
(r) $2x^3 - \frac{13}{2}x^2 + 6x + C$
(s) $5x^5 - 9x + C$
(t) $\frac{x^3}{3} - \frac{8}{5}x^{\frac{5}{2}} + 2x^2 + C$

2 (a) $\frac{2}{3}x^{\frac{3}{2}} + C$ (b) $\frac{5}{8}x^{\frac{8}{5}} + C$ (c) $2x^{\frac{1}{2}} + C$
(d) $4x^{\frac{1}{4}} + C$ (e) $2x^{\frac{5}{2}} + C$ (f) $\frac{2}{5}x^{\frac{5}{2}} + C$
(g) $-2x^{-\frac{1}{2}} + C$ (h) $3x^{\frac{4}{3}} + C$
(i) $\frac{4}{3}x^{\frac{9}{4}} + C$ (j) $2x^{-\frac{1}{4}} + C$
(k) $\frac{2\sqrt{3}}{3}x^{\frac{3}{2}} + C$ (l) $x^{\frac{1}{2}} + C$

3 (a) $-\frac{1}{3}x^{-3} + 2x + C$
(b) $x^2 + x^{-1} + C$
(c) $4x^{\frac{1}{2}} - x^{-1} + C$
(d) $-\frac{3}{2x^2} + C$
(e) $-\frac{1}{3x} + C$
(f) $-\frac{5}{24x^3} + C$
(g) $-9x^{-1} - 6x + \frac{x^3}{3} + C$
(h) $2x^2 - x^{-1} + C$
(i) $\frac{3}{2}x + 2x^{-1} + C$
(j) $\frac{12}{5}x^{\frac{5}{2}} + \frac{10}{3}x^{\frac{3}{2}} - 12x^{\frac{1}{2}} + C$

Exercise 6B

1 (a) $y = x^2 - x + 1$ (b) $y = \frac{x^4}{4} + \frac{x^2}{2} + 1$
(c) $y = \frac{x^5}{5} + x^2 + 1$ (d) $y = -\frac{8}{7}x^{\frac{7}{4}} + 1$
(e) $y = 3x^3 - 12x^2 + 16x + 1$
(f) $y = x^3 - \frac{5}{2}x^2 + 2x + 1$

2 (a) 8 (b) $\frac{1}{4}$ (c) $\frac{14}{3}$
(d) $\frac{14}{3}$ (e) $-\frac{1}{10}$ (f) $\frac{9}{10}$

4 $-\frac{1}{6}$, **5** $31\frac{1}{3}$ **6** −18

7 $10 - \frac{16}{3}\sqrt{2}$ **8** $\frac{2}{7}$, 0, $\frac{2}{9}$, 0

9 $y = \frac{1}{4}x^4 - \frac{1}{2}x^{-2} + \frac{17}{4}$; $\frac{65}{8}$

10 $5\frac{1}{2}$

11 (a) $130\frac{4}{5}$ (b) $-\frac{62}{81}$

12 2, 8

Exercise 6C

1 (a) 42 (b) $36\frac{2}{7}$
(c) $1\frac{1}{3}$ (d) 6.75

2 (a) $22\frac{2}{3}$ (b) $\frac{74}{3}$
(c) $\frac{49}{3}$ (d) 2

3 $\frac{4}{3}$ **4** 4 **5** $4\frac{1}{2}$

6 21; sign positive, indicating region is above x-axis; −21

7 $4\frac{1}{2}$ **8** $-\frac{1}{4}$, $\frac{1}{4}$, 0 **9** (1, 0), (4, 0)

10 $\frac{1}{3}$ **11** $\frac{1}{6}$ **12** $\frac{32}{3}$ **13** $78\frac{2}{3}$

14 $8\frac{1}{3}$ **15** $48\frac{1}{3}$ **16** 16 **17** 8.93

18 (a) (0, −4) minimum, (−2, 0) maximum
(b) $2\frac{3}{4}$

19 (a) 4 (b) −4

20 $\frac{4}{3}$ **21** $\frac{1}{6}$

22 $y = 2x + 1 - x^{-1}$; 0

23 (a) $\frac{5}{2}$ (b) $\frac{7}{12}$

24 $y = x + 5$; $\frac{4}{3}$

Exercise 7A

2 34.8°, 145.2°

5 (a) $\cos x = \pm\frac{4}{5}$ (b) $x = \pm 2 (x \in \mathbb{R})$
(c) $x = -2$ or 7

6 (a) false (b) true (c) true

7 (a) true (b) false (c) false

8 12 units^2 **10** $a^2 \geqslant 4b$

13 either $k > 0$ or $k < -5$; necessary and sufficient

15 $(x - a)^2(x + 2a)$ **16** $1 \pm \sqrt{3}$

20 $\frac{7}{2} - \frac{3}{2}\sqrt{5}$ **21** $5 + 2\sqrt{6}$

22 $x = 1$, $x = 2$ **23** 15 cm^2

24 $5x - 2y - 29 = 0$

25 (b) 8 (c) $3x - 4$

26 Normal is $x + 1 = 0$

27 60°, 300° **28** 13 467

Review exercise 2

1 1.5, 6.25

2 $-\frac{1}{3}$, 2, −12, 72, −432

3 $\frac{9}{16}$

4 $4\frac{1}{2}$

5 (a) 1 (b) $\frac{2}{3}$ (c) 54

6 $-4\frac{1}{2}$

7 (a) (i) 148 (ii) $161\frac{1}{3}$ (b) 12, 14

8 4 m, 2 m

9 $85\frac{1}{3}$

10 (a) 14.9 (b) 689

11 (a) 5, −30; 13 (b) 49.8

12 $-40x^3 + 3x^2 + 6x$

13 $57\frac{1}{6}$

14 (a) $\dfrac{10\,100}{3}$ (b) $\frac{7}{9}$, $\frac{k}{9}$

15 (a) 320, $\frac{1}{2}$

16 $A = \pi(5t^2 + 4)^2$
(a) 30 cm s^{-1} (b) 2940π cm^2 s^{-1}

17 6.75

18 (a) 18.5 m (b) 27.7 m (c) 32 m

19 $A = 2\pi x^2 + \dfrac{32\pi}{x}$; 2, 24π cm^2

20 $2\frac{2}{3}$

22 24.3 cm^3 (3 s.f.)

23 (a) $A = 6$, $B = 9$
(b) $\frac{3}{5}x^{\frac{5}{3}} + \frac{9}{2}x^{\frac{4}{3}} + 9 + C$
(c) 149.1

24 (a) 26 733 (b) 53 467

25 $23\frac{19}{27}$

26 $\frac{1}{216}$

27 57.7 km h^{-1}, 83.2 km h^{-1}

28 (b) £12 079.98

29 $\dfrac{dy}{dx} = \frac{9}{2}x^{\frac{1}{2}} - \frac{7}{2}x^{-\frac{1}{2}} + 3x^{-\frac{3}{2}}$

$\dfrac{d^2y}{dx^2} = \frac{9}{4}x^{-\frac{1}{2}} + \frac{7}{4}x^{-\frac{3}{2}} - \frac{9}{2}x^{-\frac{5}{2}}$

30 (a) (2, 20), (4, 16) (b) 36

31 (b) 200 (c) $\frac{1000}{3}$ (d) 8.9×10^{-4}

32 (a) £13 780 (b) £42 198

33 (a) (1, 0) min; (−1, 4) max;
(b) (1, 0) min; (−3, 256) max

34 (b) $(x + 2)(x - 1)(x - 6)$
(c) $(-\frac{2}{3}, 14\frac{22}{27})$, (4, −36)
(d) $20\frac{2}{3}$

35 (i) (a) $3\frac{1}{2}$
(b) $-\frac{1}{2}$, $5\frac{1}{2}$, $11\frac{1}{2}$, $17\frac{1}{2}$
(ii) (a) $-\frac{1}{2}$, 8
(b) $-4\frac{1}{2}$, $1\frac{1}{2}$, $-\frac{1}{2}$, $\frac{1}{6}$; 4, 10, 25, 62.5
(c) $-\frac{1}{3}$, $\frac{5}{2}$ (d) $-3\frac{3}{8}$

36 (a) 2, 3 (b) 5

37 (b) 30 000, 100

38 £3216

39 (i) (a) −1, 1, 3 (b) 2400 (c) 25
(ii) (a) 12, −8; 3, $\frac{1}{2}$ (b) −32

40 (b) $2x - 250x^{-2}$ (c) $x = 5$, $S = 75$

41 (a) (−1, 1), (3, 5) (b) $5\frac{1}{3}$

42 (a) 1, $\frac{1}{3}$, $-\frac{1}{9}$ (b) 5.986 …

43 (a) (i) −10 (ii) $2\frac{1}{2}$ (iii) 0
(b) (i) $\frac{1}{2}$ (ii) 12 (iii) 24 570

44 (b) $64 - 64x + 12x^2$ (c) $1\frac{1}{3}$
(e) $37\frac{25}{27}$

45 (a) $\frac{3}{2}x^{-\frac{1}{2}} + 2x^{-\frac{3}{2}}$ (b) $2x^{\frac{3}{2}} - 8x^{\frac{1}{2}} + C$
(c) $6 - 2\sqrt{3}$

46 (i) $-3, -1, 1, 3$; $n^2 - 4n$, 15
(ii) $-5, -4, \frac{1}{2}$; -8

47 (a) $-\frac{1}{2}$ (b) $\frac{3}{4}, -2$
(c) 14 (d) 867.62

48 (c) 5.43 (e) 139

49 $32\sqrt{2}$, $\frac{512}{5}\sqrt{2}$

50 (a) (i) £3370 (ii) £5679
(b) £50 423

51 (i) (a) $7, 1\frac{1}{2}$ (b) 462
(ii) (a) $\frac{1}{2}$ (b) 3 (c) 364.5

52 (b) 25 (c) 2 (d) 625

53 (a) $2y + x = 0$ (b) $y = -1\frac{1}{4}$
(c) $\frac{125}{48}$

54 (i) (a) $5, 3\frac{1}{2}$ (b) 16
(ii) (a) $-\frac{1}{5}, 5$ (b) 3
(c) 118 096

55 (b) $V = 100x - \frac{4}{3}x^3$
(c) $333\frac{1}{3}\text{ cm}^3$

56 (i) 4
(ii) (a) 97 (c) 2
(d) $\frac{5}{4}$ (e) 4110

57 (b) (4, 16) (c) 133 (3 s.f.)

58 (a) (6, 12) (b) $13\frac{1}{3}$

59 (a) 120 (b) 3836.25
(c) 427.5 (d) 4192.5

60 (a) (i) $1\frac{1}{2}$ (ii) -3 (iii) 21
(b) $\frac{1}{4}$

61 (b) $0, \pm\sqrt{8}$ (c) 3

62 (a) (1, 0), (5, 0), (6, 5)
(b) $10\frac{1}{6}$

63 (b) $(\frac{5}{3}, 9\frac{13}{27})$, $(-1, 0)$ (c) $21\frac{1}{3}$

64 (a) $(0, -14)$
(b) $(1, -11)$, $(\frac{7}{3}, -12.2)$
(c) $1 < x < \frac{7}{3}$

65 (a) 5 (b) 4
(c) $5 + 4(p - 1) > 100$
(d) $p = 25$

66 (b) 10 (c) 5 cm

67 (a) 1 (b) $y = x + 1$
(c) $A(-1, 0)$, $B(5, 6)$
(d) $17\frac{1}{2}$ (e) $33\frac{1}{3}$

68 (i) (a) 9 (b) 85, -4 (c) 940
(ii) (a) $-\frac{1}{5}$ (b) -20.8

69 (c) 1200

70 (i) (a) -3 (b) 48 (c) 11
(ii) (a) $\frac{2}{3}$ (b) 81 (c) 166 000

71 (b) $\frac{10}{3}$ (d) $\dfrac{2300\pi}{27}$

72 (a) -2 (b) (6, 17) (c) $\frac{4}{3}$

Exercise 8A

1 x^4 **2** a^6 **3** $4a^3$ **4** $30a^2b^2$
5 x^{10} **6** y^{12} **7** $12y^7$ **8** $24y^8$
9 $60a^5b^4$ **10** $36x^7y^4$ **11** $96p^{10}q^4$
12 $48a^5b^5$ **13** $30p^5q^9$ **14** $30x^3y^2z^3$
15 $48a^4b^3c^7$ **16** $72a^3b^2$ **17** $360a^5b^7c^6$
18 $84p^4q^4$ **19** $6a^4b^4pq^3$ **20** $\frac{1}{10}a^6p^4q^3$
21 a^{30} **22** x^{14} **23** a^8b^{12}
24 x^4 **25** x^4 **26** a^8
27 p^{-3} **28** $5p^3$ **29** $3a^{-1}$
30 $12a^7$ **31** $2a^8$ **32** 4
33 $4x^4$ **34** $5x^7$ **35** $3ab^9$
36 $2a^2bc^{-2}$ **37** $\frac{5}{3}a^4b$ **38** $36abc^2$
39 $6a^2b^{-2}c^2$ **40** $3x^7y^3z$
41 3.19×10^3 **42** 4.59×10^3
43 1.044×10^6 **44** 9×10^0
45 2.025×10^{-1} **46** $\frac{1}{2}$
47 $\frac{4}{9}$ **48** $\frac{125}{8}$
49 $-\frac{1}{12}$ **50** $\frac{256}{3}$
51 (a) $y = \dfrac{2}{x^2}$ (b) $\pm\frac{1}{5}$
52 100, $\frac{100}{9}$ **53** 17 : 3
54 (a) 78.4 m (b) 100 km h^{-1}
55 (a) 20 (b) 51.84
56 $y = 24x^{-3} - 2$
57 144 : 49; 3.02 kg **58** 31 cm
59 (a) 5 cm (b) 189 cm^3
60 (a) 1.5625×10^{13} (b) 4×10^{-5}

Exercise 8B

1 3 **2** 36 **3** −12 **4** $-6\frac{5}{12}$

5 $-2\frac{1}{4}$ **6** $\dfrac{x+5}{6}$ **7** $\dfrac{9x-2}{35}$ **8** $\dfrac{15q-2p}{15}$

9 $\dfrac{1}{x^2(x+1)}$ **10** $\dfrac{7x+2}{12}$ **11** $4(1+4y)$

12 $abc(b-c)$ **13** $5x(x-3)$

14 $(x-4)(x+4)$ **15** $x(x+20)$

16 $2(x-2)(x+2)$ **17** $6(2-y)(2+y)$

18 $x^2(x^2+3x+1)$ **19** $(y+1)^2$

20 $(3x-1)(x+1)$ **21** 6

22 −1.8 **23** 2 **24** 3 **25** 1

26 2 **27** 30 **28** $\frac{5}{6}$ **29** $-\frac{1}{4}$

30 7 **31** −23 **32** 6 **33** (7, 4)

34 (3, −1) **35** (−2, 3) **36** (−4, −5)

37 $(-\frac{1}{2}, 1\frac{1}{2})$ **38** (1, −2) **39** $(\frac{1}{2}, \frac{1}{2})$

40 (−3, −2) **41** (−2, +6) **42** (−14, −2)

43 (2, −3) **44** $(2, \frac{1}{2})$ **45** (−3, −9)

46 $(-1, -2\frac{1}{2})$ **47** $(-\frac{2}{3}, \frac{3}{4})$ **48** $\dfrac{5-y}{y}$

49 $\dfrac{y}{1-y}$ **50** $\dfrac{3y+2}{1-y}$ **51** $\dfrac{y}{3y-1}$

52 $\dfrac{2}{3y-1}$ **53** $t=\dfrac{bc-ad}{a-b-c+d}$

54 $a=\dfrac{2S-dn(n-1)}{2n}$ **55** $F=\dfrac{9C+160}{5}$

Exercise 8C

1 2 **2** −1 **3** 3 **4** $-\frac{8}{5}$

5 3 **6** −3 **7** $7\frac{1}{2}$ **8** 2

9 $\frac{13}{5}$ **10** 3 **11** $-\frac{17}{9}$ **12** $-\frac{1}{5}$

13 3 **14** −1

15 AB $\frac{8}{5}$, BC $-\frac{4}{9}$, CA −3

16 AB $\frac{2}{3}$, BC 2, CA −2

17 AB $-\frac{7}{2}$, BC −1, CA $-\frac{4}{9}$

18 both have gradient $-\frac{1}{2}$, so A, B and C lie on a straight line

19 both have gradient $\frac{3}{2}$ and are on same line

21 (a) 5 (b) −7 (c) −2 (d) −1
(e) 1.5 (f) $\frac{7}{3}$ (g) $\frac{2}{3}$ (h) $\frac{3}{4}$
(i) $-\frac{3}{2}$ (j) $-\frac{4}{3}$ (k) $\frac{5}{2}$ (l) $-\frac{6}{5}$

22 (a) −4 (b) 6 (c) 3 (d) −2
(e) $\frac{7}{2}$ (f) $\frac{5}{3}$ (g) 3 (h) $\frac{7}{5}$
(i) $\frac{7}{2}$ (j) $\frac{8}{5}$ (k) $-\frac{7}{2}$ (l) −4

23 (a) (0, 4) (b) (0, −7) (c) $(0, \frac{3}{2})$
(d) $(0, -\frac{2}{3})$ (e) $(0, \frac{5}{2})$ (f) (0, 3)
(g) $(0, \frac{7}{3})$ (h) $(0, \frac{3}{2})$

24 $y=2x+3$ **25** $y=5x-4$

26 $y=2-x$ **27** $y=-\frac{2}{3}x+1$

28 $y=-\frac{1}{2}x-7$ **29** $y=2x+1$

30 $y=-3x+3$ **31** $y=\frac{1}{2}x+5$

32 $y=-\frac{4}{5}x-2$ **33** $y=-7x+6$

34 $y=3x-\frac{1}{2}$ **35** $\sqrt{10}$

36 $\sqrt{34}$ **37** $\sqrt{122}$ **38** $5\sqrt{5}$

39 13 **40** $\sqrt{13}$ **41** $2\sqrt{13}$

42 $2\sqrt{37}$ **43** $2\sqrt{53}$ **44** $\sqrt{29}$

45 $5\sqrt{10}$ **46** $\sqrt{13}$

49 $PQ = 6.32$, $QR = 5.39$, $PR = 4.12$

50 (a) 14.4 (b) 10.2

Exercise 8D

1 (a) 0.471 (b) 0.451 (c) 1.56
(d) 0.854 (e) 1.13 (f) 0.271
(g) 0.441 (h) 0.714 (i) 0.220
(j) 0.946

2 (a) 20.0° (b) 83.6° (c) 56.7°
(d) 70.7° (e) 61.9° (f) 7.55°
(g) 33.5° (h) 51.4° (i) 28.0°
(j) 52.0°

3 (a) 0.962 (b) 0.703 (c) 1.34
(d) 0.671 (e) 0.701 (f) 0.823
(g) 0.978 (h) 0.456 (i) 0.355
(j) 0.752

4 6.79 cm **5** 6.37 cm **6** 5.89 cm

7 9.80 cm **8** 19.7° **9** 70.6°

10 18.6° **11** 15.5 cm **12** 29.3 cm

13 21.3 cm **14** 8.02 cm **15** 5.07 cm

16 (a) 4.95 cm (b) 38.6 cm^2

17 (a) 17.6 m (b) 35.6 m

18 (a) 2.82 cm (b) 5.30 cm (c) 43.2°

19 (a) 51.3° (b) 12.8 m
(c) 23.0 m (d) 350 m^2

20 (a) 7.5 cm (b) 6.73 cm
(c) 14.0 cm (d) 31.2 cm
(e) 38.2 cm^2

21 1880 cm^2

22 (a) 5.79 cm (b) 8.34 cm
(c) 3.06 cm; 98.6 cm^2

23 (a) 33.0 m (b) 12.0 m

24 (a) (i) 2.94 km (ii) 4.05 km
(b) (i) 8.29 km (ii) 1.90 km
(c) (i) 283° (ii) 8.51 km

Exercise 8E

1 $AC = 4.5$ cm, $BC = 5.7$ cm, $C = 91°$

2 $C = 67.4°$, $AB = 12.2$ cm, $BC = 9.11$ cm

3 $A = 69.5°$, $AB = 22.5$ cm, $BC = 24.0$ cm

4 $B = 79.3°$, $AB = 17.2$ cm, $AC = 17.9$ cm

5 $B = 50°$ $AB = 6.7$ cm, $BC = 5.7$ cm

6 $B = 56.2°$, $AB = 7.6$ cm, $BC = 5.9$ cm

7 $C = 33.3°$, $B = 99.7°$, $AC = 11.3$ cm

8 $A = 35.3°$, $B = 80.7°$, $AC = 21.2$ cm

9 $A = 33.3°$, $C = 73.7°$, $AB = 12.7$ cm

10 $A = 30.2°$, $C = 96.6°$, $AB = 11.7$ cm

11 $A = 56.3°$, $B = 55.0°$, $BC = 15.5$ cm

12 $A = 34.6°$, $B = 73.1°$, $AC = 21.4$ cm

13 $B = 44.8°$, $C = 99.0°$, $AB = 19.1$ cm *or*
$B = 135.2°$, $C = 8.6°$, $AB = 2.9$ cm

14 $B = 75.6°$, $A = 52.1°$, $BC = 15.6$ cm *or*
$B = 104.4°$, $A = 23.3°$, $BC = 7.8$ cm

15 $C = 30°$, $A = 132.6°$, $BC = 36.2$ cm *or*
$C = 150°$, $A = 12.6°$, $BC = 10.7$ cm

16 $B = 59°$, $A = 73.7°$, $BC = 20.4$ cm *or*
$B = 121°$, $A = 11.7°$, $BC = 4.3$ cm

17 4.39 cm **18** 11.2 cm **19** 6.42 cm

20 7.46 cm **21** 8.00 cm **22** 9.82 cm

23 16.1 cm **24** 13.9 cm **25** 44.0°

26 60° **27** 109.5° **28** 82.8°

29 141.4° **30** 57.1° **31** 15.9°

32 22.6°

33 $AB = 2.45$ cm, $A = 60.2°$, $B = 76.8°$,
area 2.39 cm^2

34 $BC = 2.87$ cm $B = 88.1°$ $C = 56.9°$,
area 5.74 cm^2

35 $AB = 18.8$ cm, $A = 22.1°$ $B = 17.9°$,
area 31.8 cm^2

36 $AC = 12.6$ cm, $A = 21.9°$, $C = 48.1°$,
area 23.5 cm^2

37 $AC = 4.70$ cm, $A = 33.2°$, $C = 106.8°$,
area 9.00 cm^2

38 2.62 naut. miles, 341°

39 6.40 km **40** 224°

41 (a) 6.53 cm (b) 9.10 cm (c) 13.7 cm

42 1094.9 cm^2

43 (a) 14.4 cm (b) 6.43 cm

Examination style paper P1

2 (2, −3), (−2, 3)

3 (b) 0.36

4 (a) $\frac{\pi}{3}, \frac{5\pi}{3}$
(b) 1.05, 1.82, 4.46, 5.24

5 (a) $44\frac{26}{27}$ (b) 128

6 (a) $x + 4y - 9 = 0$ (b) $4x - y - 9 = 0$
(c) $2x + 8y - 7 = 0$ (d) (3.5, 0)

7 (a) −735
(b) (i) 84.5 (ii) 31 terms

8 (b) $150\pi\text{ cm}^2$

List of symbols and notation

The following notation will be used in all Edexcel examinations.

$\in$	is an element of
$\notin$	is not an element of
$\{x_1, x_2, \ldots\}$	the set with elements $x_1, x_2, \ldots$
$\{x : \ldots\}$	the set of all x such that ...
$\mathrm{n}(A)$	the number of elements in set A
$\emptyset$	the empty set
e	the universal set
A'	the complement of the set A
$\mathbb{N}$	the set of natural numbers, $\{1, 2, 3, \ldots\}$
$\mathbb{Z}$	the set of integers, $\{0, \pm 1, \pm 2, \pm 3, \ldots\}$
$\mathbb{Z}^+$	the set of positive integers, $\{1, 2, 3, \ldots\}$
$\mathbb{Z}_n$	the set of integers modulo n, $\{0, 1, 2, \ldots, n-1\}$
$\mathbb{Q}$	the set of rational numbers $\left\{\frac{p}{q} : p \in \mathbb{Z}, q \in \mathbb{Z}^+\right\}$
$\mathbb{Q}^+$	the set of positive rational numbers, $\{x \in \mathbb{Q} : x > 0\}$
$\mathbb{Q}_0^+$	the set of positive rational numbers and zero, $\{x \in \mathbb{Q} : x \geqslant 0\}$
$\mathbb{R}$	the set of real numbers
$\mathbb{R}^+$	the set of positive real numbers, $\{x \in \mathbb{R} : x > 0\}$
$\mathbb{R}_0^+$	the set of positive real numbers and zero, $\{x \in \mathbb{R} : x \geqslant 0\}$
$\mathbb{C}$	the set of complex numbers
(x, y)	the ordered pair x, y
$A \times B$	the cartesian product of sets A and B, $A \times B = \{(a, b) : a \in A, b \in B\}$
$\subseteq$	is a subset of
$\subset$	is a proper subset of
$\cup$	union
$\cap$	intersection
$[a, b]$	the closed interval, $\{x \in \mathbb{R} : a \leqslant x \leqslant b\}$
$[a, b)$	the interval $\{x \in \mathbb{R} : a \leqslant x < b\}$
$(a, b]$	the interval $\{x \in \mathbb{R} : a < x \leqslant b\}$
(a, b)	the open interval $\{x \in \mathbb{R} : a < x < b\}$
$y R x$	y is related to x by the relation R
$y \sim x$	y is equivalent to x, in the context of some equivalence relation
$=$	is equal to
$\neq$	is not equal to
$\equiv$	is identical to *or* is congruent to

$\approx$	is approximately equal to
$\cong$	is isomorphic to
$\propto$	is proportional to
$<$	is less than
$\leqslant, \ngtr$	is less than or equal to, is not greater than
$>$	is greater than
$\geqslant, \nless$	is greater than or equal to, is not less than
∞	infinity
$p \wedge q$	p and q
$p \vee q$	p or q (or both)
$\sim p$	not p
$p \Rightarrow q$	p implies q (if p then q)
$p \Leftarrow q$	p is implied by q (if q then p)
$p \Leftrightarrow q$	p implies and is implied by q (p is equivalent to q)
$\exists$	there exists
$\forall$	for all
$a+b$	a plus b
$a-b$	a minus b
$a \times b,\ ab,\ a.b$	a multiplied by b
$a \div b,\ \frac{a}{b},\ a/b$	a divided by b
$\sum_{i=1}^{n} a_i$	$a_1 + a_2 + \ldots + a_n$
$\prod_{i=1}^{n} a_i$	$a_1 \times a_2 \times \ldots \times a_n$
$\sqrt{a}$	the positive square root of a
$\lvert a \rvert$	the modulus of a
$n!$	n factorial
$\binom{n}{r}$	the binomial coefficient $\frac{n!}{r!(n-r)!}$ for $n \in \mathbb{Z}^+$ $\frac{n(n-1)\ldots(n-r+1)}{r!}$ for $n \in \mathbb{Q}$
$\mathrm{f}(x)$	the value of the function f at x
$\mathrm{f}: A \to B$	f is a function under which each element of set A has an image in set B
$\mathrm{f}: x \mapsto y$	the function f maps the element x to the element y
f^{-1}	the inverse function of the function f
$\mathrm{g} \circ \mathrm{f},\ \mathrm{gf}$	the composite function of f and g which is defined by $(\mathrm{g} \circ \mathrm{f})(x)$ or $\mathrm{gf}(x) = \mathrm{g}(\mathrm{f}(x))$
$\lim_{x \to a} \mathrm{f}(x)$	the limit of $\mathrm{f}(x)$ as x tends to a
$\Delta x,\ \delta x$	an increment of x
$\frac{\mathrm{d}y}{\mathrm{d}x}$	the derivative of y with respect to x
$\frac{\mathrm{d}^n y}{\mathrm{d}x^n}$	the nth derivative of y with respect to x

$f'(x), f''(x), \ldots f^{(n)}(x)$	the first, second, ... nth derivatives of $f(x)$ with respect to x
$\int y\,dx$	the indefinite integral of y with respect to x
$\int_a^b y\,dx$	the definite integral of y with respect to x between the limits $x = a$ and $x = b$
$\frac{\partial V}{\partial x}$	the partial derivative of V with respect to x
$\dot{x}, \ddot{x}, \ldots$	the first, second, . . . derivatives of x with respect to t
e	base of natural logarithms
e^x, $\exp x$	exponential function of x
$\log_a x$	logarithm to the base a of x
$\ln x$, $\log_e x$	natural logarithm of x
$\lg x$, $\log_{10} x$	logarithm to the base 10 of x
sin, cos, tan cosec, sec, cot	the circular functions
arcsin, arccos, arctan arccosec, arcsec, arccot	the inverse circular functions
sinh, cosh, tanh cosech, sech, coth	the hyperbolic functions
arsinh, arcosh, artanh, arcosech, arsech, arcoth	the inverse hyperbolic functions
i	square root of -1
z	a complex number, $z = x + iy$
Re z	the real part of z, Re $z = x$
Im z	the imaginary part of z, Im $z = y$
$\|z\|$	the modulus of z, $\|z\| = \sqrt{(x^2 + y^2)}$
arg z	the argument of z, $\arg z = \arctan \frac{y}{x}$
z^*	the complex conjugate of z, $x - iy$
$\mathbf{M}$	a matrix $\mathbf{M}$
$\mathbf{M}^{-1}$	the inverse of the matrix $\mathbf{M}$
$\mathbf{M}^T$	the transpose of the matrix $\mathbf{M}$
det $\mathbf{M}$, $\|\mathbf{M}\|$	the determinant of the square matrix $\mathbf{M}$
$\mathbf{a}$	the vector $\mathbf{a}$
$\overrightarrow{AB}$	the vector represented in magnitude and direction by the directed line segment AB
$\hat{\mathbf{a}}$	a unit vector in the direction of $\mathbf{a}$
$\mathbf{i}, \mathbf{j}, \mathbf{k}$	unit vectors in the directions of the cartesian coordinate axes
$\|\mathbf{a}\|$, a	the magnitude of $\mathbf{a}$
$\|\overrightarrow{AB}\|$, AB	the magnitude of $\overrightarrow{AB}$
$\mathbf{a} . \mathbf{b}$	the scalar product of $\mathbf{a}$ and $\mathbf{b}$
$\mathbf{a} \times \mathbf{b}$	the vector product of $\mathbf{a}$ and $\mathbf{b}$

A, B, C, etc	events
$A \cup B$	union of the events A and B
$A \cap B$	intersection of the events A and B
$\mathrm{P}(A)$	probability of the event A
A'	complement of the event A
$\mathrm{P}(A\|B)$	probability of the event A conditional on the event B
X, Y, R, etc.	random variables
x, y, r, etc.	values of the random variables X, Y, R, etc
$x_1, x_2 \ldots$	observations
$f_1, f_2, \ldots$	frequencies with which the observations $x_1, x_2, \ldots$ occur
$\mathrm{p}(x)$	probability function $\mathrm{P}(X = x)$ of the discrete random variable X
$p_1, p_2, \ldots$	probabilities of the values $x_1, x_2, \ldots$ of the discrete random variable X
$\mathrm{f}(x)$, $\mathrm{g}(x), \ldots$	the value of the probability density function of a continuous random variable X
$\mathrm{F}(x), \mathrm{G}(x), \ldots$	the value of the (cumulative) distribution function $\mathrm{P}(X \leqslant x)$ of a continuous random variable X
$\mathrm{E}(X)$	expectation of the random variable X
$\mathrm{E}[\mathrm{g}(X)]$	expectation of $\mathrm{g}(X)$
$\mathrm{Var}(X)$	variance of the random variable X
$\mathrm{G}(t)$	probability generating function for a random variable which takes the values 0, 1, 2, ...
$\mathrm{B}(n, p)$	binomial distribution with parameters n and p
$\mathrm{N}(\mu, \sigma^2)$	normal distribution with mean μ and variance σ^2
μ	population mean
σ^2	population variance
σ	population standard deviation
$\bar{x}$, m	sample mean
s^2, $\hat{\sigma}^2$	unbiased estimate of population variance from a sample, $s^2 = \frac{1}{n-1}\sum(x_i - \bar{x})^2$
ϕ	probability density function of the standardised normal variable with distribution N(0, 1)
Φ	corresponding cumulative distribution function
ρ	product-moment correlation coefficient for a population
r	product-moment correlation coefficient for a sample
$\mathrm{Cov}(X, Y)$	covariance of X and Y

Index